L2.5

AF841765

UNIVERSITY OF SUNDERLAND
WITHDRAWN

UNIVERSITY OF SUNDERLAND
UNIVERSITY LIBRARY

The D-2 Digital Video Recorder

For Howard

The D-2 Digital Video Recorder

John Watkinson

Focal Press
London & Boston

Focal Press
is an imprint of Butterworth–Heinemann
PART OF REED INTERNATIONAL P.L.C.

All rights reserved. No part of this publication may be reproduced in any material form (including photocopying or storing it in any medium by electronic means and whether or not transiently or incidentally to some other use of this publication) without the written permission of the copyright owner except in accordance with the provisions of the Copyright, Designs and Patents Act 1988 or under the terms of a licence issued by the Copyright Licensing Agency Ltd, 33–34 Alfred Place, London, England WC1E 7DP. Applications for the copyright owner's written permission to reproduce any part of this publication should be addressed to the Publishers.

Warning: The doing of an unauthorized act in relation to a copyright work may result in both a civil claim for damages and criminal prosecution.

This book is sold subject to the Standard Conditions of Sale of Net Books and may not be re-sold in the UK below the net price given by the Publishers in their current price list.

First published 1990

© John Watkinson, 1990

British Library Cataloguing in Publication Data

Watkinson, John
The D-2 digital video recorder.
1. Digital videorecording equipment
I. Title
621.38833

ISBN 0-240-51302-9

Library of Congress Cataloging-in-Publication Data

Watkinson, John.
The D-2 digital video recorder/John Watkinson.
p. cm.
Includes bibliographical references and index.
ISBN 0-240-51302-9 :
1. Digital television. 2. Video tape recorders and recording.
I. Title.
TK6678.W383 1990
621.388'33–dc20 90-2539

UNIVERSITY OF SUNDERLAND
LIBRARY
Class No. 621.3889 W08
Access No. 1110968716 8
Order No. E06240/5 42413

Composition by Genesis Typesetting, Laser Quay, Rochester, Kent
Printed and bound in Great Britain by Courier International Ltd., Tiptree, Essex

Preface

Composite video was developed solely to provide a way of broadcasting colour television in a way which would be compatible with existing monochrome receivers. It requires no more bandwidth than a monochrome signal, and so full bandwidth monochrome video recorders could be simply adapted to record colour, and the composite video recorder was born.

Whilst the composite signal has advantages for broadcasting, the same cannot be said for processing and recording, and the development of units which work with component signals has overcome these problems at a stroke.

Some years ago, claims were made that the future lay solely in components, and that composite video would be restricted to terrestrial broadcasting, but this has turned out to be idealistic.

If one has the luxury of starting with a clean sheet of paper, the indisputable advantages of component working make it the obvious choice on quality and flexibility grounds.

In the real world, broadcasters have invested heavily in composite signal handling systems, and in these days of greater financial constraints, it simply makes no sense to scrap the lot. Generally the weakest link in a signal chain has been the recorder, and this is particularly true of composite signals which are sensitive to timebase error. For this reason the use of digital techniques in a composite recorder gives a greater tangible improvement in quality than in component recorders. This is the thinking behind the D-2 format which is the subject of this book. It is not the purpose of this book to compare composite and component video, but simply to document as clearly as possible how the D-2 format works and to describe the subsystems necessary to support the format. Interfacing both audio and composite video in the digital domain is fully detailed.

Digital recording uses techniques such as error correction which have not been seen in analog recorders, and all of these new subjects are treated from first principles so that no great prior knowledge is necessary to obtain full benefit from the information here.

This is not just a theory book, because it also contains numerous examples of current practice; nor is it just an introductory book, because the detailed descriptions omit nothing of consequence, and there are references for those interested in deeper study.

Acknowledgements

The D-2 format was jointly proposed by Ampex and Sony, and the author is indebted to innumerable employees of both companies who have given their valuable time to discuss technical details and to find relevant documents. Without their active assistance this book would not have been possible.

John Watkinson
Burghfield Common, England

Contents

Chapter 1

Introduction to D-2

1.1 The advantages of digital video

The first methods used in television transmission and recording were understandably analog, and the signal formats were essentially determined by the requirements of the cathode ray tube as a display, so that the receiver might be as simple as possible, and be constructed with a minimum number of vacuum tubes. Following the development of magnetic audio recording during the Second World War, a need for television recording was perceived. This was due to the number of time zones across the United States. Without recording, popular television programmes had to be performed live several times so they could be seen at peak viewing time in each time zone. Ampex eventually succeeded in recording monochrome video in the 1950s, and the fundamentals of the Quadruplex machine which combined a rotary head with frequency modulation were so soundly based that modern analog video recorders only refine the process.[1] Study of every shortcoming in the analog process has led to the development of some measure to reduce it, and current machines are capable of excellent performance. The point to appreciate, however, is that analog recording has become a mature technology, and has almost reached limits determined by the laws of physics. The process of refinement produces increasingly small returns.

As there is now another video technology, known as digital, the previous technology has to be referred to as analog. It is appropriate to begin with a comparison of the fundamental differences between these technologies.

In an analog system, information is conveyed by the infinite variation of some continuous parameter, such as the voltage on a wire or the strength of flux. When it comes to recording, distance along the medium is a further analog of time. However much the signal is magnified, more and more detail will be revealed until a point is reached where the actual value is uncertain because of noise. A parameter can only be a true analog of the original if the conversion process is linear, otherwise harmonic distortion is introduced. If the speed of the medium is not constant there will not be a true analog of time and the result is known as timebase error.

It is a characteristic of an analog system that the degradations at the output incorporate all the degradations introduced in each stage through which the signal has passed. This sets a limit to the number of stages a signal can pass through before it becomes too impaired to be worth watching. Down at signal level, all impairments can be reduced to the addition of some unwanted signal such as noise or distortion, and timing instability such as group-delay effects and chroma phase errors. In an analog system, such effects can never be separated from the original signal; in the digital domain they can be eliminated.

In a digital video system, the information is in binary form. The signals sent have only two states and change at predetermined times according to a stable clock. If the binary signal is degraded by noise, this will be rejected at the receiver, as the signal is judged solely on whether it is above or below some threshold. However, the signal will be conveyed with finite bandwidth, and this will restrict the rate at which the voltage changes. Superimposed noise can move the point at which the receiver judges that there has been a change of state. Time instability has this effect too. This instability is also rejected because, on receipt, the signal is reclocked by the stable clock, and all changes in the system will take place at the edges of that clock. Figure 1.1 shows that however many stages a binary signal passes through, it still comes out the same, only later. It is possible to convey an analog waveform

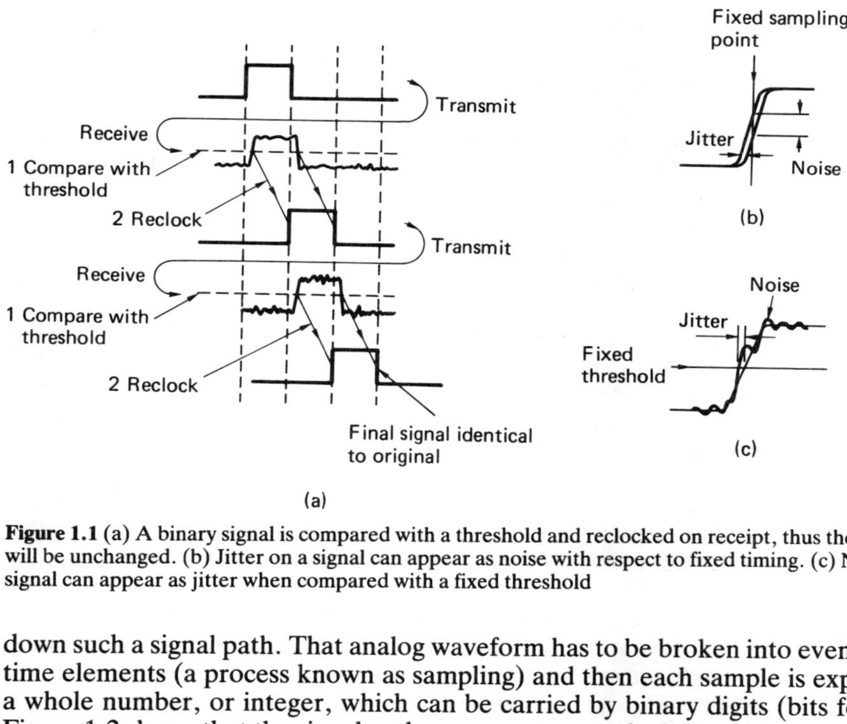

Figure 1.1 (a) A binary signal is compared with a threshold and reclocked on receipt, thus the meaning will be unchanged. (b) Jitter on a signal can appear as noise with respect to fixed timing. (c) Noise on a signal can appear as jitter when compared with a fixed threshold

down such a signal path. That analog waveform has to be broken into evenly spaced time elements (a process known as sampling) and then each sample is expressed as a whole number, or integer, which can be carried by binary digits (bits for short). Figure 1.2 shows that the signal path may convey sample values either in parallel on several wires, where each wire carries a binary signal representing a different power of two, or serially in one channel, at higher speed, a process called pulse code modulation (PCM). The only drawback of this scheme is that a single high-quality video channel requires between 1 and 200 million bits per second. Digital video only became viable when advances in high-density recording made such a data rate available at reasonable cost.

In simple terms, the signal waveform in a digital recorder is conveyed as if someone had measured the instantaneous voltage at regular intervals with a digital

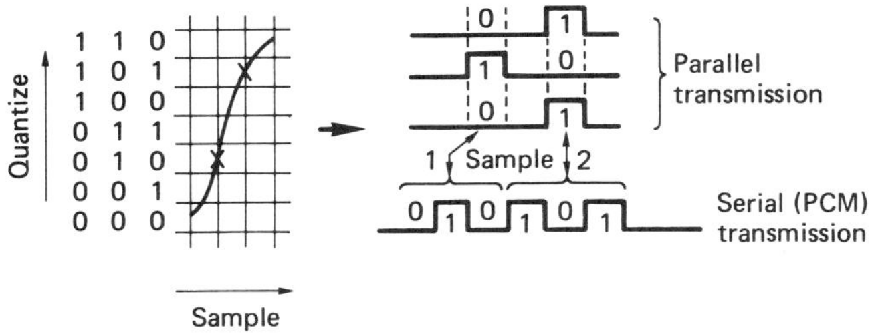

Figure 1.2 When a signal is carried in numerical form, either parallel or serial, the mechanisms of Figure 1.1 ensure that the only degradation is in the conversion processes

voltmeter and written the readings in binary on a roll of paper. The rate at which these measurements are made and the accuracy of the meter now wholly determine the quality of the system, because once a parameter is expressed as discrete numbers, those numbers can be conveyed unchanged through a recording process. This dependence on the quality of conversion is the price paid to make quality independent of the signal path. As it is so critical to picture quality, a whole chapter of this book is dedicated to sampling and conversion. Since a digital system faithfully conveys the original analog waveform, it is perfectly feasible to digitize composite video signals. Unless there is a defect in the conversion stage, there is complete freedom from differential phase and gain error in a composite digital recorder.

A magnetic head cannot know the meaning of signals which are passed through it, so there is no distinction at the head/medium interface between analog and digital recording. Thus a digital signal will suffer all the degradations that beset an analog signal: particulate noise, distortion, dropout, modulation noise, print-through, crosstalk and so on. However, there is a difference in the effect of these degradations on the meaning of the signals. As stated, digital recording uses a binary code, and the presence or absence of a flux change is the only item of interest. Provided that flux changes can generate a playback pulse which is sensibly bigger than the noise, the numerical meaning will be unchanged by reasonable distortions of the waveform. In other words, a bit is still a bit, whatever its shape. This implies that the bits on the medium can be very small indeed, and can be packed very close together; hence the required data rate is achievable. If the trivial example of the paper tape recording is pursued further, suppose that the tape upon which the voltages were written became crumpled up. If it were smoothed out, the numbers would still be legible, and could be copied without error to a new piece of paper. By comparison, if a photograph is crumpled up, it will look like a crumpled-up photograph for evermore.

Large disturbances of the recording, such as dropout or severe interference, may cause flux changes to be missed, or simulate ones which did not exist. The result is that some of the numbers recorded will be incorrect. In numerical systems, provision of an error-correction system is feasible; in analog systems it is not. In PCM systems, corruption of high-order bits can cause severe disturbance of the video waveform, and a properly engineered error-correction system is absolutely essential to return the corrupted numbers to their original value. It is probably true to say that, without error-correction systems, digital video recording would not be technically feasible.

In the digital domain, signals can be easily conveyed and stored in electronic circuitry. Unavoidable speed variations in recorders cause the numbers to appear at a fluctuating rate. The use of a temporary store allows those numbers to be read out at constant rate, a process known as timebase correction. In this way, timebase errors and chroma phase errors can be eliminated. The rock-solid vector display of a composite digital recorder is uncanny at first. The ease with which delay is provided in the digital domain also allows linear phase filters and interpolators to be created in a straightforward manner, because the group-delay problems of analog filters cannot occur. Filters can be constructed which work with mathematical precision and freedom from component drift and with a response that can be easily changed. These filters will be useful for constructing oversampling video converters and in the colour processors necessary to allow composite recordings to be played at variable speeds.

The main advantages of digital video recording can be summarized as follows.

- The quality of a digital video recording is independent of the head and medium in a properly engineered system. Frequency response, linearity and noise are determined only by the quality of the conversion processes. There is complete freedom from moiré and other analog recording artifacts. The independence of the quality from the medium also means that a recorder will not display different picture or audio quality if different brands of tape are used, provided that they all have acceptable error rates.
- A digital recording is no more than a series of numbers, and hence can be copied through an indefinite number of generations without degradation. This implies that the life of a recording can be truly indefinite, because even if the medium begins to decay physically, the sample values can be copied to a new medium with no loss of information.
- The use of error-correction techniques eliminates the effects of dropout whereas analog machines can only use concealment.
- The use of timebase correction on replay eliminates timebase error and with it chroma phase error, and can be further used to synchronize more than one machine to sample accuracy, eliminating the need for lengthy timing-in processes.
- The use of digital recording and error correction allows the signal-to-noise ratio of the recorded tracks to be relatively poor. The tracks can be narrow and hence achieve a saving in tape consumption despite the greater bandwidth. Professional analog recorders must use wide tape tracks to give extremely good first-generation quality which then allows several generations of dubbing.
- The advantages listed also apply to the audio channels of video recorders. It is natural to digitize both audio and video, so that much common circuitry can be used. The audio quality possible in a DVTR then exceeds that of a professional analog audio recorder, which could not be said for analog VTRs. The rigid timebase control prevents phase errors between channels, which is particularly important with the advent of stereo audio in television.

1.2 History of D-2

The D-2 format began life as a development by Ampex. Some years ago, most manufacturers were building experimental digital recorders,[2] and many of these were composite machines. It was then argued that component recording was superior to composite recording, which is not in doubt, and that the immediate need was for a digital format which recorded components, which does not actually follow. This led to the development of the D-1 format, which has been described elsewhere.[3,4]

At the same time, Ampex were faced with the need to find a replacement for the venerable ACR-25 cartridge machine. This unit allows automated broadcasting of commercials and short spots using cassettes mechanically retrieved and automatically threaded. The Quadruplex format is used, on 2 inch tape, now well obsolete. The newer C-format could not be adapted to cassette operation because the extended scanner wrap needed to give field per scan virtually precluded mechanical threading.

It was clear that the replacement would have to be digital, but it was also clear that the purpose of the cart machine is to play back to air in a broadcast environment, and this demands composite recording in the majority of instances.

The composite digital recording research at Ampex was to turn into the ACR-225 automated broadcasting system.

Although the D-1 format was capable of higher technical quality, this would remain academic for those who could not afford to replace their entire installation with component equipment. What was needed by broadcasters was essentially a digital replacement for the 1 inch analog composite recorders.

The format of the Ampex cart machine joined forces with the composite digital research which had been taking place at Sony and grew into that role, becoming known as D-2.[5]

There is a world of difference between a laboratory curiosity which records digital video to prove a principle and a machine which has the facilities necessary to be used in a television system. Now that users have become accustomed to the facilities offered by C-format, no proposed replacement will succeed if it does not also offer all of those facilities. Even if it does so, it still will not succeed unless it offers a tangible advantage over any predecessor. Sheer technical quality is obviously an advantage of digital recording, but the C-format has been refined to such a high standard that it is virtually impossible for the eye to distinguish a first-generation C-format recording on a top-flight machine from a D-2 recording. In successive generations, the digital recorder will maintain the same quality, whereas the analog quality will inevitably fall.

If it is to succeed, D-2 will do so partly because it takes nothing away operationally and gives a small quality advantage, but mainly because it will cost less to run and maintain.

D-1 gave a quality advantage, but it took away operational features and cost more to run. It should be realized that D-2 is a more recent technology than D-1 and is in some respects developed from it. The application is quite different, and so a direct comparison is neither fair nor relevant. Both are tremendous technical achievements.

1.3 A typical D-2 machine outlined

An outline of a representative D-2 recorder follows to illustrate some general principles, and to put the major chapters of this book into perspective.

Figure 1.3 shows how a digital recording of composite video can be made, starting with the sampling and quantizing processes explained in Chapter 2. Once in

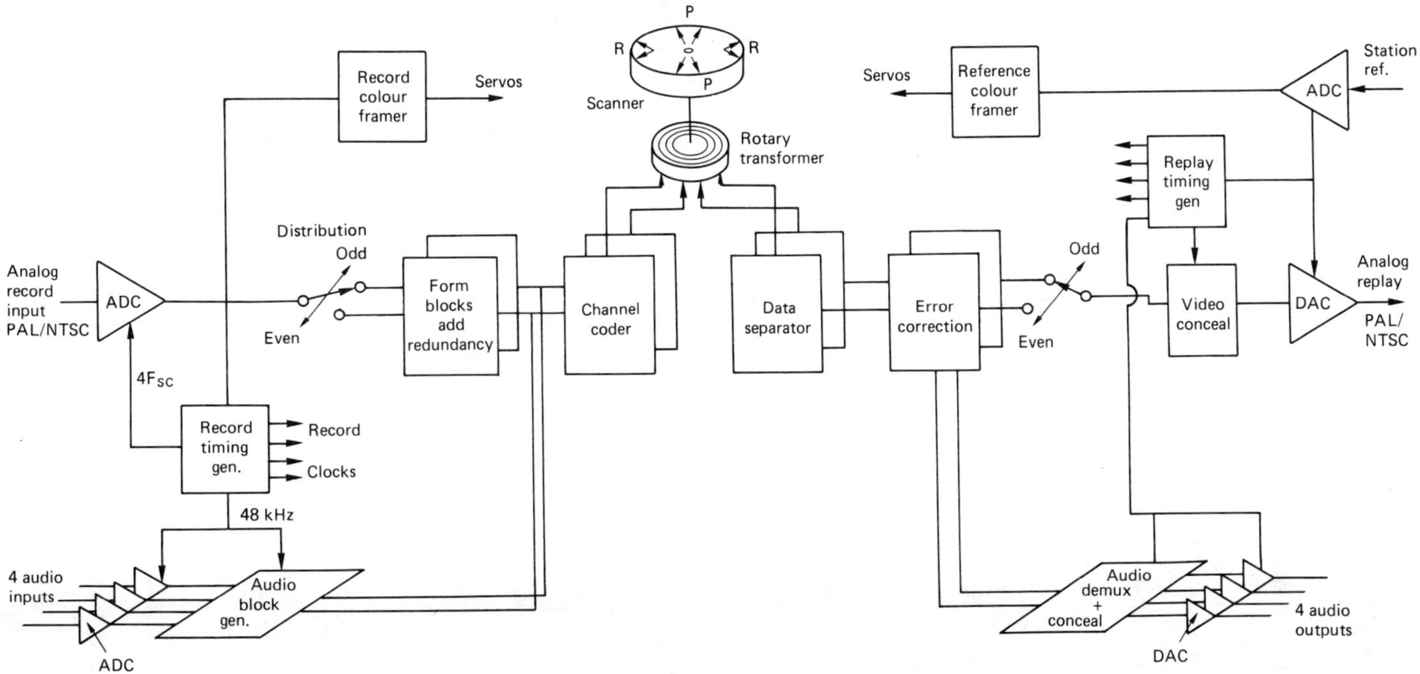

Figure 1.3 Analog video is sampled at $4F_{sc}$, and audio at 48 kHz. Samples are distributed between two channels, formed into blocks, and coded for recording. The audio blocks are timeshared with video through a common tape channel consisting of a channel coder, the rotating head assembly and the data separator. On replay, timing is generated by station reference. Errors which cannot be corrected are concealed before conversion

digital form, the data are formed into blocks, and the encoding section of the error-correction system supplies additional bits designed to protect the data against errors. These blocks are then converted into some form of channel code which combines the data with clock information so that it is possible to identify how many bits were recorded even if several adjacent bits are identical. Chapter 4 deals with the subject of digital recording.

The coded data are recorded on tape by a rotary head. Upon replaying the recording of the hypothetical machine of Figure 1.3, the errors caused by various mechanisms will be detected and corrected or concealed using the extra bits appended during the encoding process. Chapter 5 treats the subject of error correction as comprehensively as possible without becoming lost in the mathematics.

The replayed digital recording may be output directly if the recorder is playing back colour framed to reference, but if this is not the case, a colour processor will be necessary to restore the subcarrier sequence. Manipulations of digital video, such as gain control, colour processing and oversampling filters, are introduced in Chapter 3.

In addition to analog video connections, the digital recorder will also have digital inputs and outputs, and it is necessary to provide some standard interconnection to allow connection between units of different manufacture. Chapter 3 also details digital communications between equipment.

Chapter 6 details the mechanisms necessary for rotary-head tape recording and contrasts the different approaches used in Ampex and Sony equipment. Chapter 7 deals with the signal systems necessary to realize the D-2 format.

Last but not least, the digital audio recording stages of D-2 are treated in some detail in Chapter 8.

1.4 Disadvantages

In this comparison of digital and analog technologies, the emphasis so far has been on the advantages of digital video. In the interests of fairness and truth we must also look at some of the problems of digital video.

(1) Digital recorders must use thin tape with a fine surface finish to allow the very-short-wavelength recordings needed. Contamination must be avoided at all costs. The tape tracks are very narrow, and greater mechanical stability is demanded from tape and transport to achieve reliable interchange. Digital recorders may be at a disadvantage in harsh environments.

(2) The signal quality of a digital format is chosen irrevocably at the design stage, and no improvement is then possible without fundamental change. This contrasts with the process of gradual refinement observed in analog recorders.

(3) Inevitably, digital recorders are more complex than analog recorders, and this has implications for repair. Faultfinding with traditional video techniques is impossible, and digital machines need computer-related approaches using diagnostic programs and signature analysis. It should not be inferred that digital machines are less reliable than analog machines; indeed digital circuitry will have fewer if any adjustments, reducing the need for periodic alignments.

(4) Digital filters are totally repeatable, since there are no adjustments. If a filter contains a small ripple in its response, every filter to that design will have exactly

the same ripple, at exactly the same frequency. Upon passing the signal through the same unit over several generations, the ripple will build up. Filter design must take this into account. Although a digital recorder may introduce no degradations to the signal, this is not quite true of the digital processing it may contain.

References

1. GINSBURG, C. P. Comprehensive description of the Ampex video tape recorder. *SMPTE J.*, **66**, 177–182 (1957)
2. BALDWIN, J. L. E. Digital television recording – history and background. *SMPTE J.*, **95**, 1206–1214 (1986)
3. WATKINSON, J. R. *The Art of Digital Video*, Chapter 10. Sevenoaks: Focal Press (1990)
4. GREGORY, S. *Introduction to the 4:2:2 Digital Video Tape Recorder.* London: Pentech Press (1988)
5. Ampex Corporation, Ampex digital format for video and audio tape recording of composite video signals using 19 mm type D-2 cassette. Redwood City, CA (1988)

Chapter 2

Composite digital conversion

The strength of a digital recorder lies in storage without degradation. If video is to take advantage of digital techniques, then means must be found to express the colour and movement of a real-life scene as a series of numbers. Once in the digital domain, degradations can be controlled, and so the conversion processes to and from the digital domain become the main sources of degradation. In the interests of quality, these processes deserve a great deal of attention, and will form the main subjects of this chapter.

2.1 The discrete picture

At the moment it is not possible to convert directly from an image to the digital domain or vice versa. Generally cameras and displays are analog devices, and there is always an intermediate analog video stage at each end of the digital video chain. In the case of D-2, the digital domain is entered via an analog process which is a broadcast standard, namely PAL, PAL-M or NTSC, and the fundamentals of these must be explored here.

To a human observer, a real scene appears continuous, whereas digital systems are, by definition, discrete, as are many aspects of analog television systems, particularly CCD cameras. At its most fundamental, a monochrome digital image is an array of points, generally rectangular, at which the brightness is stored as a number. The points are known as picture elements, generally abbreviated to pixels, although sometimes the abbreviation is more savage and they are known as pels. As shown in Figure 2.1, the array will generally be arranged with a regular spacing between pixels, which are in rows and columns. By placing the pixels close together, it is hoped that the observer will perceive a continuous image. Obviously the finer the pixel spacing, the greater the resolution of the picture will be, but the amount of digital data needed to store one picture will increase as the square of the resolution, and with it the costs. In this chapter the theoretical connection between resolution and pixel spacing will be examined, along with the reasons why it is seldom achieved in practice.

If it is desired to convey a coloured image, then each pixel is no longer a scalar brightness value, but a vector which describes in some way the brightness, hue and saturation of that point in the picture. The method which is in almost universal use at the camera is to use one sensor for each primary colour. Each sensor is fed by light of a different colour from an optical splitter block. Again the amount of data

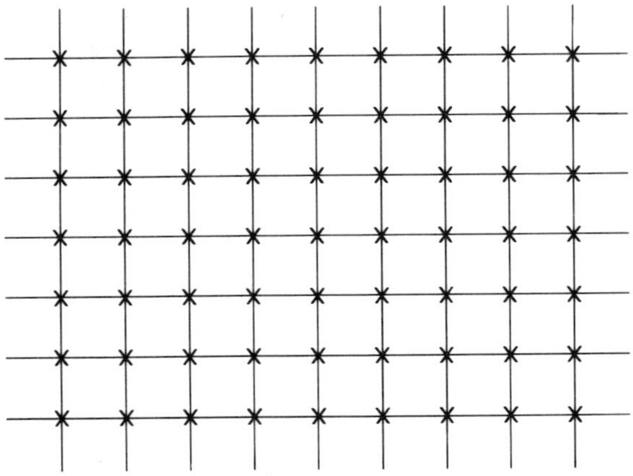

Figure 2.1 A picture can be stored digitally by representing the brightness at each of the above points by a binary number. For a colour picture each point becomes a vector and has to describe the brightness, hue and saturation of that part of the picture. Samples are usually but not always formed into regular arrays of rows and columns, and it is most efficient if the horizontal and vertical spacing are the same

has increased, and with it the difficulty of implementation. It is not possible to represent accurately every colour which occurs naturally, owing to the current lack of suitable display technology, and so there are further shortcomings in the reproduced image.

In order to produce moving pictures, the current approach is simply to provide a mechanism where the value of every pixel can be updated periodically. This effectively results in a three-dimensional array, where two of the axes are spatial and the third is temporal.

2.2 Sampling

Sampling is a way of expressing a continuous phenomenon by means of a periodic measurement. In television systems the input image which falls on the camera sensor will be continuous in time and continuous in two spatial dimensions corresponding to the height and width of the sensor. All three of these continuous dimensions will be sampled, as has been seen. There is a direct connection between the concept of temporal sampling, where the input changes with respect to time at some frequency and is sampled at some other frequency, and spatial sampling, where an image changes a given number of times per unit distance and is sampled at some other number of times per unit distance. The connection between the two is the process of scanning. Temporal frequency can be obtained by multiplying spatial frequency by the speed of the scan. Figure 2.2 shows a hypothetical image sensor which has 1000 discrete sensors across a width of 1 centimetre. The spatial sampling rate of this sensor is thus 1000 per centimetre. If the sensors are measured sequentially during a scan which takes 1 millisecond to go across the 1 centimetre width, the result will be a temporal sampling rate of 1 MHz.

Figure 2.3(a) shows a waveform which has been sampled at a relatively high frequency, and intuitively it is clear that at this sampling rate the waveform has

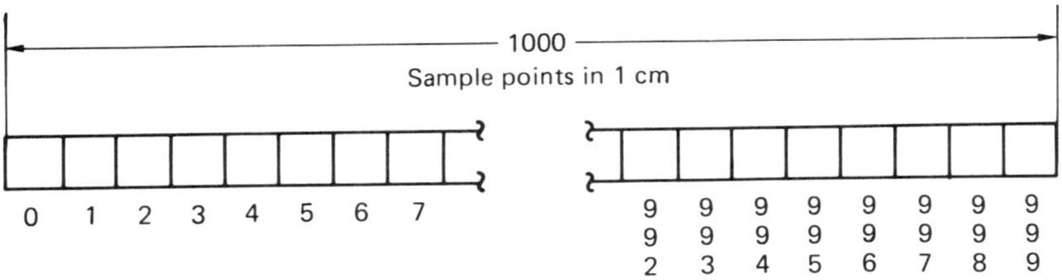

Figure 2.2 If the above spatial sampling arrangement of 1000 points per cm is scanned in one millisecond, the sampling rate will become 1 megahertz

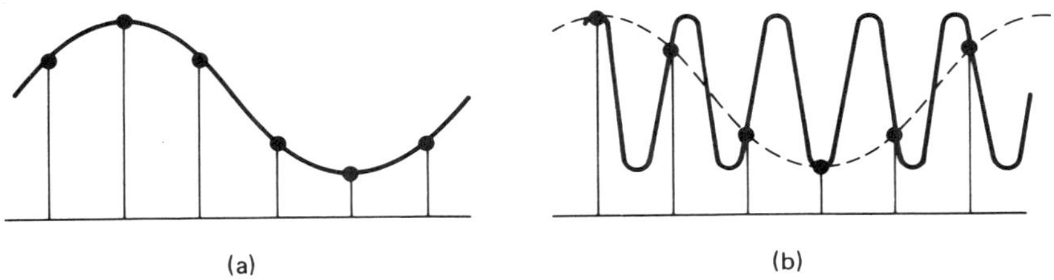

Figure 2.3 At (a), the sampling is adequate to reconstruct the original signal. At (b) the sampling rate is inadequate, and reconstruction produces the wrong waveform (dotted). Aliasing has taken place

been preserved. Figure 2.3(b) shows an example where the sampling rate is clearly inadequate. A phenomenon called *aliasing* has taken place, where a signal of a new frequency has replaced the original. Aliasing is commonly seen in film and television material, particularly with such subjects as stagecoach wheels. The spoke passing frequency of the coach wheels exceeds the ability of the frame rate of the film camera (24 Hz) to follow them, and so the wheels can appear to be turning slowly or even backwards.

Obviously aliasing can be prevented by sampling at a high frequency, but in practical systems bandwidth has to be paid for, and sampling at an unnecessarily high rate will cause problems when the information needs to be stored or transmitted. It is necessary to establish the precise connection between sampling rate and bandwidth.

Figure 2.4 shows that sampling is essentially a process of modulation. The sampling waveform, which is a steady stream of narrow impulses, is amplitude modulated, or multiplied, by the baseband signal. Since the sampling waveform is impulsive rather than sinusoidal, its spectrum consists of the sampling rate and all of its harmonics. Multiplication causes sidebands, and these can be seen above and below the sampling frequency and each of the harmonics. In order to return to the continuous baseband signal it is necessary to use a low-pass filter which rejects everything but the original spectrum.

Figure 2.4 also shows the result when an excessive baseband spectrum enters a sampled system. The lower sideband of the sampling rate now overlaps the baseband, and no filter can determine whether energy in this area belongs to the baseband or the sideband. In particular, frequencies above half the sampling rate are heterodyned down to a new position the same distance below.

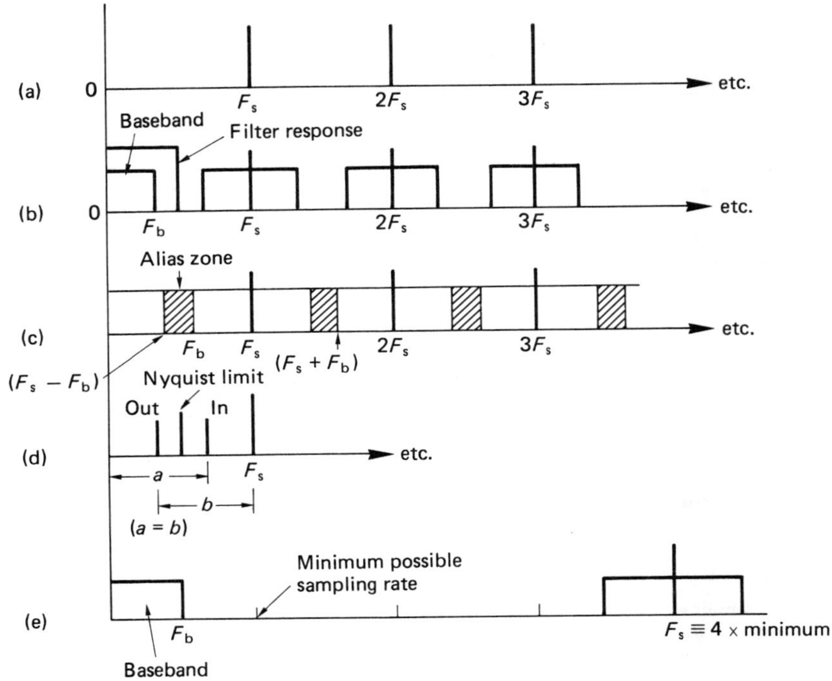

Figure 2.4 (a) Spectrum of sampling pulses. (b) Spectrum of samples. (c) Aliasing due to sideband overlap. (d) Beat-frequency production. (e) Oversampling

Figure 2.4 shows that in the extreme case, a rectangular baseband spectrum of one-half the sampling rate could be accommodated, and this is the conclusion of Shannon's sampling theory, also discovered in the Soviet Union by Kotelnikov.

Figure 2.5 shows how it is possible to return to the continuous analog domain using a low-pass filter. The impulse response of an LPF which cuts off at one-half the sampling frequency is a $\sin x/x$ curve where the zero crossings are separated by the sampling period. This ensures that samples do not interfere with each other, since at the location of a given sample, the contributions from all other samples will be zero. The $\sin x/x$ impulses from all of the samples combine to produce a continuous waveform.

In sampled systems a low-pass filter is necessary before the sampling stage to prevent frequencies above half of the sampling rate entering the system. This will be called an *anti-aliasing* filter. A second identical filter will also be needed upon returning to the continuous domain, in order to reject the unwanted sidebands of the sampling. This will be called a *reconstruction* or *anti-image* filter.

The eye has a temporal filtering effect which is known as persistence of vision, and the phosphors of CRTs continue to emit light after the electron beam has passed. It might be thought that a temporal sampling rate (frame rate in progressive scan systems) which was high enough to make flicker invisible would be adequate, but in fact this is not the case.

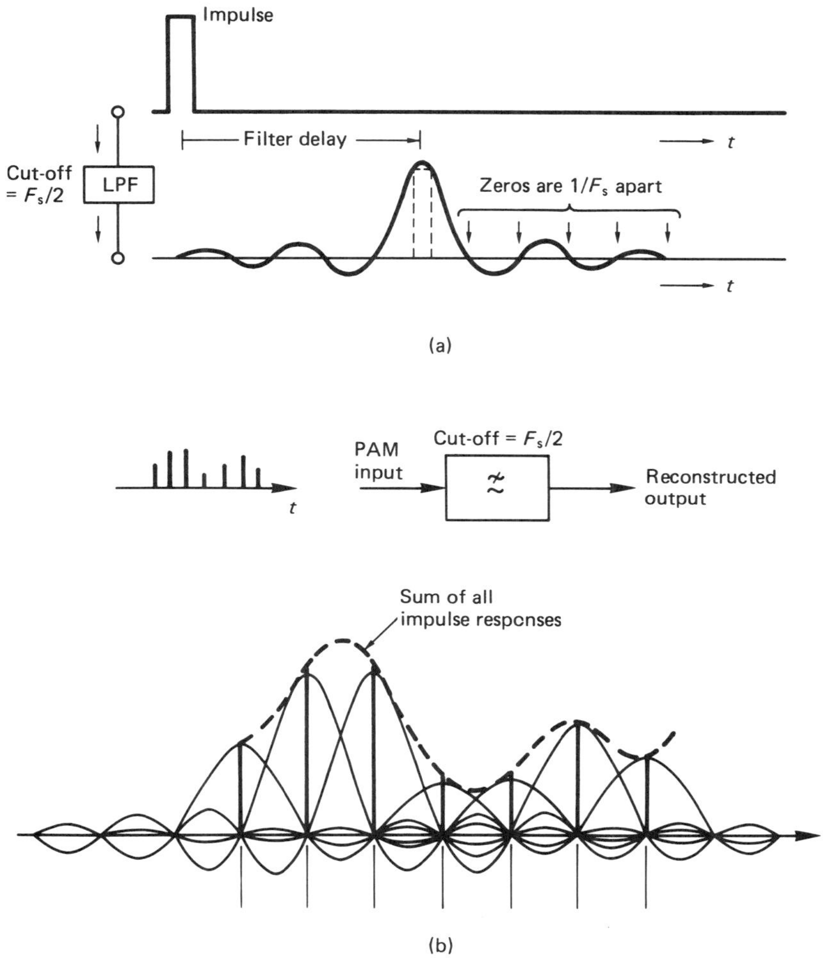

Figure 2.5 The impulse response of a low-pass filter which cuts off at $F_s/2$ has zeros at $1/F_s$ spacing which correspond to the position of adjacent samples, as shown in (b). The output will be a signal which has the value of each sample at the sample instant, but with smooth transitions from sample to sample

Since conventional analog video systems sample temporally and vertically, there will be two opportunities for aliasing to take place. Digital video is a three-dimensional sampling system, so aliasing can potentially occur in all three.

Whilst the maximum information capacity predicted by theory would be reached with filters which cut off at exactly half the sampling rate, such filters cannot be realized.

If an electrical filter is being used on a temporal signal, as the slope of the filter approaches infinity, so the group delay of the filter also approaches infinity. Realizable filters will have a finite slope, and so the sampling rate will always need to be raised above twice the baseband frequency by some factor which depends on the characteristics of the filter.

Where the input is a spatial spectrum, the filter must also be spatial rather than temporal, and so it cannot readily be implemented with electronics. However, the same restrictions apply to the achievable cut-off rate.

2.3 Aperture effect

The foregoing has discussed the effect of infinitely small sampling pulses, which are never achieved in practice. Whenever the sampling pulses at the input of a system, or the output pulses, have finite duration, the aperture effect is observed. This will also be seen in spatial sampling, where the sampling point has finite dimensions.

Figure 2.6(a) shows a sampling system where the impulses have grown until they are as wide as the sample period. They can be considered to consist of an infinite number of narrow pulses. When the sample polarity alternates as shown, many of the positive pulses are cancelled by the negative pulses, and a loss occurs which clearly will not be observed at low frequencies. The resultant frequency response is shown in Figure 2.6(b) where it will be seen to be a $\sin x/x$ curve with the first null at the sampling rate. This corresponds to a signal level of 0.64 of the LF level at half the sampling rate. Figure 2.6(c) shows that there will be a family of responses according to the impulse width, becoming more linear as the impulse narrows.

The aperture effect will show up in many aspects of television. Lenses have finite modulation transfer functions, such that a very small object becomes spread in the image. The image sensor will also have a finite aperture function.

In tube cameras, the beam will have a finite radius, and will not necessarily have a uniform energy distribution across its diameter. In CCD cameras, the sensor is split into elements which may almost touch in some cases. The element integrates light falling on its surface, and so will have a rectangular aperture. In both cases there will be a roll-off of higher spatial frequencies.

It is highly desirable to prevent spatial aliasing, since the result is visually irritating. In tube cameras the aliasing will be in the vertical dimension only, since the horizontal dimension is continuously scanned. Such cameras seldom attempt to prevent vertical aliasing. CCD sensors can, however, alias in both horizontal and vertical dimensions, and so an anti-aliasing optical filter is generally fitted between the lens and the sensor. This takes the form of a plate which diffuses the image formed by the lens. Such a device can never have a sharp cut-off nor will the aperture be rectangular. The aperture of the anti-aliasing plate is in series with the aperture effect of the CCD elements, and the combination of the two effectively prevents spatial aliasing, and generally gives a good balance between horizontal and vertical resolution, allowing the picture a natural appearance.

Conventional tube cameras generally have better horizontal resolution, and produce vertical aliasing which has a similar spatial frequency to real picture information, but which lacks realism. With a conventional approach, there are effectively two choices. If aliasing is permitted, the theoretical information rate of the system can be approached. If aliasing is prevented, realizable anti-aliasing filters cannot sharp cut, and the information conveyed is below system capacity.

These considerations also apply at the television display. The display must filter out spatial frequencies above one-half the sampling rate. In a conventional CRT this means that an optical filter should be fitted in front of the screen to render the raster invisible. Again the aperture of a simply realizable filter would attenuate too much of the wanted spectrum, and so the technique is not used. The technique of

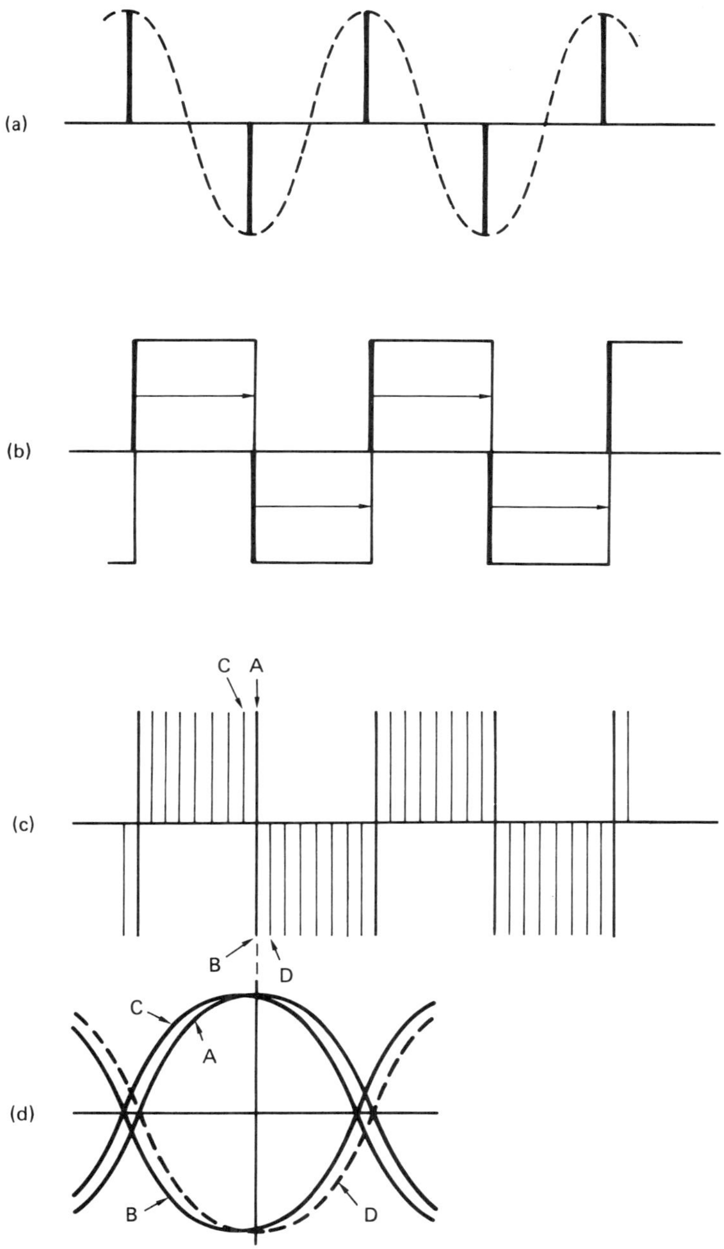

Figure 2.6(a) The mechanism of aperture effect which causes high-frequency losses when impulse has finite period. (a) Maximum-frequency sine wave in samples of zero duration. (b) Zero-order hold version of above (100% aperture). (c) Consider rectangular pulse as infinite number of delta functions. (d) Result of low-pass filtering some of above impulses. A cancels B completely. C cancels D partially

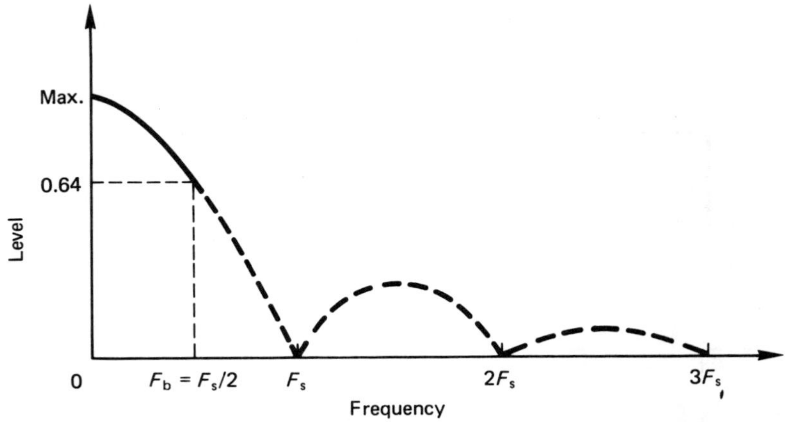

Figure 2.6(b) Frequency response with 100% aperture nulls at multiples of sampling rate. Area of interest is up to half sampling rate

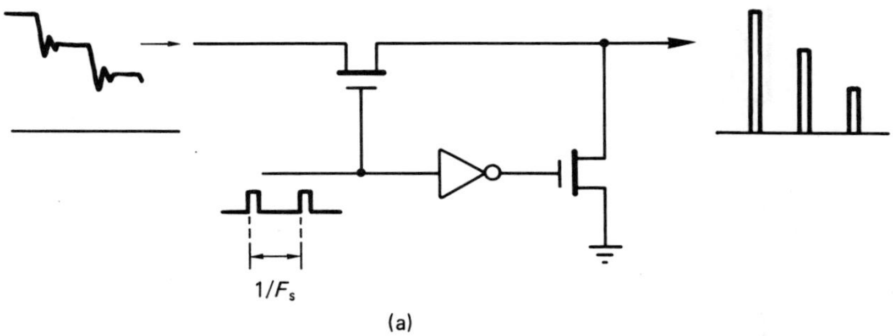

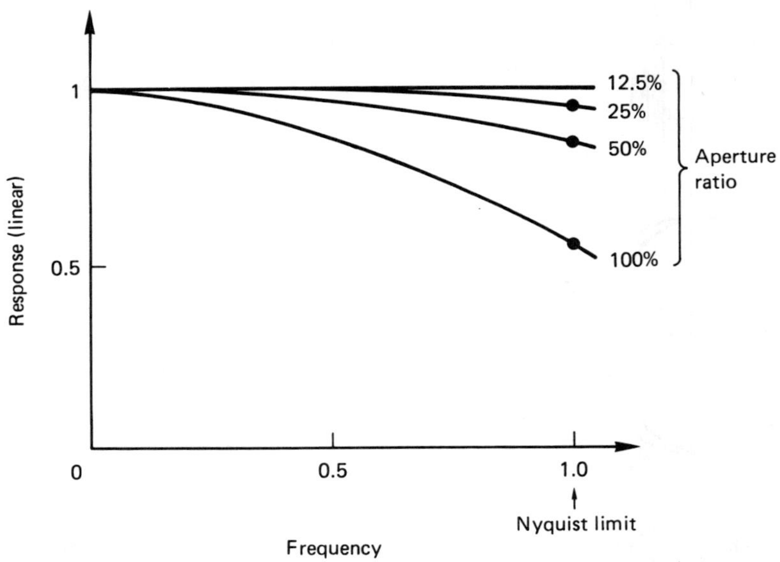

Figure 2.6(c) (a) Resampling circuit eliminates transients and reduces aperture ratio. (b) Response of various aperture ratios

spot wobble was most effective in reducing raster visibility in monochrome television sets without affecting horizontal resolution, and its neglect remains a mystery.

2.4 Kell effect

As noted, in conventional tube cameras and CRTs the horizontal dimension is continuous, whereas the vertical dimension is sampled. The aperture effect means that the vertical resolution in real systems will be less than sampling theory permits, and to obtain equal horizontal and vertical resolutions a greater number of lines is necessary. The magnitude of the increase is described by the so-called Kell factor,[1] although the term factor is a misnomer since it can have a range of values depending on the apertures in use and the methods used to measure resolution.[2] In digital video, sampling takes place in horizontal and vertical dimensions, and the Kell parameter becomes unnecessary. The outputs of digital systems will, however, be displayed on raster scan CRTs, and the Kell parameter of the display will then be effectively in series with the other system constraints.

2.5 Interlace

The terrestrial broadcast standards PAL and NTSC use 2:1 interlace. Figure 2.7(a) shows that, in such a system, there is an odd number of lines in a frame, and the frame is split into two fields. The first field begins with a whole line and ends with a half line, and the second field begins with a half line, which allows it to interleave

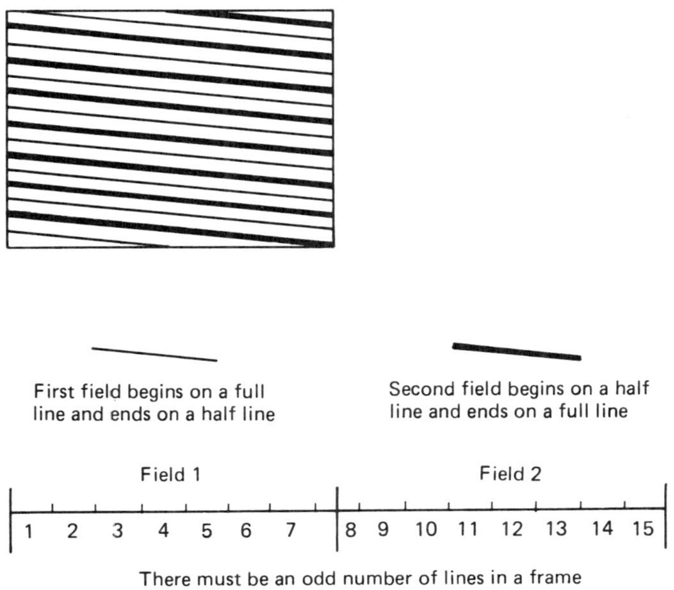

Figure 2.7(a) 2:1 Interlace

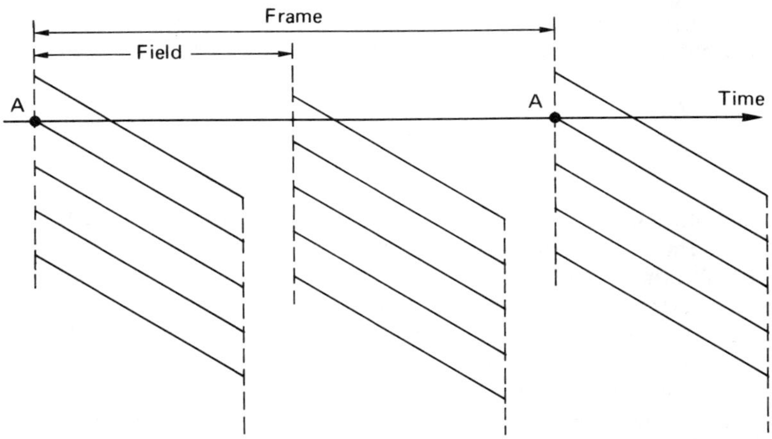

Figure 2.7(b) In an interlaced system, a given point A is only refreshed at frame rate, causing twitter on fine vertical detail

spatially with the first field. The field rate is intended to determine the flicker frequency, whereas the frame rate determines the bandwidth needed, which is thus halved along with the information rate. Information theory tells us that halving the information rate must reduce quality, and so the saving in bandwidth is accompanied by a variety of effects. Figure 2.7(b) shows the spatial/temporal sampling points in a 2:1 interlaced system. If an object has a sharp horizontal edge, it will be present in one field but not in the next. The refresh rate of the edge will be reduced to frame rate, 25 Hz or 30 Hz, and becomes visible as twitter. Whilst the vertical resolution of a test card is maintained with interlace, apart from the twitter noted, the ability of an interlaced standard to convey motion is halved. Tests[3] have shown that, all other things being equal, a 2:1 interlaced picture has to be viewed from almost twice as far away as a progressive scan picture in order to render the artifacts of scanning invisible. Unfortunately PAL and NTSC were all designed to have equal vertical and horizontal resolution using a Kell parameter which was only valid for progressive scan. The horizontal resolution is wasted and efficiency in the information theory sense is poor. In the light of what is now known, interlace causes degradation roughly proportional to bandwidth reduction, and so should not be considered for any future standards.

2.6 Types of video

Whilst it is easy to theorize about optimal digital systems, having nice, neat arrays of pixels, the D-2 recorder simply has to accept the standard PAL or NTSC signals from the real, imperfect, world of analog video.

Figure 2.8 shows some of the basic types of analog colour video. Each of these types can, of course, exist in a variety of line standards. Since practical colour cameras generally have three separate sensors, one for each primary colour, an RGB system will exist at some stage in the internal workings of the camera, even if it does not emerge in that form. RGB consists of three parallel signals each having

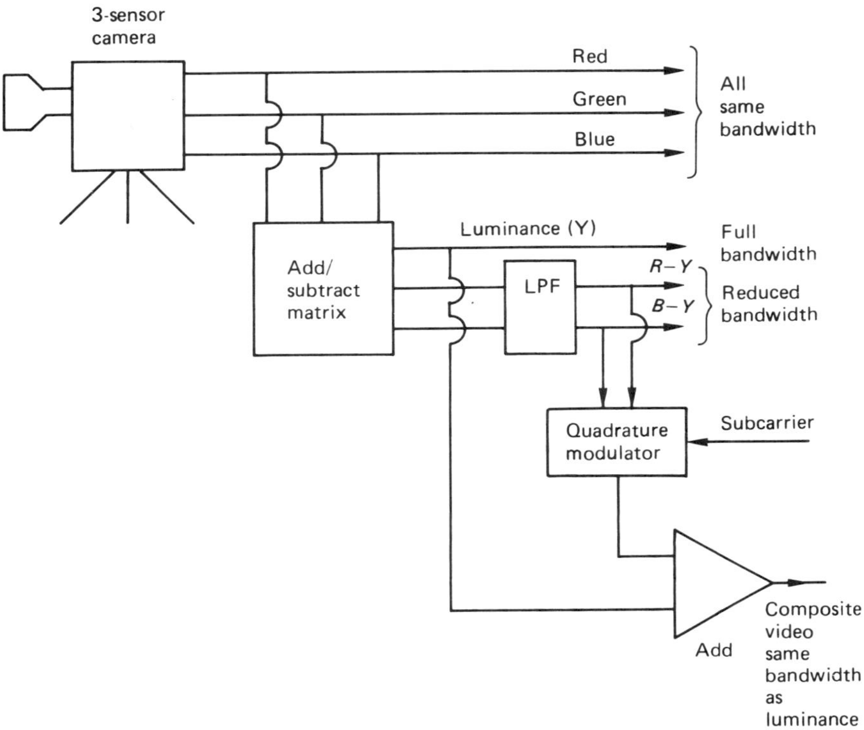

Figure 2.8 The major types of analog video. Red, green and blue signals emerge from the camera sensors, needing full bandwidth. If a luminance signal is obtained by a weighted sum of *R, G* and *B*, it will need full bandwidth, but the colour difference signals *R–Y* and *B–Y* need less bandwidth. Combining *R–Y* and *B–Y* into a subcarrier modulation scheme allows colour transmission in the same bandwidth as monochrome

the same spectrum and is used where the highest accuracy is needed, often for production of still pictures. Examples of this are paint systems and in computer-aided design (CAD) displays. RGB is seldom used for real-time video recording, and there is no standard RGB recording format for post-production or broadcast, although the IBA did build an experimental analog RGB recorder. As the red, green and blue signals directly represent part of the image, this approach is known as component video.

Some saving of bandwidth can be obtained by using colour difference working. The human eye relies on brightness to convey detail, and much less resolution is needed in the colour information. *R, G* and *B* are matrixed together to form a luminance (and monochrome-compatible) signal *Y* which has full bandwidth. The matrix also produces two colour difference signals, $R-Y$ and $B-Y$, but these do not need the same bandwidth as *Y*: one-half or one-quarter will do depending on the application. Analog colour difference recorders such as Betacam and M II record these signals separately. The D-1 format records 525/60 or 625/50 colour difference signals digitally. In casual parlance, colour difference formats are often called component formats to distinguish them from composite formats.

For colour television broadcast in a single channel, the PAL and NTSC systems interleave into the spectrum of a monochrome signal a subcarrier which carries two colour difference signals of restricted bandwidth. The subcarrier is intended to be invisible on the screen of a monochrome television set. A subcarrier-based colour system is generally referred to as composite video, and the modulated subcarrier is called chroma.

PAL uses the scanning standard of 625/50, whereas NTSC scans at 525/59.94, although the two colour systems have a few things in common. Analog composite recorders include the B- and C-formats, and the D-2 format records PAL or NTSC digitally. A D-2 machine having analog inputs and outputs can replace a C-format machine directly.

2.7 Spectra of composite video

In video, sampling takes place in more than one dimension at once, and the resulting spectra can be complicated. It is well worth looking at the sampling spectra of analog video before the additional horizontal sampling of digital video takes place.

Some examples will now be given for different standards. Figure 2.9(a) shows the spectrum due to sampling images at the US frame rate of 30 Hz. If the camera integrates incoming light over the frame period, as most do, there will be a temporal aperture effect which slopes off the spectrum. This spectrum is then itself sampled by the line scan process at 15 750 Hz, resulting in the combined spectrum of Figure 2.9(b). This spectrum will repeat up to the allowable bandwidth of the video signal, as shown in Figure 2.9(c). For component working, a camera will produce three signals having the same spectrum. In colour difference working one wideband luminance signal having the same spectrum as the components is accompanied by two colour difference signals which have the same spectral structure, but whose overall bandwidth is less, as in Figure 2.9(d).

For monochrome-compatible colour broadcasting, advantage is taken of the gaps in the spectra. In NTSC (National Television Systems Committee), the two colour difference signals are used to generate reduced bandwidth signals which alternately phase-modulate a subcarrier. The subcarrier is suppressed to produce a chroma signal.

In order to allow the maximum perceived resolution with minimum bandwidth, the modulator phase was shifted slightly to give a pair of signals known as I and Q (in-phase and quadrature). This placed the Q signal on the colour diagram axis to which the eye is least sensitive.

The chroma is given by:

$$I \sin F_{sc} + Q \cos F_{sc}$$

such that, at 90° intervals, the chroma waveform represents I, Q, $-I$, $-Q$ repeatedly. If the receiver samples the chroma at the same intervals, the two baseband colour difference signals can be restored.

The chroma modulation process takes the spectrum of the colour difference signals and produces upper and lower sidebands around the frequency of subcarrier. Since both colour and luminance signals have gaps in their spectra at multiples of line rate, it follows that if the subcarrier frequency is chosen to fall

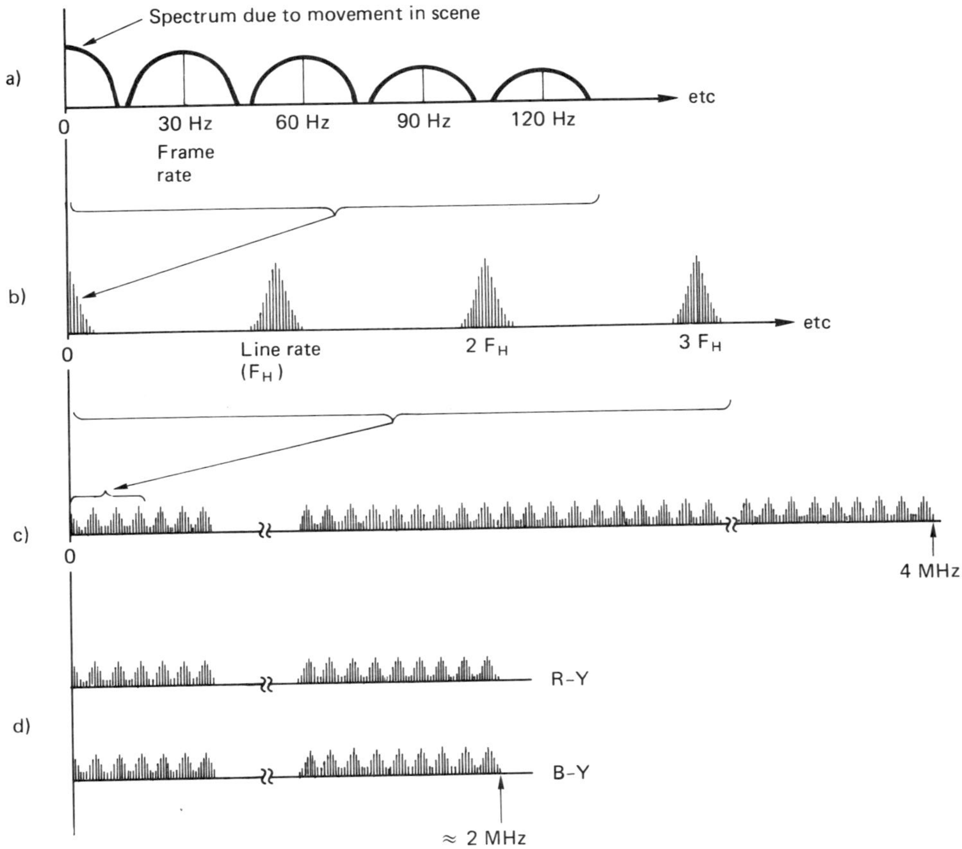

Figure 2.9 At (a) the spectrum of the scene movement is repeated in the upper and lower sidebands of the frame sampling and its harmonics. At (b) the spectrum of (a) is sampled vertically by the line rate and appears as lines with frame-rate spacing around multiples of line rate. This spectrum continues to fill the system bandwidth as at (c) which could be a luminance, *R, G,* or *B* signal. In colour difference working, the luminance signal at (c) will be accompanied by two colour difference signals of approximately half the bandwidth, as at (d)

half-way between two of the harmonics of line rate, i.e. it is an odd multiple of half line rate, then the two spectra will interleave and share the same spectrum. In NTSC the subcarrier frequency is 227.5 times line rate. Figure 2.10 shows that this frequency means that on successive lines the subcarrier will be phase inverted. The effect of the chroma added to luminance is to make the luminance alternately too dark or too bright. The phase inversion causes this effect to cancel over pairs of lines.

The existence of line pairs means that two frames or four fields must elapse before the same relationship between line pairs and frame sync repeats. When editing NTSC recordings, this four-field sequence must not be broken. If this rule is not followed, it is possible for a sudden inversion of subcarrier phase to occur at the edit point, which renders the signal unbroadcastable.

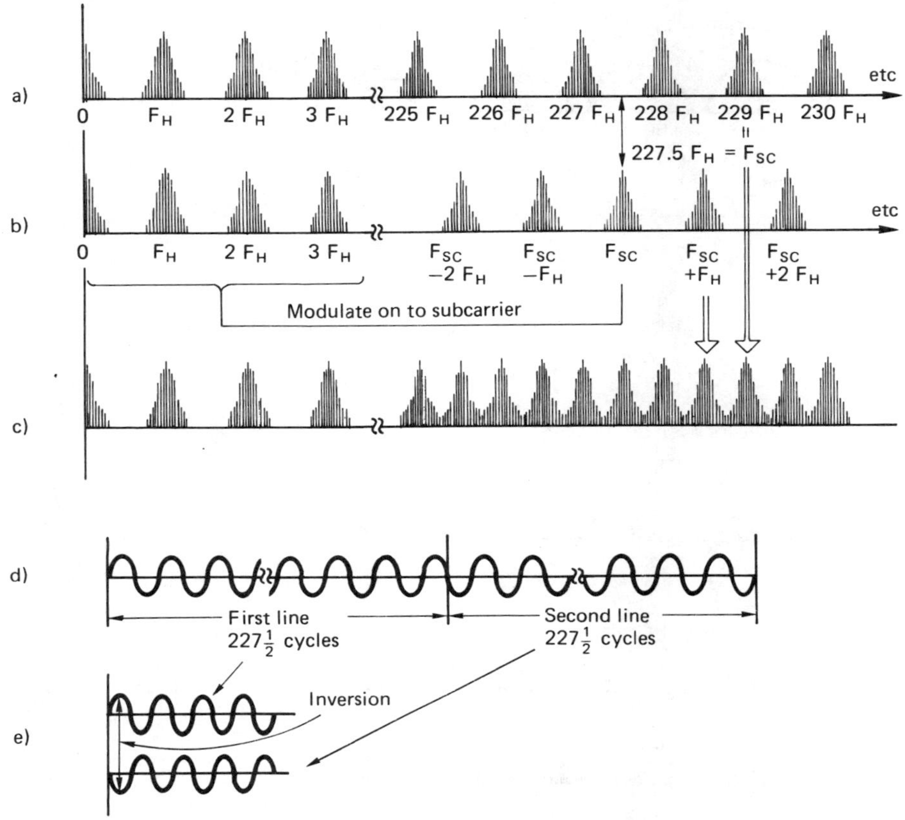

Figure 2.10 In NTSC, the luminance or monochrome spectrum is shown at (a). In (b) the colour difference spectrum is shifted up by modulating a subcarrier at 227.5 times line frequency. The spectra of the resulting chroma and the original luminance will then mesh as at (c). The half-angle offset in F_{sc} results in a two-line sequence (d) where the first line ends on a half cycle, and the second line chroma will be inverted with respect to the first as at (e)

As the chroma of NTSC is suppressed carrier, it is necessary for the receiver to regenerate the carrier frequency so that it knows when to sample the chroma. This is done by broadcasting a portion of subcarrier, known as the burst, between the sync pulse and the active line. A phase-locked loop in the receiver will freewheel between bursts. The half line offset of NTSC subcarrier means that if an oscilloscope is triggered from H-sync, normal and inverted bursts will appear superimposed on the trace.

Whilst the spectral interleaving of NTSC works fine for the video, there was a problem with the sound carrier. To prevent vision interference with the sound, the sound carrier of the 525/60 monochrome system had itself been placed at an odd multiple of half line rate, and so the addition of the chroma signal meant that chroma and sound could mutually interfere. The audio subcarrier frequency was standardized, and any change would have meant modification to every monochrome television set in the United States on the introduction of colour. The

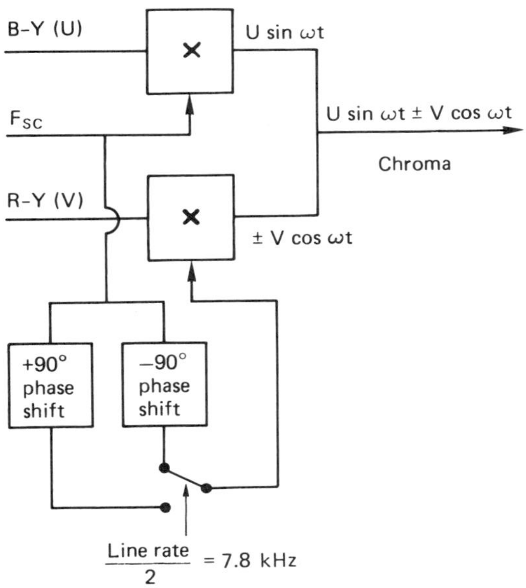

Figure 2.11(a) In PAL the phase of the feed to the *V* modulator is reversed on alternate lines

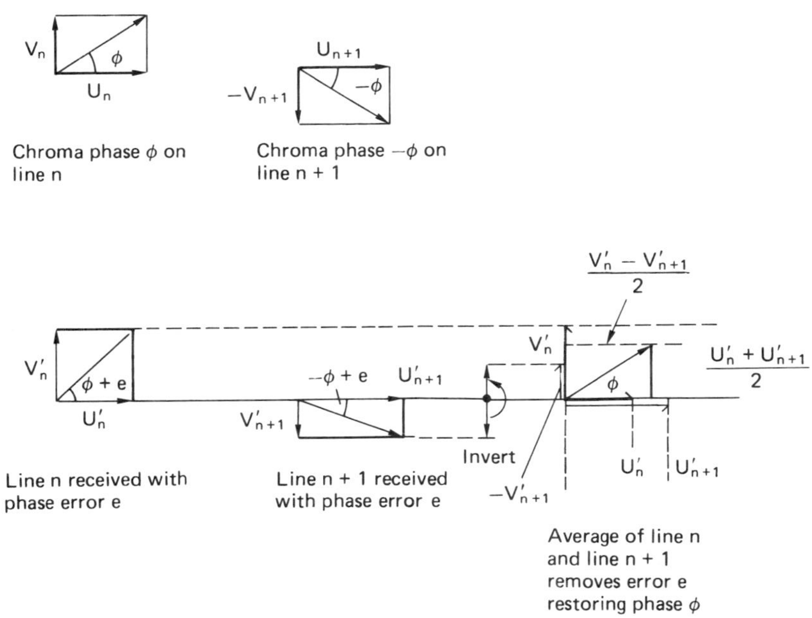

Figure 2.11(b) Top, chroma as transmitted. Bottom, chroma received with phase error. *V* is reversed on second line and the two lines are averaged, eliminating phase error and giving a small saturation error

solution adopted was to contract the spectrum of the video slightly by reducing the frame rate to 29.97 Hz. This made the line rate 15 734.25 Hz and the subcarrier frequency 3.5795 MHz. The small disparity between field rate and 60 Hz was of little consequence when it was introduced, but the development of the video recorder revealed a difficulty in synchronizing NTSC recordings to real time, which was alleviated by the invention of drop-frame timecode. This allows a single count to count real-time seconds and NTSC rate frames at the same time by having dropped frames, which do not exist in the video, but which make the seconds count correct. The use of 29 February in the calendar has a similar effect.

The dependence of NTSC on the phase of subcarrier to convey colour accurately led to some difficulties in multipath reception conditions in areas where tall buildings are found, and to the somewhat unfair redefinition of the system title as Never Twice the Same Colour. In fact NTSC works well in the majority of other locations.

When the PAL (Phase Alternating Line) system was being developed, it was decided that immunity to received phase errors should be a goal. Figure 2.11(a) shows how this was achieved. The two colour difference signals (U and V) are used to quadrature-modulate a subcarrier in a similar way as for NTSC, except that the phase of the V signal is reversed on alternate lines. The receiver must then reinvert the V signal in sympathy. If a phase error occurs in transmission, it will cause the phase of V to lead and lag alternately, as shown in Figure 2.11(b). If the colour difference signals are averaged over two lines, the phase error is eliminated and replaced with a small saturation error which is subjectively much less visible. This does, however, have a fundamental effect on the spectrum.

Figure 2.12(a) shows the spectrum due to sampling at a frame rate of 25 Hz. The line rate sampling of 15 625 Hz then gives the spectrum of Figure 2.12(b). The colour difference signals will have the same spectral structure, but less bandwidth. The $B-Y$ signal becomes U, but the $R-Y$ signal will be inverted on alternate lines to become $\pm V$. The inversion signal is a square wave at half line rate, and this effectively modulates the V signal at half line rate so that the V spectrum is divided by two, and so contains spectral entries at half line rate spacing. When U and V modulate the subcarrier, the spectrum of the chroma also contains energy at multiples of half line rate, and so spectral interleaving with a half-cycle offset of subcarrier frequency will not work, as Figure 2.12(c) shows. The solution is to adopt a subcarrier frequency with a quarter cycle per line offset. Multiplying the line rate by 283¾ allows the luminance and chrominance spectra to mesh as in Figure 2.12(d).

The quarter-cycle offset is fundamental to the elimination of phase errors, which PAL achieves, but it does cause some other effects which raise the complexity of implementation, leading to the retaliatory name of Problems Are Lurking. The receiver needs to know whether or not to invert V on a particular line, and this information is conveyed by swinging the phase of the burst $\pm 135°$ on alternate lines. A damped PLL in the receiver will run at the average phase, i.e. subcarrier phase, but it will develop a half line rate phase error whose polarity determines the sense of V-switch.

The quarter-cycle offset means that there are now line quartets instead of line pairs, and four frames or eight fields have to elapse before the same relationship of subcarrier to frame timing repeats. This restricts the way in which PAL recordings can be edited.

The quarter-cycle offset also means that the line pair cancellation of NTSC is

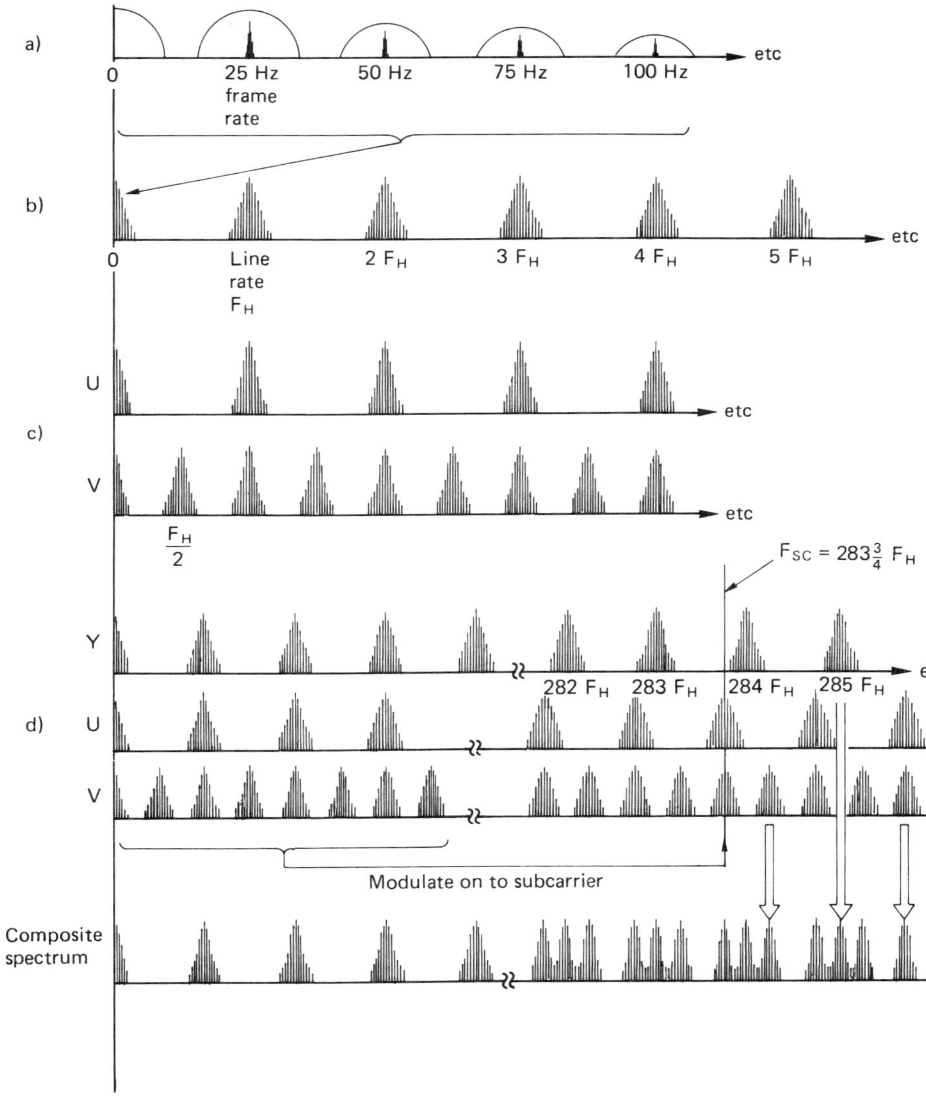

Figure 2.12 PAL spectra. At (a) is the spectrum due to sampling the image spectrum at the frame rate. This spectrum is in turn sampled at line rate, as in (b) which continues up to about 5.5 MHz. The *V*-switch process effectively reverses the polarity of *V* on alternate lines, so that two lines elapse before a cycle of *V* elapses instead of one line. This halves the spectral frequencies, as (c) shows. Clearly the half-cycle offset of NTSC cannot be used in the presence of the half line rate energy in the *V* signal. Instead the subcarrier has a three-quarter cycle offset, as (d) shows, and this allows the luminance and chroma to mesh as shown

absent, and another means has to be found to achieve visibility reduction. This is done by adding half frame rate to subcarrier frequency, such that an inversion in subcarrier is caused from one field to the next. Since in an interlaced system lines one field apart are adjacent on the screen, cancellation is achieved. The penalty of this approach is that subcarrier phase creeps forward with respect to H-sync at one cycle per frame. The eight-field sequence contains 2500 unique lines all having the subcarrier in a slightly different position. Observing burst on an H-triggered oscilloscope shows a stable envelope with a blurred interior. Measuring the phase of subcarrier with respect to sync requires specialist equipment or a great deal of determination. This is of little consequence for broadcasting, but it does raise the complexity of D-2 converters.

PAL-M and -N are unholy mixtures of PAL and NTSC which offer some of the advantages of PAL, but it is not necessary to detail them here since the digital implementation of both can be deduced from that of other standards.

2.8 Choice of sampling rate

When composite video is to be digitized, the input will be a single waveform having spectrally interleaved luminance and chroma. Any sampling rate which allows sufficient bandwidth would convey composite video from one point to another. However, if processing in the digital domain is contemplated, there will be less choice.

In the D-2 colour processor it will be necessary to decode the composite signal which will require some kind of digital filter. Whilst it is possible to construct filters with any desired response, it is a fact that a digital filter whose response is simply related to the sampling rate will be much less complex to implement. This is the reasoning which has led to the near universal use of four times subcarrier sampling rate. Figure 2.13 shows the spectra of PAL and NTSC sampled at $4 \times F_{sc}$. It will be evident that there is a considerable space between the edge of the baseband and the lower sideband. This allows the anti-aliasing and reconstruction filters to have a

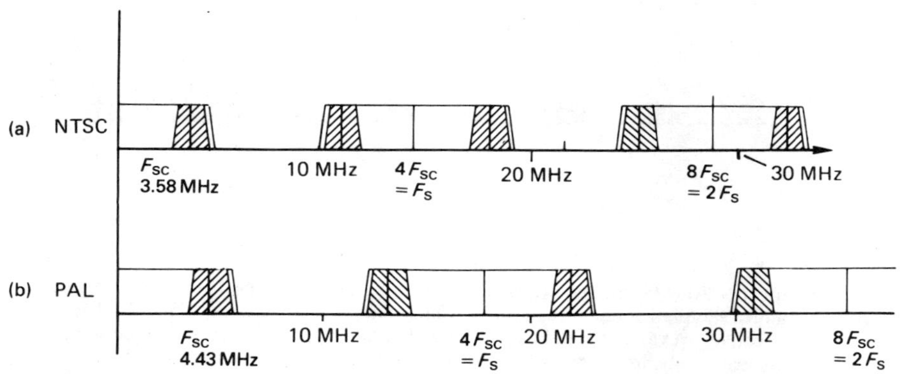

Figure 2.13 The spectra of NTSC at (a) and of PAL at (b) where both are sampled at four times the frequency of their respective subcarriers. This high sampling rate is unnecessary to satisfy sampling theory, and so both are oversampled systems. The advantages are in the large spectral gap between baseband and sideband which allows a more gentle filter slope to be employed, and in the relative ease of colour processing at a sampling rate related to subcarrier

more gradual cut-off, so that ripple in the passband can be reduced. This is particularly important for D-2 recorders, since they are digital devices in an analog environment, and signals may have been converted to and from the digital domain many times in the course of production.

2.9 Filter design

The discussion so far has assumed that perfect anti-aliasing and reconstruction filters are used. Perfect filters are not available, of course, and because designers must use devices with finite slope and rejection, aliasing can occur, but it can be made less significant by raising the sampling frequency slightly. It is not easy to specify such filters, particularly the amount of stopband rejection needed. The amount of aliasing resulting would depend on, among other things, the amount of out-of-band energy in the input signal. Very little is known about the energy in typical source material outside the usual frequency range. As a further complication, an out-of-band signal will be attenuated by the response of the anti-aliasing filter to that frequency, but the residual signal will then alias, and the reconstruction filter will reject it according to its attenuation at the new frequency to which it has aliased. To take the opposite extreme, if a camera were used which had no response at all above the video band, no anti-aliasing filter would be needed.

It would also be acceptable to bypass one of the filters involved in a copy from one digital machine to another via the analog domain, although a digital transfer is of course to be preferred.

The nature of the filters used has a great bearing on the subjective quality of the system. Entire books have been written about analog filters, and they will only be treated briefly here.

Figure 2.14 shows the terminology used to describe the common, elliptic low-pass filter. These are popular because they can be realized with fewer components than other filters of similar response. It is a characteristic of these

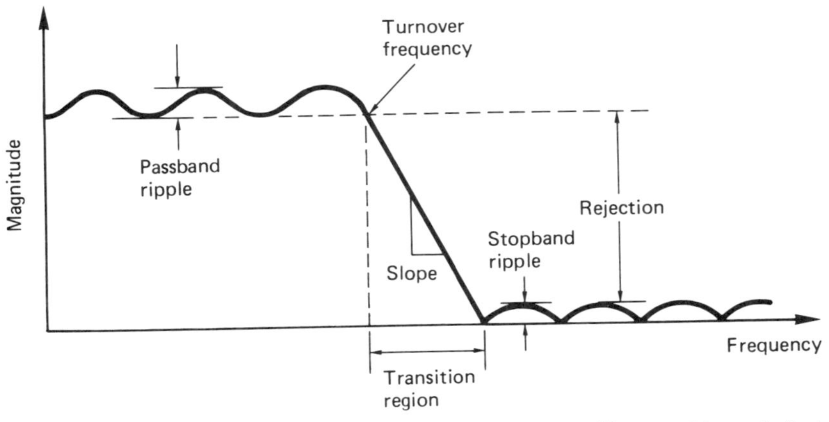

Figure 2.14 The important features and terminology of low-pass filters used for anti-aliasing and reconstruction

elliptic filters that there are ripples in the passband and stopband. In much equipment the anti-aliasing filter and the reconstruction filter will have the same specification, so that the passband ripple is doubled. Sometimes slightly different filters are used to reduce the effect.

Active filters can simulate inductors using op-amp techniques, but they tend to suffer non-linearity at high frequencies where the falling open-loop gain reduces the effect of feedback. Active filters also can contribute noise, but this is not necessarily a bad thing in controlled amounts, since it can act as a dither source for the subsequent quantizing stage.

For video applications, the phase response of such filters must be linear. Since a sharp cut-off is generally achieved by cascading many filter sections which cut at a similar frequency, the phase responses of these sections will accumulate. The phase may start to leave linearity at only a half of the passband frequency, and near the cut-off frequency the phase error may be severe. Effective group-delay equalization is necessary.

It is possible to construct a ripple-free phase-linear filter with the required stopband rejection, but it may be expensive of design effort and component complexity, and it might drift out of specification as components age. The money may be better spent in avoiding the need for such a filter. Much effort can be saved in analog filter design by using oversampling. Reference to Chapter 3 will show that digital filters are inherently phase linear and, using LSI, can be inexpensive to construct. The technical superiority of oversampling converters along with economics means that they will be increasingly used in the future, which is why the subject is more prominent here than the treatment of filter design.

2.10 Oversampling

Oversampling is a technique which allows sampled systems to approach theoretical performance limits more closely. It has become virtually universal in digital audio, since at the relatively low frequencies of audio it is easy to implement. It is only a matter of time before oversampling becomes common in video.

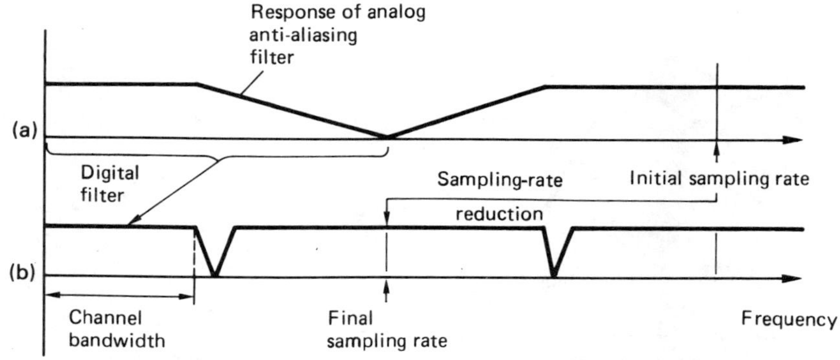

Figure 2.15 In a temporal oversampling system, the sampling rate is considerably higher than normal, and a gentle slope is adequate for the response of the anti-aliasing filter as shown at (a). Following sampling, the samples are digitally low-pass filtered in a phase-linear filter to the target channel bandwidth. The sampling rate can then be reduced without aliasing taking place. It is most convenient to use sampling rates which are integer multiples of the final sampling rate

In the purest sense of the word, oversampling simply means using a sampling rate which is in excess of that required by sampling theory. By using a high sampling rate, the need for a steep-cut analog anti-aliasing filter is eliminated, but the higher sampling rate results in an unnecessarily high data rate. This is reduced to the desired sampling rate by a digital filter which can be designed for a sharper cut-off than an analog filter, whilst retaining phase linearity. The resulting sampling rate carries an information bandwidth which is closer to the theoretical limit.

Figure 2.15(a) shows an oversampling system, where the high sampling rate opens a wide gap between the baseband and the lower sideband. The analog anti-aliasing filter need only fall to the stopband response at half of the high sampling rate to prevent aliasing, and at the band edge of the baseband signal, attenuation will be minimal. Figure 2.15(b) shows that the sampling rate is then reduced by a digital filter. In order to prevent aliasing at the lower sampling rate the filter must incorporate a low-pass response. In the digital domain an optimal aperture function is more readily implemented, and so the resultant frequency response will be flatter.

Oversampling can be used for either direction of conversion. Figure 2.16(a) shows the circuit diagram of a typical analog reconstruction filter following a DAC, which would typically be used to return from the digital domain to raster scanned analog. Following the low-pass filter is a phase-correction network which is as complex as the filter itself. Figure 2.16(b) shows an oversampling converter which increases the sampling rate in a digital interpolator before the DAC.[4] The resultant analog filter need not be so complex, and will need less adjustment. Digital filters need no alignment and do not drift. As the cost of logic circuits continues to fall, it may become cheaper to implement oversampling converters, in addition to the improved performance gained.

2.11 Quantizing

Figure 2.17 shows that the process of quantizing divides the voltage range up into quantizing intervals Q. These may be of differing size for particular applications, e.g. telephony, but in this case it becomes very difficult to process sample values arithmetically. Hence in most digital video and audio equipment all the quantizing intervals are the same, and the term *uniform quantizing* is applied. The term 'linear quantizing' will also be found, but this is a contradiction in terms.

Whatever the exact voltage of the input signal, the quantizer will express it as the number of the interval in which it falls. When that number arrives at the DAC, it will create the voltage corresponding to the centre of the interval. Quantizing thus makes errors, which cannot exceed $\pm\frac{1}{2}Q$. If the output of the DAC is resampled to avoid aperture effect, the impulses from the DAC can be compared with the impulses at the input sampling stage. The difference between the two will be an impulse train (Figure 2.18) which can be thought of as an unwanted signal added by the quantizing process to a perfect signal. This is the quantizing error signal; it warrants considerable study since it has some unexpected characteristics.

Where the input signal exercises the range of the quantizer and has a complex waveform such as that produced by a contrasty, detailed picture, the size of the quantizing error will be anywhere between $-\frac{1}{2}Q$ and $+\frac{1}{2}Q$ with uniform probability, as in Figure 2.19. This probability density function should be contrasted with that of thermal noise in electronic components, which has a

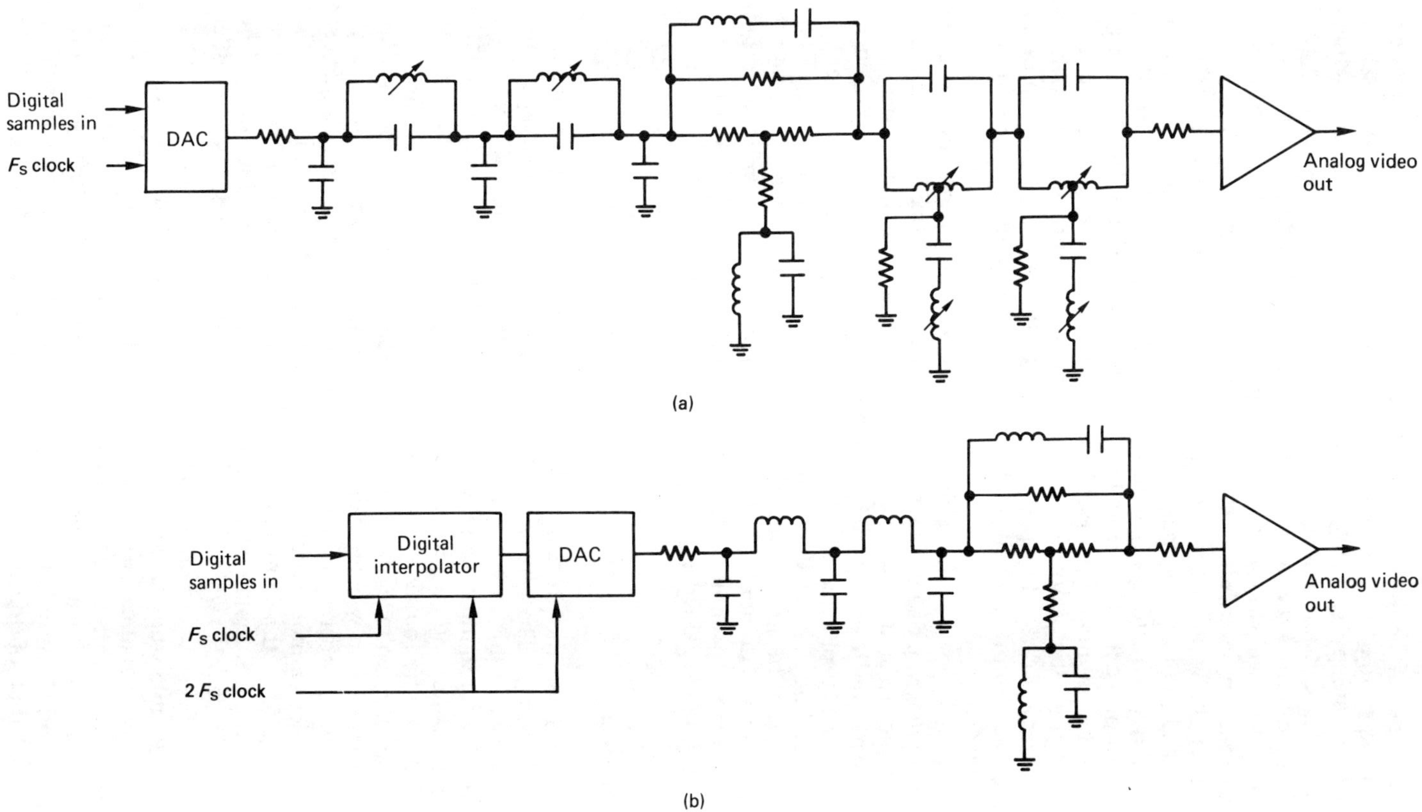

Figure 2.16 A conventional DAC and analog filter is shown at (a). It is complex and requires adjustment. At (b) a digital interpolator doubles the sampling rate supplied to the DAC so only a simple analog filter is required

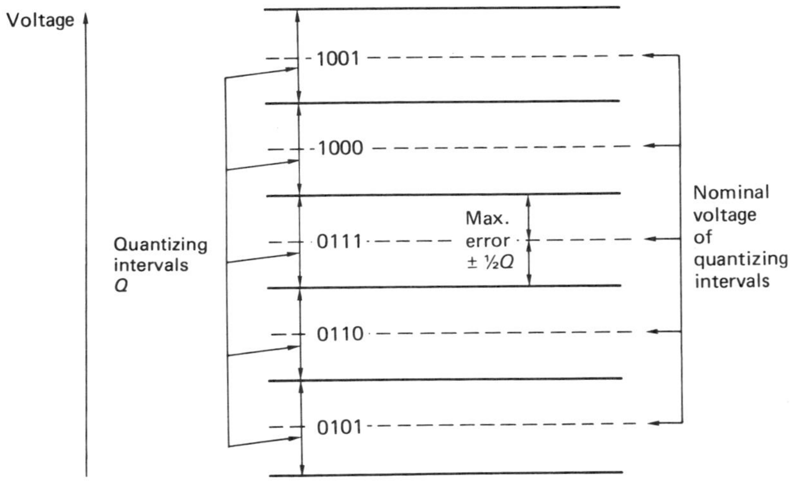

Figure 2.17 Quantizing. All voltages within a particular quantizing interval are assigned the same number, which corresponds to the voltage of the centre of the interval. The maximum quantizing error cannot exceed $\pm \frac{1}{2}Q$

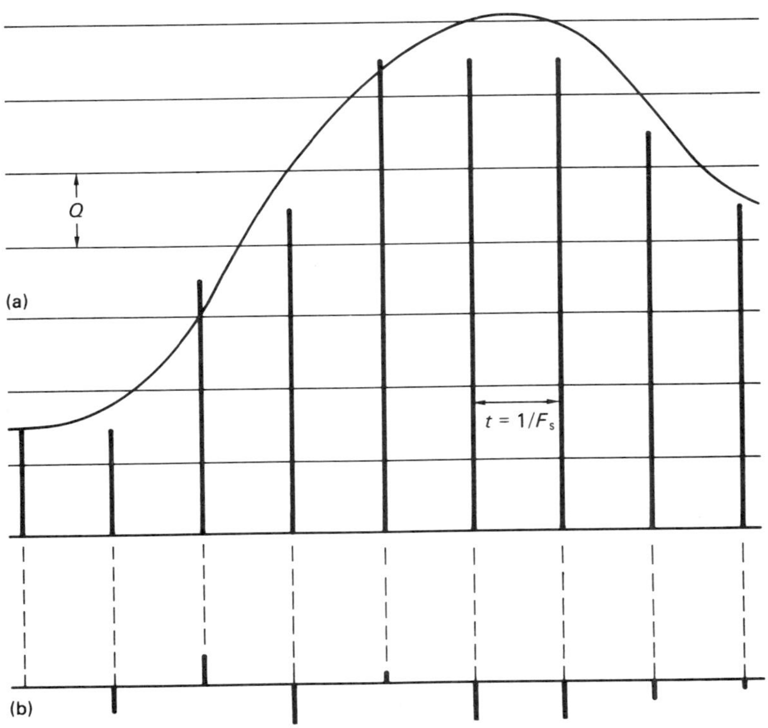

Figure 2.18 At (a) an arbitrary signal is represented to finite accuracy by PAM needles whose peaks are at the centre of the quantizing intervals. The errors caused can be thought of as an unwanted signal (b) added to the original

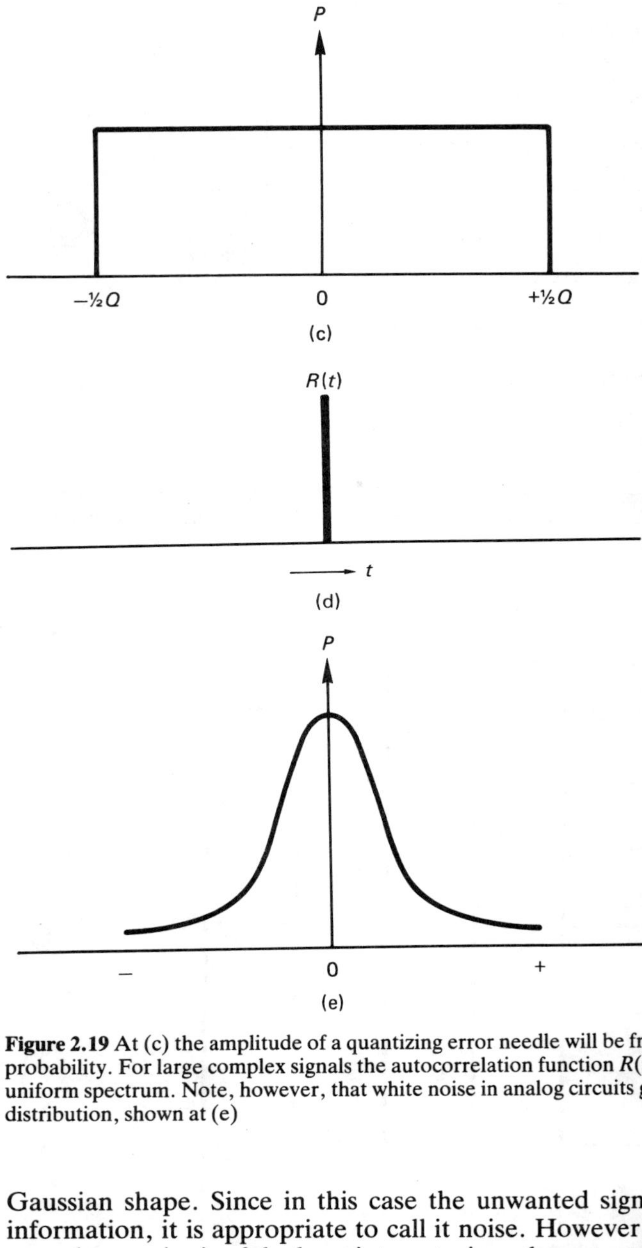

Figure 2.19 At (c) the amplitude of a quantizing error needle will be from $-\frac{1}{2}Q$ to $+\frac{1}{2}Q$ with equal probability. For large complex signals the autocorrelation function $R(t)$ has one spike as at (d), giving a uniform spectrum. Note, however, that white noise in analog circuits generally has Gaussian amplitude distribution, shown at (e)

Gaussian shape. Since in this case the unwanted signal is uncorrelated with the information, it is appropriate to call it noise. However, the large-signal case is the one where noise is of the least interest, since the presence of the signal masks noise. Nevertheless many treatments of quantizing proceed to make the simple connection between the number of bits in the word, n, and the signal-to-noise ratio, namely $6.02n + 1.76$ dB. There are two shortcomings to this simplistic approach. First, the noise power calculated has an infinite spectrum, and no account is taken of the effect of the reconstruction filter on this spectrum. Secondly, and much more

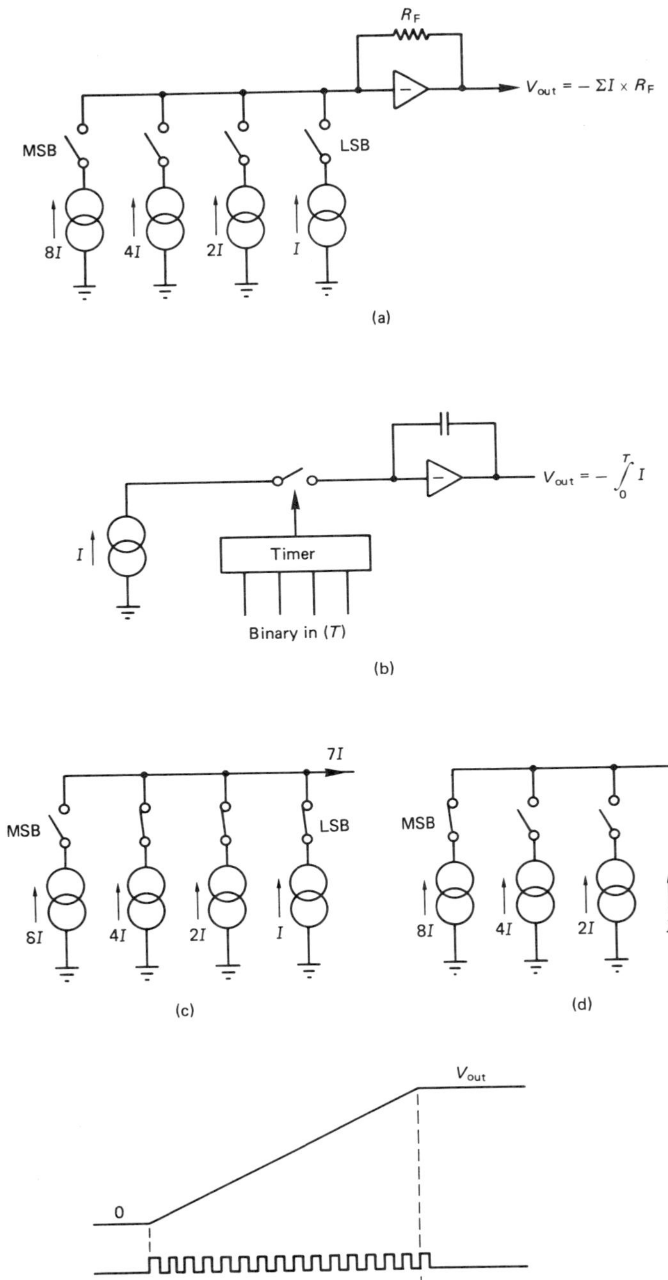

Figure 2.20 Elementary conversion: (a) weighted current DAC; (b) current flow with 0111 input; (c) current flow with 1000 input

importantly, the mathematics only holds if the probability density function of the quantizing error is uniform. At low levels, and particularly with spectrally pure or simple signals, the quantizing error ceases to be random and becomes a function of the input signal. Once an unwanted signal becomes a deterministic function of the wanted signal, it has to be referred to as distortion rather than noise. As the level of the analog input is reduced, the quantizing error becomes less random, and noise modulation occurs. Where more than one frequency is present in the input there will be intermodulation products.

The harmonics caused by these non-linear processes can alias with the sampling rate to produce anharmonic frequencies in the passband spectrum. Where the sampling rate is a multiple of the signal frequency, the effect is harmonic distortion. It is easy to demonstrate this phenomenon graphically (Figure 2.20). The quantizing error has been determined from the signal, and low-pass filtered by the reconstruction filter. Strong correlations between quantizing errors can be seen and the resultant harmonics of the signal are visible. The effect is visible as contouring on the picture. This result is true for a perfect quantizer, and the effect can be demonstrated easily by temporarily disabling some of the low-order bits in a system.

In practice, it is difficult to obtain a perfectly deterministic quantizing error because of noise in the analog input signal.[5] This has a randomizing effect on the quantizing error, and reduces the distortion accordingly.[6] Figure 2.21 shows that the effect of dither is to smear the transfer function of the quantizer horizontally. Where the RMS noise voltage is one-third of a quantizing interval,[7] the quantizing error will be as random as the noise, and the quantizing process becomes perfectly linear. Another way of describing the effect is to say that the voltage of the input between quantizing intervals is conveyed in the duty cycle of the binary switching.

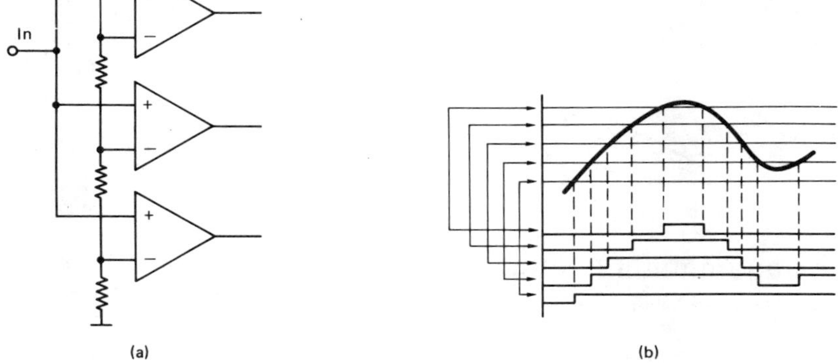

Figure 2.21 The flash converter. At (a) each quantizing interval has its own comparator, resulting in waveforms of (b). A priority encoder is necessary to convert the comparator outputs to a binary code

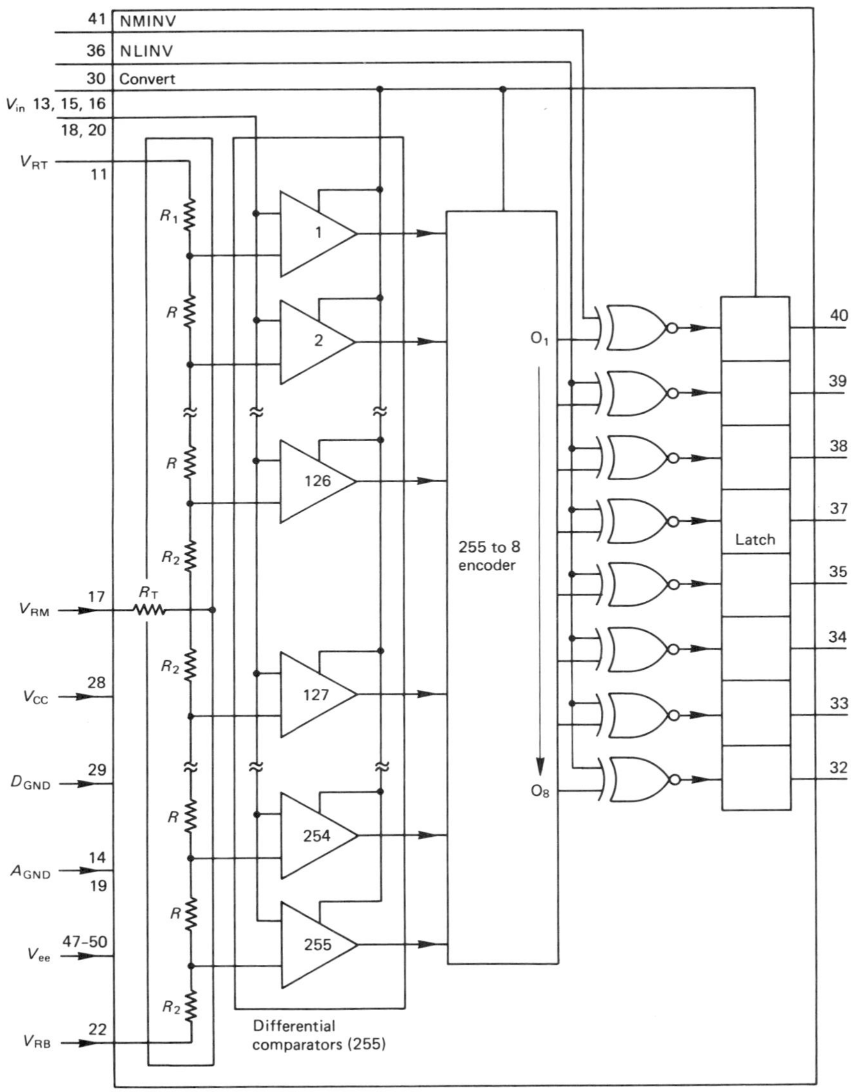

(c)

Figure 2.21 (*Continued*). Shown at (c) is a typical 8 bit flash converter primarily intended for video applications (Courtesy TRW)

This randomizing technique is known as *dither*; in addition to linearizing the system it also defines the signal-to-noise ratio. If sufficient noise is not already superimposed on the input signal, a noise source can be built into the converter. A diode can be used as a source of Gaussian noise. The probability density function of the dither is important. If rectangular PDF noise is created by connecting a digitally generated pseudo-random sequence to a DAC, the probability of the resultant quantizing error will be the combined probabilities of the dither and the undithered quantizing error. Since the pseudo-random generator output cannot be correlated with the undithered quantizing error, the combination of the two results in a triangular probability density function. The result will be slight noise modulation: the amplitude of the noise changes with signal amplitude, although the linearizing effect is unimpaired. If sufficient Gaussian dither is used to produce a linear system, namely ⅓Q RMS, the signal-to-noise ratio can easily be shown to be 6.02n dB. For most practical purposes, 6 dB per bit is an adequate assessment of signal to noise, but only if suitable dither is employed. In a composite signal, the presence of the chroma will tend to dither the luminance, but only in coloured regions of the frame.

In an extension of the application of dither,[8] digitally generated dither is converted to the analog domain and added to the input signal prior to quantizing. That same digital dither is then subtracted from the digital quantizer output. The effect is that the transfer function of the quantizer is smeared diagonally (Figure 2.22). The significance of this diagonal smearing is that the amplitude of the dither is not critical. However much dither is employed, the noise amplitude will remain the same. If dither of several quantizing intervals is used, it has the effect of making all the quantizing intervals in an imperfect converter appear to have the same size.

The importance of correctly dithering a quantizer cannot be emphasized enough, since failure to dither irrevocably distorts the converted signal: there can be no process which will subsequently remove that distortion.

In principle a sample can be quantized into any number range desired, but it is convenient to use binary circuitry to handle and store the numbers, and accordingly quantizing ranges will always be some power of two. The appropriate range is hard to define since, as with any signal which will be assessed by a human being, results are subjective. It is tempting to take the signal-to-noise ratio of an equivalent analog system and divide by six to obtain the number of bits. Initially digital video machines used 8 bits, as do the D-1 and D-2 formats, but there is a trend today to go to 10 bits for some applications.

2.12 Basic digital-to-analog conversion

There is only one way of obtaining an analog signal from PCM data at the frequencies of digital video. Essentially the input sample controls binary-weighted currents and these are summed as shown in Figure 2.20. In Figure 2.20(b), the binary code is about to have a major overflow, and all the low-order currents are flowing. In Figure 2.20(c), the binary input has increased by one, and only the most significant current flows. This current must equal the sum of all the others plus one least significant current to an accuracy of rather better than one least significant current. In this simple 4 bit example, the necessary accuracy is only one part in 16, but for an 8 bit system it would become one part in 256, or about 0.4%. This degree of accuracy can be achieved with care, and it must be maintained in the presence of ageing and temperature change.

To prevent interaction between the stages in weighted-current converters, the currents must be switched to ground or into the virtual earth by change-over switches. The on resistance of these switches is a source of error, particularly the MSB, which passes most current. A solution in monolithic converters is to fabricate

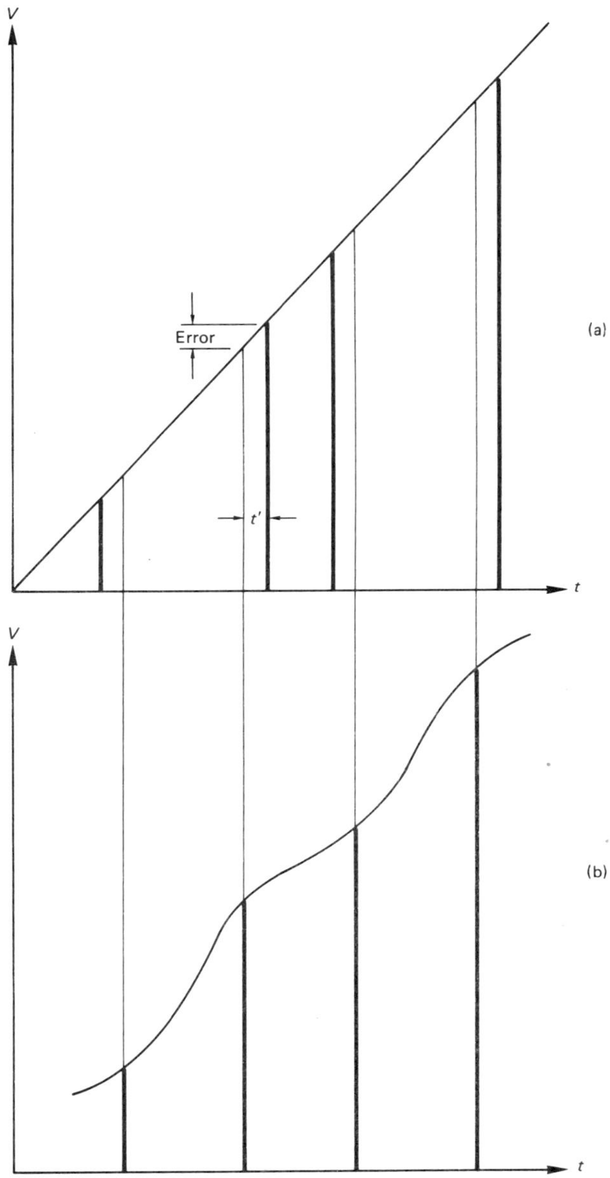

Figure 2.22 The effect of sampling timing jitter on noise, and calculation of the required accuracy for a 16 bit system. (a) Ramp sampled with jitter has error proportional to slope. (b) When jitter is removed by later circuits, error appears as noise added to samples

switches whose area is proportional to the weighted current, so that the voltage drops of all the switches are the same. The error can then be removed with a suitable offset. The layout of such a device is dominated by the MSB switch since, by definition, it is as big as all the others put together.

2.13 Basic analog-to-digital conversion

The general principle of a quantizer is that different quantized voltages are compared with the unknown analog input until the closest quantized voltage is found. The code corresponding to this becomes the output.

The flash converter is probably the simplest technique available for PCM video conversion. The principle is shown in Figure 2.21. The threshold voltage of every quantizing interval is provided by a resistor chain which is fed by a reference voltage. This reference voltage can be varied to determine the sensitivity of the input. There is one voltage comparator connected to every reference voltage, and the other input of all of these is connected to the analog input. The input voltage determines how many of the comparators will have a true output. As one comparator is necessary for each quantizing interval, then, for example, in an 8 bit system there will be 255 binary comparator outputs, and it is necessary to use a priority encoder to convert these to a binary code. Although the device is simple in principle, it contains a lot of circuitry and can only be practicably implemented on a chip. The analog signal has to drive a lot of inputs, and a low-impedance driver is essential to avoid restricting the slewing rate of the input. The extreme speed of a flash converter is a distinct advantage in oversampling. Because computation of all bits is performed simultaneously, no analog track/hold circuit is required, and droop is eliminated. In fact the sampling process takes place on the binary comparator outputs after quantizing in a flash converter. The clock which takes these samples must have very low jitter, with minimum skew between bits, otherwise noise will be introduced. The mechanism is illustrated in Figure 2.22, where a changing signal is being sampled. The mistiming due to clock jitter causes the wrong voltage to be sampled. The effect clearly rises with the rate of change of input voltage; therefore the amount of noise caused will depend on the spectrum of the input, making it a form of modulation noise. It is difficult to specify the clock jitter tolerance, since the worst-case assumptions demand an accuracy within about 100 ps!

2.14 Imperfections of converters

ADCs and DACs have the same transfer function, since they are only distinguished by the direction of operation, and therefore the same terminology can be used to classify the shortcomings of both.

Figure 2.23 shows the transfer functions resulting from the main types of converter error.

(a) *Offset error*. A constant appears to have been added to the digital signal. This has a serious effect in video systems, since it causes black-level shifts. Offset error is sometimes cancelled by digitally sampling the converter output during blanking and feeding it back to the analog input as a small control voltage.

(b) *Gain error*. The slope of the transfer function is incorrect. Since converters are referred to one end of the range, gain error causes an offset error. Severe gain errors will cause clipping.

(c) *Linearity*. (Also known as *integral linearity*.) The deviation of the transfer function from a straight line (ignoring the quantizing steps). It has exactly the same significance as linearity in analog circuits, since if it is inadequate, harmonic

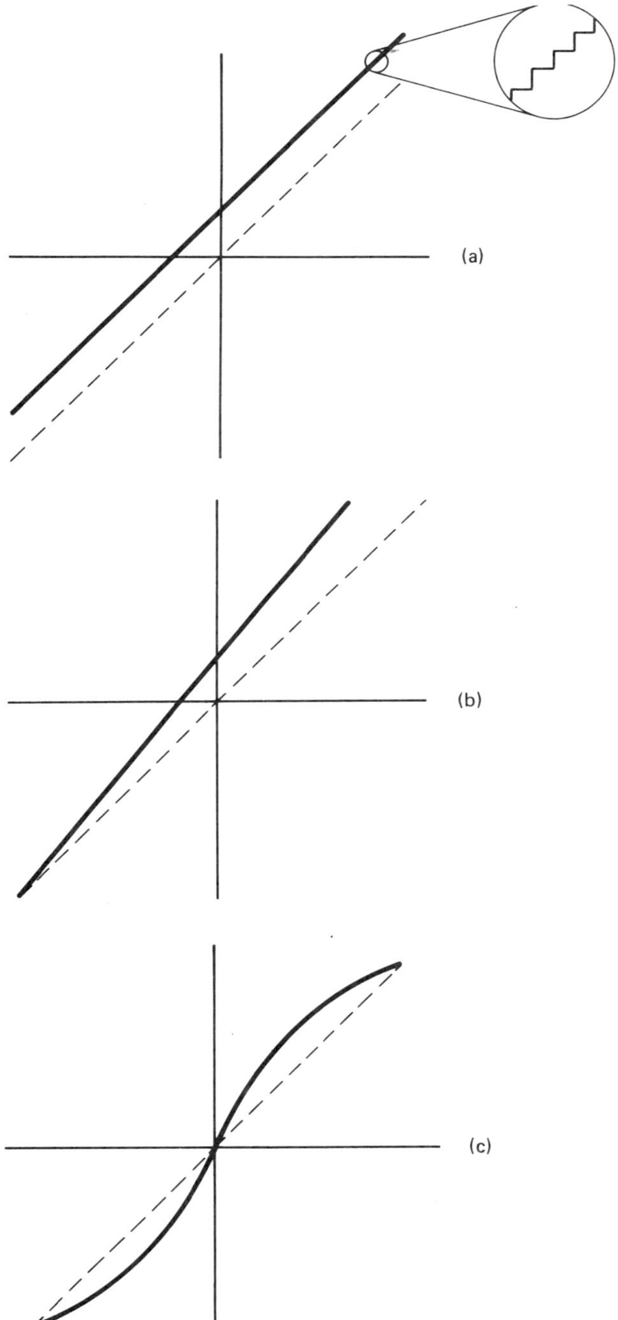

Figure 2.23 Main converter errors (solid line) compared with perfect transfer function (dotted line). These graphs hold for ADCs and DACs, and the axes are interchangeable; if one is chosen to be analog, the other will be digital

distortion will be caused. Subclassifications of linearity which are used are *differential non-linearity*, which is the amount by which adjacent quantizing intervals differ in size, and *monotonicity*, which is a special case of differential non-linearity. Non-monotonicity means that the output does not increase for an increase in input. Figure 2.24 shows how this can happen. With a converter input code of 01111111 (127 decimal), the seven low-order current sources of the converter will be on. The next code is 10000000 (128 decimal), where only the eighth current source is operating. If the current it supplies is in error on the low side, the analog output for 128 may be less than that for 127. If a device has better than ½Q linearity it must be monotonic.

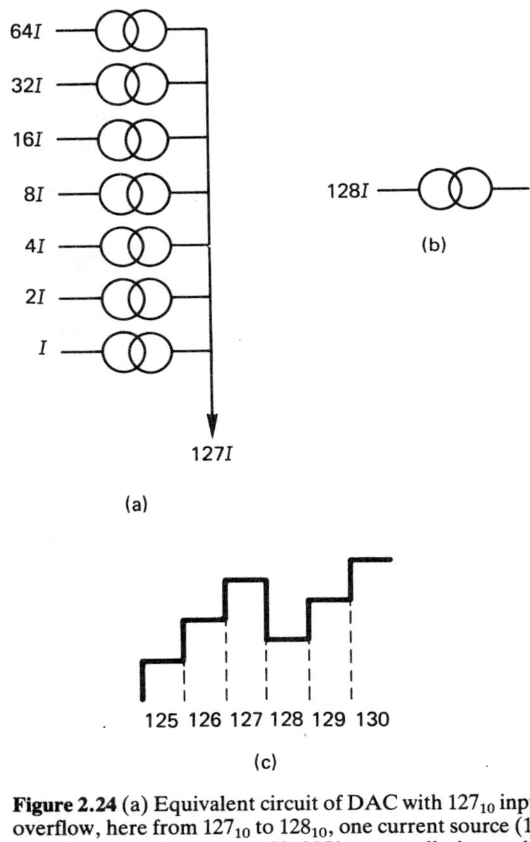

Figure 2.24 (a) Equivalent circuit of DAC with 127_{10} input. (b) DAC with 128_{10} input. On a major overflow, here from 127_{10} to 128_{10}, one current source (128I) must be precisely I greater than the sum of all the lower-order sources. If 128I is too small, the result shown at (c) will occur. This is non-monotonicity

(d) *Absolute accuracy*. The difference between actual and ideal output for a given input. For video this is rather less important than linearity. For example, if all the current sources in a converter have good thermal tracking, linearity will be maintained, even though the absolute accuracy drifts.

References

1. KELL, R., BEDFORD, A. and TRAINER, M. A. An experimental television system, Part 2. *Proc. IRE*, **22**, 1246–1265 (1934)
2. HSU, S. The Kell factor: past and present. *SMPTE J.*, **95**, 206–214 (1986)
3. JESTY, L. C. The relationship between picture size, viewing distance and picture quality. *Proc. IEE*, **105B**, 425–439 (1958)
4. WILLIAMS, R. and PRATER, J. Digital filtering: the right stuff for video. *Electron. Syst. Des. Mag.*, January, 91–96
5. BENNETT, W. R. Spectra of quantized signals. *Bell Syst. Tech. J.*, **27**, 446–472 (1948)
6. GOODALL, W. M. Television by pulse code modulation. *Bell Syst. Tech. J.*, **30**, 33–49 (1951)
7. VANDERKOOY, J. and LIPSHITZ, S. Resolution below the least significant bit in digital systems with dither. *J. Audio Eng. Soc.*, **32**, 106–113 (1984)
8. ROBERTS, L. G. Picture coding using pseudo-random noise. *IRE Trans. Inf. Theory*, **IT-8**, 145–154 (1962)

Chapter 3

Composite digital video coding and processing

The conversion process expresses the analog input as a binary code. In addition to recording and replaying a binary code, a DVTR will also carry out signal processing which will be applied to the sample stream in various places to perform in the digital domain functions such as timebase correction and colour processing.

These subjects are detailed here, with a brief introduction to binary coding and arithmetic.

3.1 Introduction to logic

The strength of binary logic is that the signal has only two states, and considerable noise and distortion can be tolerated before the state becomes uncertain. At every logical element, the signal is compared with a threshold and can thus pass through any number of stages without being degraded. The two states of the signal when measured with an oscilloscope are simply two voltages, usually referred to as high and low. The actual voltage levels will depend on the type of logic chips in use and on the supply voltage used. Within logic, these levels are not of much consequence, and it is only necessary to know them when interfacing between different families or when driving external devices. The pure logic designer is not interested at all in these voltages, only in their meaning. Just as the electrical waveform from a camera represents picture brightness, so the waveform in a logic circuit represents the truth of some statement. As there are only two states, there can only be *true* or *false* meanings. The true state of the signal can be assigned by the designer to either voltage state. When a high voltage represents a true logic condition and a low voltage represents a false condition, the system is known as *positive logic*, or *high true* logic. This is the usual system, but sometimes the low voltage represents the true condition and the high voltage represents the false condition. This is known as *negative logic* or *low true* logic. Provided that everyone is aware of the logic convention in use, both work equally well.

Negative logic is often found in the TTL logic family, because in this technology it is easier to sink current to ground than to source it from the power supply. Figure 3.1 shows that if it is necessary to connect several logic elements to a common bus so that any one can communicate with any other, an open collector system is used, where high levels are provided by pull-up resistors and the logic elements only pull the common line down. If positive logic were used, when no device was operating the pull-up resistors would cause the common line to take on an absurd true state;

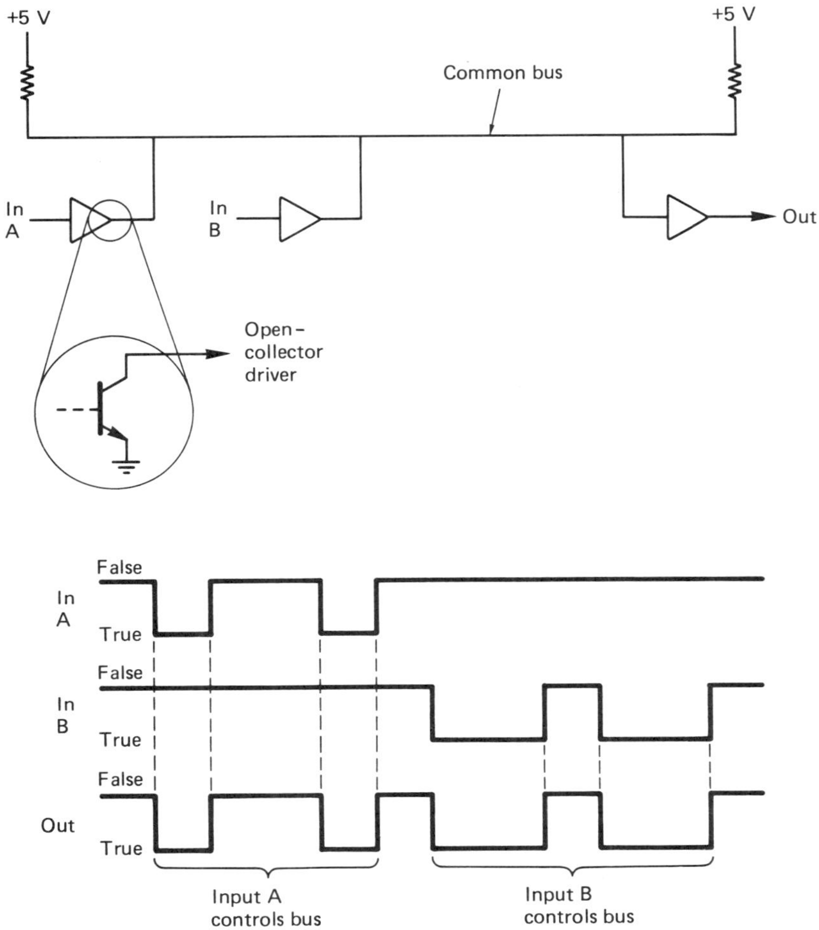

Figure 3.1 Using open-collector drive, several signal sources can share one common bus. If negative logic is used, the bus drivers turn off their output transistors with a false input, allowing another driver to control the bus. This will not happen with positive logic

whereas if negative logic is used, the common line pulls up to a sensible false condition when there is no device using the bus. Whilst the open collector is a simple way of obtaining a shared bus system, it is limited in frequency of operation owing to the time constant of the pull-up resistors charging the bus capacitance. In the so-called tri-state bus systems commonly used in digital video, there are both active pull-up and pull-down devices for speed (totem poles), but both devices will be disabled to a third state, where the output assumes a high impedance, allowing some other driver to determine the bus state.

In logic systems, all complex devices can be configured from combinations of a few fundamental gates. It is not profitable to spend too much time arguing which are the really fundamental ones, since most can be made from combinations of others. Table 3.1 shows the important simple gates and their derivatives, and introduces the logical expressions to describe them, which can be compared with the truth-table notation. Figure 3.1 also shows the important fact that when negative logic is used, the OR gate function interchanges with that of the AND gate. Sometimes schematics are drawn to reflect which voltage state represents the true condition. In the so-called intentional logic scheme, a negative logic signal always starts and ends at an inverting 'bubble'. If an AND function is required between two negative logic signals, it will be drawn as an AND symbol with bubbles on all the terminals, even though the component used will be a positive logic OR gate. Opinions vary on the merits of intentional logic.

Table 3.1 The basic logic gates compared

Positive logic name	*Boolean expression*	*Positive logic symbol*	*Positive logic truth table*	*Plain English*
Inverter or NOT gate	$Q = \overline{A}$	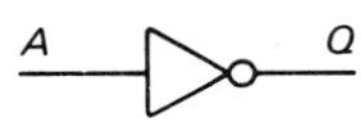	A Q 0 1 1 0	Output is opposite of input
AND gate	$Q = A \cdot B$	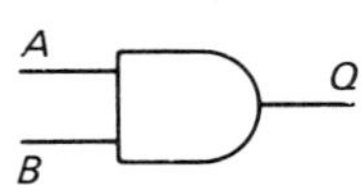	A B Q 0 0 0 0 1 0 1 0 0 1 1 1	Output true when both inputs are true only
NAND (Not AND) gate	$Q = \overline{A \cdot B}$ $= \overline{A} + \overline{B}$	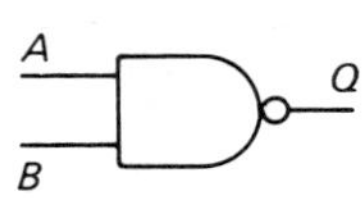	A B Q 0 0 1 0 1 1 1 0 1 1 1 0	Output false when both inputs are true only
OR gate	$Q = A + B$	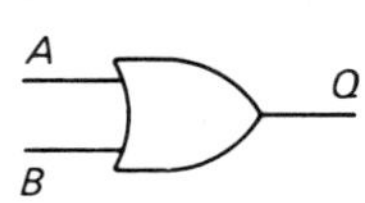	A B Q 0 0 0 0 1 1 1 0 1 1 1 1	Output true if either or both inputs true
NOR (Not OR) gate	$Q = \overline{A + B}$ $= \overline{A} \cdot \overline{B}$	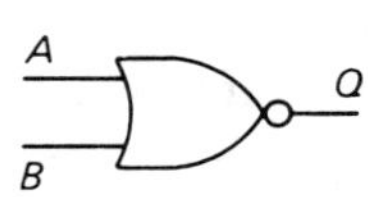	A B Q 0 0 1 0 1 0 1 0 0 1 1 0	Output false if either or both inputs true
Exclusive OR (XOR) gate	$Q = A \oplus B$	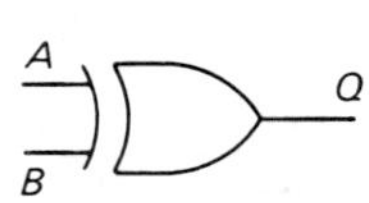	A B Q 0 0 0 0 1 1 1 0 1 1 1 0	Output true if inputs are different

If numerical quantities need to be conveyed down the two-state signal paths described here, then the only appropriate numbering system is binary, which has only two symbols, 0 and 1. Just as positive or negative logic could be used for the truth of a logical binary signal, it can also be used for a numerical binary signal. Normally, a high voltage level will represent a binary 1 and a low voltage will represent a binary 0, described as a 'high for a one' system. Clearly a 'low for a one' system is just as feasible. Decimal numbers have several columns, each of which represents a different power of ten; in binary the column position specifies the power of two. Figure 3.2 shows some binary numbers and their equivalent in decimal. The radix point has the same significance in binary: symbols to the right of it represent one-half, one-quarter and so on. Binary is convenient for electronic circuits, which do not get tired, but numbers expressed in binary become very long, and writing them is tedious and error prone. The octal and hexadecimal notations are both used for writing binary since conversion is so simple. A binary number is split into groups of three or four digits starting at the least significant end, and the groups are individually converted to octal or hexadecimal digits. Since 16 different symbols are required in hex, the letters A–F are used for the numbers above nine.

A number of binary digits or bits are needed to express a binary number. These bits can be conveyed at the same time by several signals to form a parallel system, which is most convenient inside equipment because it is fast, or one at a time down a single signal path, which is slower, but convenient for cables between pieces of equipment because the connectors require fewer pins. When a binary system is used to convey numbers in this way, it can be called a digital system.

3.2 Binary codes

There are a wide variety of waveform types encountered in video, and all of these can be sucessfully digitized. All that is needed is for the quantizing range to be a little greater than the useful voltage range of the waveform.

In monochrome, the useful waveform can only vary between black and white, since the syncs carry no information which cannot be recreated. Composite video, however, contains an embedded subcarrier, which has meaningful levels above white and below black. The quantizing range has to be extended to embrace the total possible excursion of luminance plus subcarrier. The resultant range is so close to the overall range of the signal that, in D-2, the whole signal including syncs is made to fit into the quantizing range. This is particularly useful in PAL because the sampling points are locked to subcarrier and have a complex relationship to sync.

If the quantizing range were set to exactly the video signal range, a slightly excessive gain somewhere in the analog input would cause clipping. In practice the quantizing range will be a little greater than the nominal signal range.

A format must specify the numerical values which result from reference analog levels so that all tapes will play back identical signals on all machines. Clearly black level does not correspond to digital zero. In D-2 PAL it is 64_{10}. It is as if a constant of 64 had been added to every sample, and gives rise to the term offset binary. Figure 3.3 shows how the PAL and NTSC signals fit into the quantizing range of D-2.

Whilst the composite signal recorded on tape in D-2 uses the above coding ranges, it will be necessary to handle luminance and chroma separately inside the colour processing circuits. Whilst luminance causes no difficulty, chroma can have

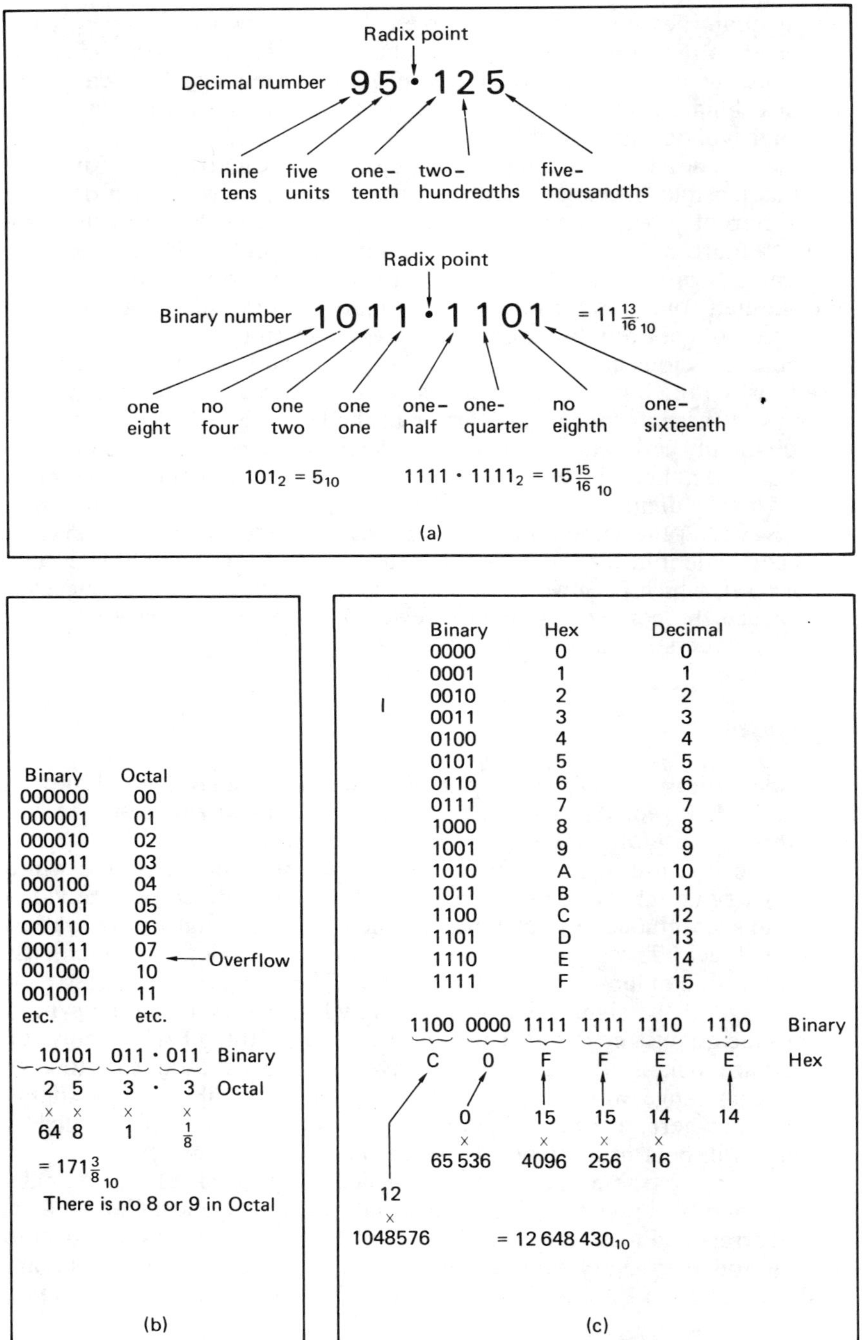

Figure 3.2 (a) Binary and decimal. (b) In octal, groups of 3 bits make one symbol 0–7. (c) In hex, groups of 4 bits make one symbol 0–F. Note how much shorter the number is in hex

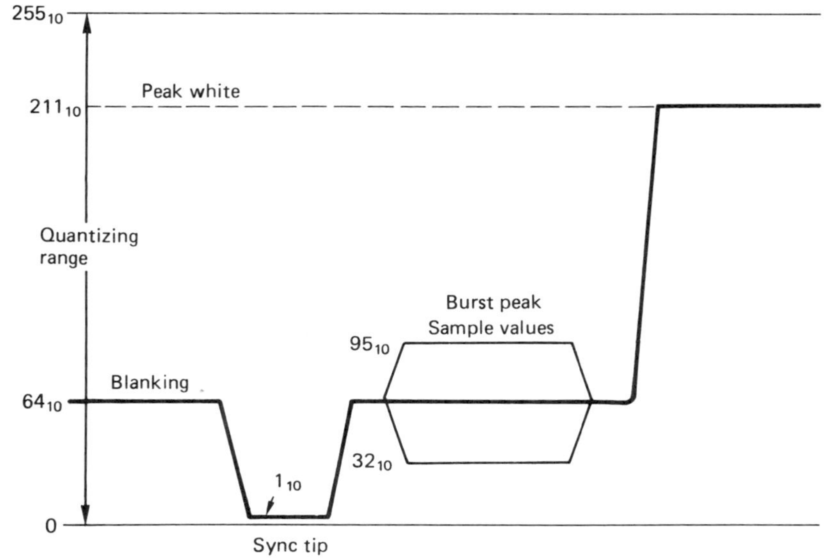

Figure 3.3 The composite PAL signal fits into the quantizing range as shown here. Note that there is sufficient range to allow the instantaneous voltage to exceed that of peak white in the presence of saturated bright colours. Values shown are decimal equivalents in a 10 or 8 bit system. In a 10 bit system the additional 2 bits increase resolution, not magnitude, so they are below the radix point and the decimal equivalent is unchanged. D-2 PAL samples in phase with burst, so the values shown are on the burst peaks and are thus also the values of the envelope

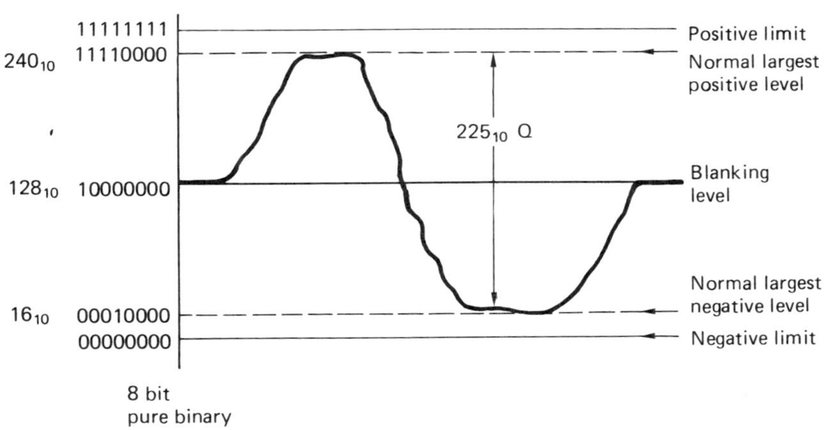

Figure 3.4 A colour difference signal can be positive or negative, but its value is always referred to blanking level, which cannot be zero in a pure binary system

positive or negative values, and a coding scheme which can accommodate this characteristic will be needed. For some purposes, offset binary can be used where the offset is made to be one-half the number range of the system. Effectively the quantizing range has been shifted to accept the positive and negative voltages in a real chroma signal. All chroma values can be expressed by binary numbers which are only positive.

The offset binary approach is perfectly acceptable where the signal has been digitized only for transmission from one place to another, but if it is required to perform signal processing in the digital domain, offset binary will be found to be unsuitable, and the data must be converted to another coding system.

Figure 3.4 shows that a chroma signal is referred to midrange. The level or saturation is measured by how far the waveform deviates from midrange, and fading, gain and mixing all take place around midrange.

At the output of a colour processor it is necessary to recombine sample values from chroma and luminance, and adding circuits assume that all bits represent the same quantizing interval so that the sum of two sample values will represent the sum of the two equivalent analog voltages.

Figure 3.5 shows that if two offset binary sample streams are added together in an attempt to perform digital summing, the result will be an offset which may lead to an overflow. Similarly, if an attempt is made to fade by, say, 6 dB by dividing all of the sample values by two, Figure 3.6 shows that a further offset results. The problem is that offset binary is referred to one end of the range. What is needed is a numbering system which operates symmetrically about the centre of the range.

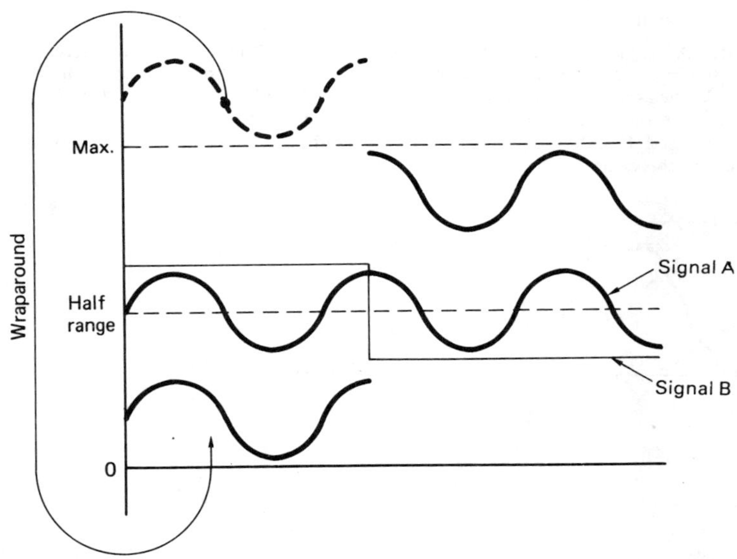

Figure 3.5 If two pure binary data streams are added to simulate mixing, offset or overflow will result

The two's complement system, which has this property, is commonly used for processing.

In two's complement, the upper half of the pure binary number range has been defined to represent negative quantities. If a pure binary counter is constantly incremented and allowed to overflow, it will produce all the numbers in the range permitted by the number of available bits, and these are shown for a 4 bit example drawn around the circle in Figure 3.7. In two's complement, however, the number range this represents does not start at zero but on the opposite side of the circle.

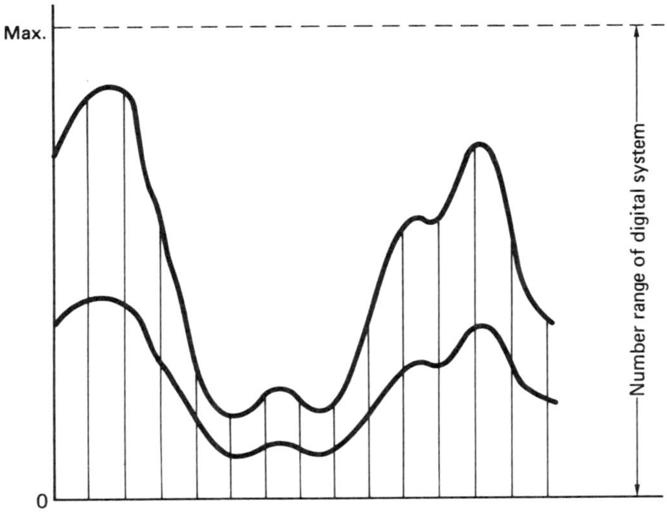

Figure 3.6 The result of an attempted attenuation in pure binary code is an offset. Pure binary cannot be used for digital video processing of colour difference signals

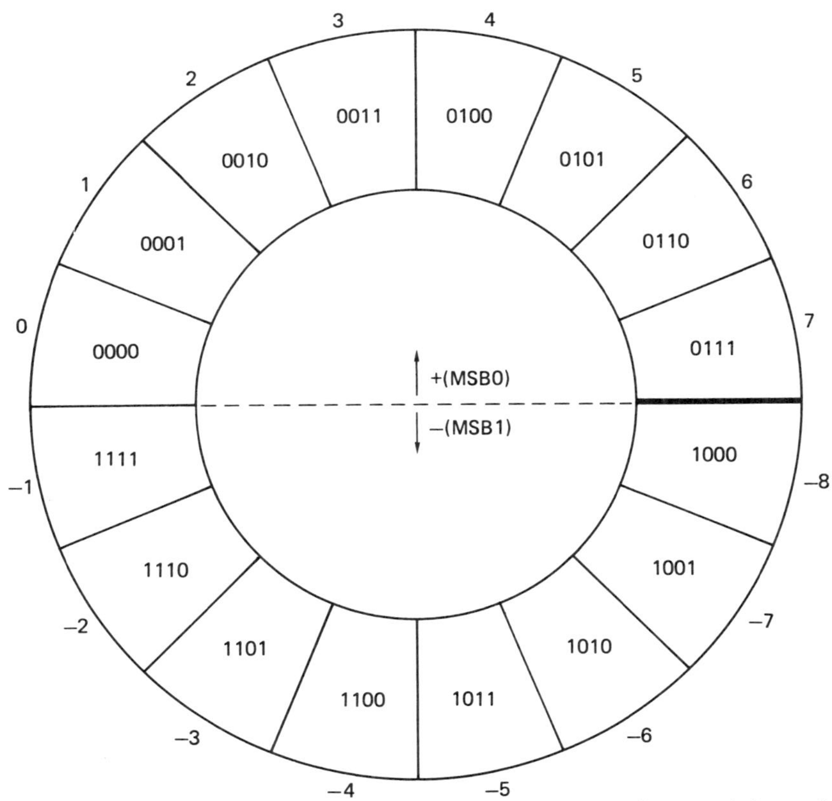

Figure 3.7 In this example of a 4 bit two's complement code, the number range is from −8 to +7. Note that the MSB determines polarity

Zero is midrange, and all numbers with the most significant bit set are considered negative. Two's complement notation can also be considered as an offset binary code with the most significant bit inverted. This system allows two sample values to be added, where the result is referred to the system midrange; this is analogous to adding analog signals in an operational amplifier. A further asset of two's complement notation is that binary subtraction can be performed using only adding logic. The two's complement is added to perform a subtraction. This permits a significant saving in hardware complexity, since only carry logic is necessary and no borrow mechanism need be supported. For these reasons, two's complement notation is in common use in digital video processing.

If a signed binary number is to be expressed as a negative number it is written down with leading zeros if necessary to occupy the wordlength of the system. All bits are then inverted, to form the one's complement, and one is added. To return to signed binary, if the most significant bit of the two's complement number is false, no action need be taken. If the most significant bit is set, the sign is negative, all bits are inverted, and one is added. Figure 3.8 shows some examples of conversion to

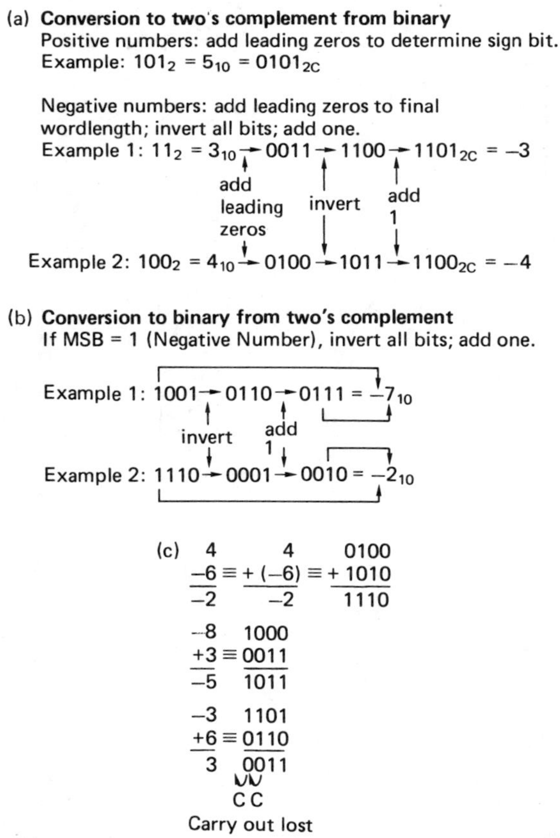

Figure 3.8 (a) Binary to two's complement conversion. (b) Two's complement to binary. (c) Some examples

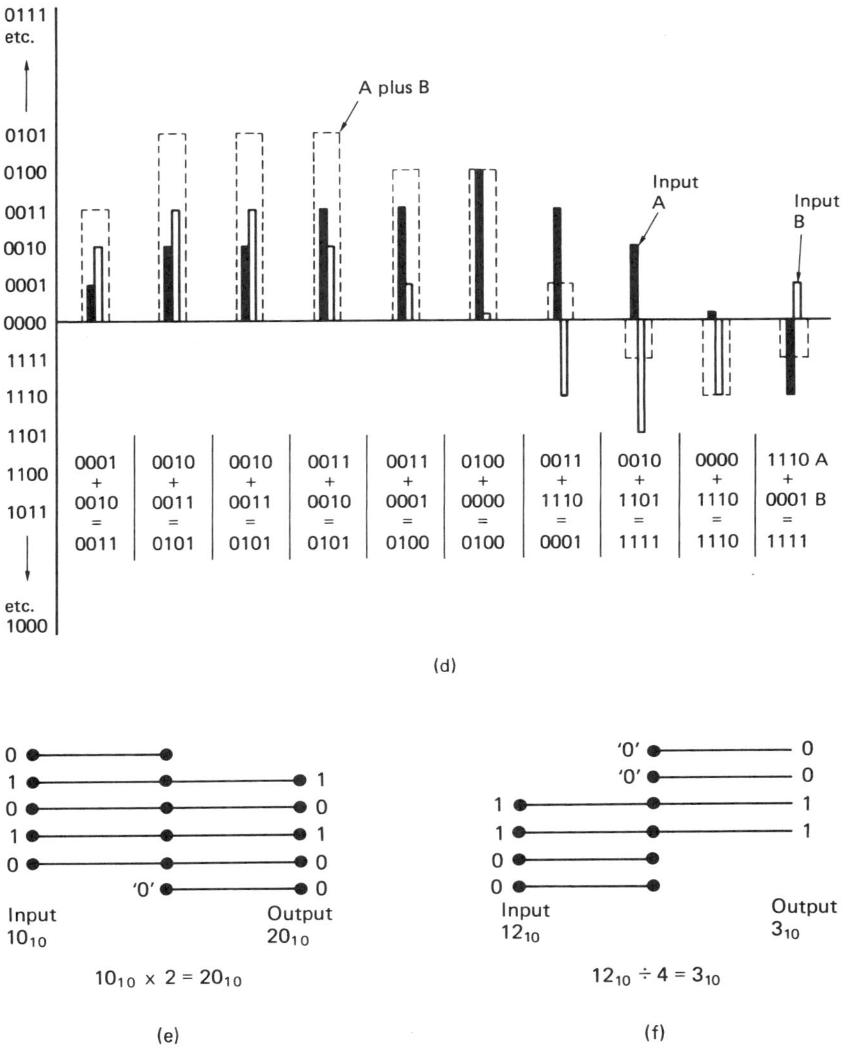

Figure 3.8 (*Continued*). (d) Using two's complement arithmetic, single values from two waveforms are added together with respect to midrange to give a correct mixing function. (e) Multiplication or (f) division by two by bit-shifting compounds to 6 dB gain change

and from two's complement, and illustrates how adding two's complement samples simulates the mixing process. Figure 3.8 also illustrates the process of attenuating by 6 dB, dividing sample values in two's complement by two by shifting one place to the right. For positive numbers, this is easy; for negative numbers, the necessary right shift must carry in a one at the left of the number, not a zero. It should also be noted that in two's complement, if a radix point exists, numbers to the right of it are added. For example, 1100.1 is not −4.5, it is −4 + 0.5 = −3.5.

1109687168

3.3 Binary adding

The circuitry necessary for adding binary numbers is shown in Figure 3.9. Addition in binary requires 2 bits to be taken at a time from the same position in each word, starting at the least significant bit. Should both be ones, the output is zero, and there is a carry-out generated. Such a circuit is called a half adder, shown in Figure

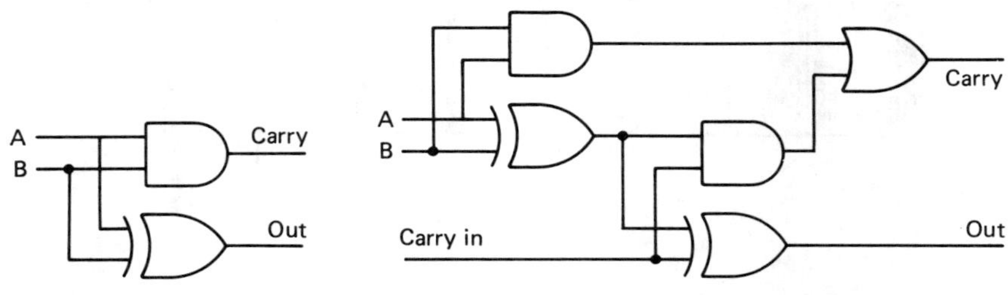

Data A	Bits B	Carry in	Out	Carry out
0	0	0	0	0
0	0	1	1	0
0	1	0	1	0
0	1	1	0	1
1	0	0	1	0
1	0	1	0	1
1	1	0	0	1
1	1	1	1	1

(b)

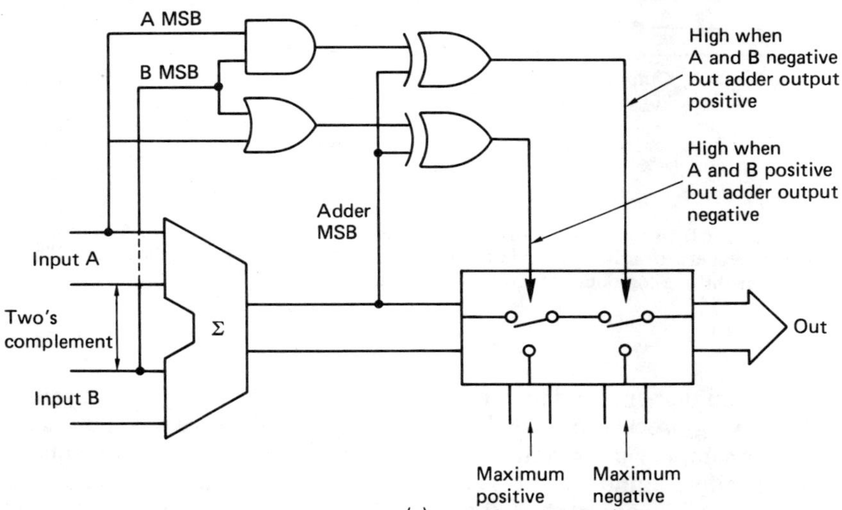

Figure 3.9 (a) Half adder; (b) full-adder circuit and truth table; (c) comparison of sign bits prevents wraparound on adder overflow by substituting clipping level

3.9(a), and is suitable for the least significant bit of the calculation. All higher stages will require a circuit which can accept a carry input as well as two data inputs. This is known as a full adder (Figure 3.9(b)). Multibit full adders are available in chip form and have carry-in and carry-out terminals to allow them to be connected in parallel to operate on long wordlengths.

When mixing by adding sample values, care has to be taken to ensure that if the sum of the two sample values exceeds the number range the result will be clipping rather than wraparound. In two's complement, the action necessary depends on the polarities of the two signals. Clearly if one positive and one negative number are added, the result cannot exceed the number range. If two positive numbers are added, the symptom of positive overflow is that the most significant bit sets, causing an erroneous negative result, whereas a negative overflow results in the most significant bit clearing. The overflow control circuit will be designed to detect these two conditions and override the adder output. If the MSB of both inputs is zero, the numbers are both positive; thus if the sum has the MSB set, the output is replaced with the maximum positive code (0111 . . .). If the MSB of both inputs is set, the numbers are both negative, and if the sum has no MSB set, the output is replaced with the maximum negative code (1000 . . .). These conditions can also be connected to warning indicators. Figure 3.9(c) shows this system in hardware. The resultant clipping on overload is sudden, and sometimes a PROM is included which translates values around and beyond maximum to soft-clipped values below or equal to maximum.

3.4 Digital fading and mixing

Fading is controlled in the digital domain by multiplying each sample value by some fixed coefficient. If that coefficient is less than one, fading will result; if it is greater than one, amplification can be obtained.

Multiplication in binary circuits is difficult. It can be performed by repeated adding, but this is too slow to be of any use. In fast multiplication, one of the inputs will be simultaneously multiplied by one, two, four, etc., by hard-wired bit shifting. Figure 3.10 shows that the other input bits will determine which of these powers will be added to produce the final sum, and which will be neglected. If multiplying by five, the process is the same as multiplying by four, multiplying by one, and adding the two products. This is achieved by adding the input to itself shifted two places. As the wordlength of such a device increases, the complexity increases exponentially, so this is a natural application for an integrated circuit. It is probably true that digital video processing would not have been viable without such chips.

In a digital vision mixer or switcher, the gain coefficients will originate in the hand-operated faders, but may also come from a pattern generator or from keying circuits.

The signal path of a simple composite digital mixer is shown in Figure 3.11. The D-2 digital input is offset binary in that it has a nominal black level of 64_{10} (PAL), and a subtraction has to be made in order that fading will take place with respect to black. On a perfect signal, subtracting 64 would achieve this, but on a slightly out of range signal, it would not. Since the digital active line is slightly longer than the analog active line, the first sample should be blanking level, and this will be the value to subtract to obtain pure binary luminance with respect to black. This is the

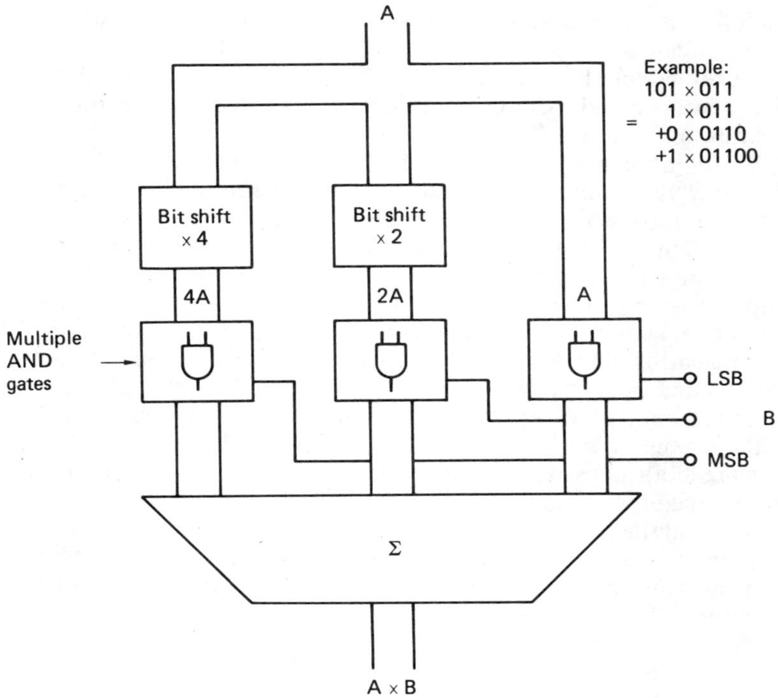

Figure 3.10 Structure of fast multiplier: The input A is multiplied by 1, 2, 4, 8, etc. by the bit-shifting technique of Figure 3.8(e). The digits of the B input then determine which multiples of A should be added together by enabling AND gates between the shifters and the adder. For long wordlengths, the number of gates required becomes enormous, and the device is best implemented in a chip

digital equivalent of black-level clamping. The two inputs are then multiplied by their respective coefficients and added together to achieve the mix. Peak limiting will be required as in Section 3.3, and then, if the output is to be to D-2, 64_{10} (PAL) must be added to each sample value to establish the correct offset. In most video applications, a crossfade will be needed, and a rearrangement of the crossfading equation allows one multiplier to be used instead of two, as shown in Figure 3.12.

The sampling rate of the two inputs must be exactly the same, and in the same phase, or the circuit will not be able to add on a sample-by-sample basis. If the two inputs have come from different sources, they must be synchronized by the same master clock, and/or timebase correction must be provided on the inputs.

Some thought must be given to the wordlength of the system. If a sample is attenuated, it will develop bits which are below the radix point. For example, if an 8 bit sample is attenuated by 24 dB, the sample value will be shifted four places down. Extra bits must be available within the mixer to accommodate this shift. Digital vision mixers with 8 bit inputs can have an internal wordlength of up to 16 bits. When several attenuated sources are added together to produce the final mix, the result will be a 16 bit sample stream. As the output will generally need to

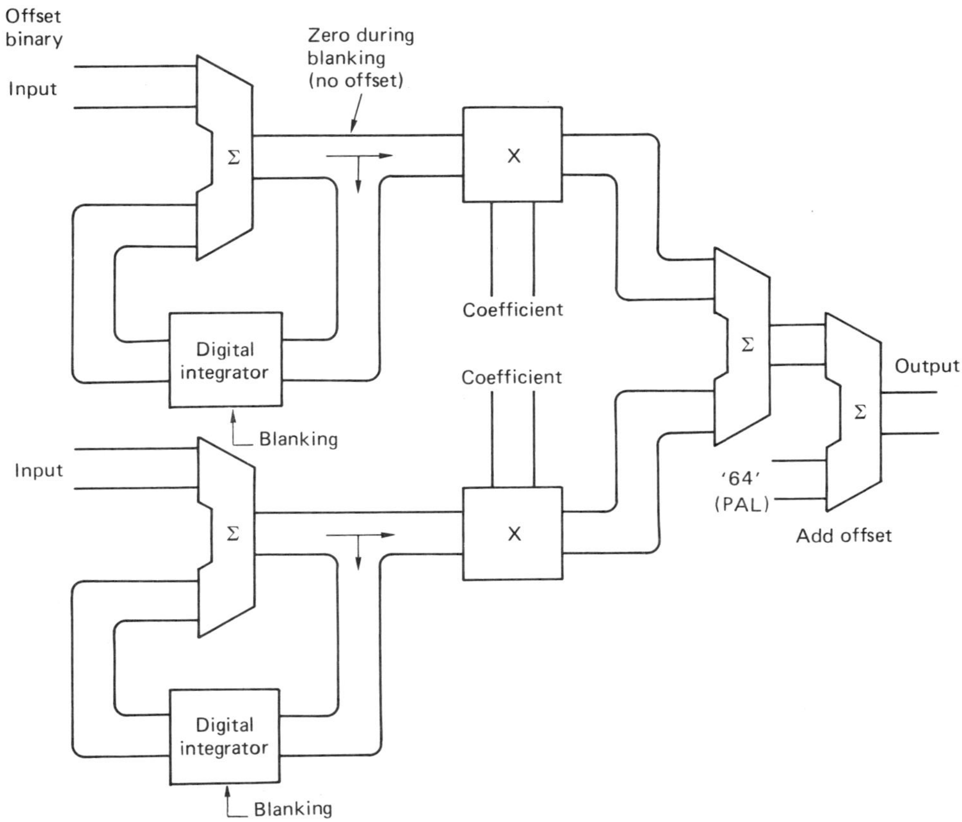

Figure 3.11 A simple digital mixer. Offset binary inputs must have the offset removed. A digital integrator will produce a counter-offset which is subtracted from every input sample. This will increase or reduce until the output of the subtractor is zero during blanking. The offset must be added back after processing if a D-2 output is required

be of the same format as the input, the wordlength must be shortened. This must be done very carefully, as will be seen in Section 3.6.

Clearly the adder which mixes the two channels can only produce a meaningful result if both input values are derived from samples which were taken at precisely the same phase relationship to subcarrier. The critical timing-in process of an analog composite switcher has been removed as far as the digital switcher is concerned, but the need for critical timing accuracy has not been eliminated – it has simply been moved to a different part of the system. A digital timebase corrector can be used to remove the timing errors between input channels, but as it can only correct to the nearest sample, which is 90° in a 4 F_{sc} system, it cannot remove errors due to the sampling clocks of the two channels having a different relationship to subcarrier. If the two channels have such a timing error, it can only be corrected in the digital domain by an interpolator of such complexity that it would be simpler to return to the analog domain for a solution. It is necessary to bear in mind that a

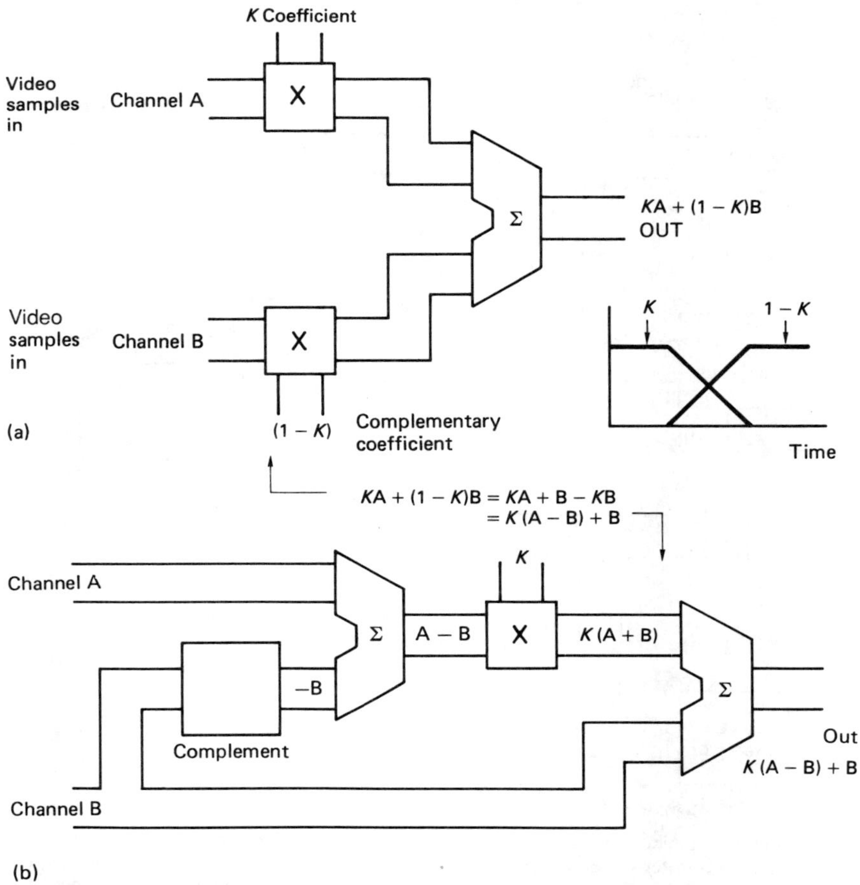

Figure 3.12 Crossfade at (a) requires two multipliers. Reconfiguration at (b) requires only one multiplier

composite digital recorder must have an input sampling clock which is accurately set to a standardized phase relationship with subcarrier, otherwise the recordings made cannot be mixed with others in the digital domain.

3.5 Blanking

It is often necessary to blank the ends of active line smoothly to prevent out-of-band signals being generated. Blanking consists of sloping off the active line by multiplying the sample values by successively smaller coefficients until blanking is reached. This is easy where the sample values have been normalized so that zero represents black, but where the usual offset of 64_{10} (PAL) is present, multiplication by descending coefficients will cause a black-level shift. The solution is to use a correction PROM which can be seen in Figure 3.13. This is addressed by the multiplier coefficient and adds a suitable constant to the multiplier output. If the multiplier were to have a gain of one-half, this would shift the black level by 32

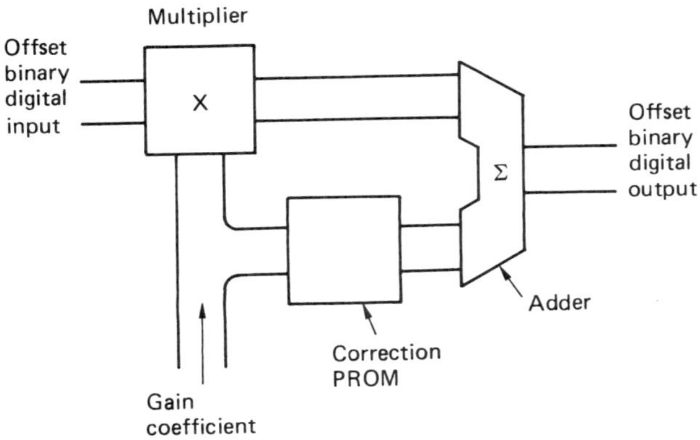

Figure 3.13 In order to fade an offset binary signal, a correction term from a PROM can be added to remove the level shift caused by fading

quantizing intervals, and so the correction PROM would add 32 to the output. Where the multiplier has fully blanked, the output will be zero, and the correction PROM has to add 64_{10} to the output.

3.6 Digital dither

When a sample value is attenuated, the extra low-order bits made available preserve the resolution of the signal and the dither in the least significant bit which

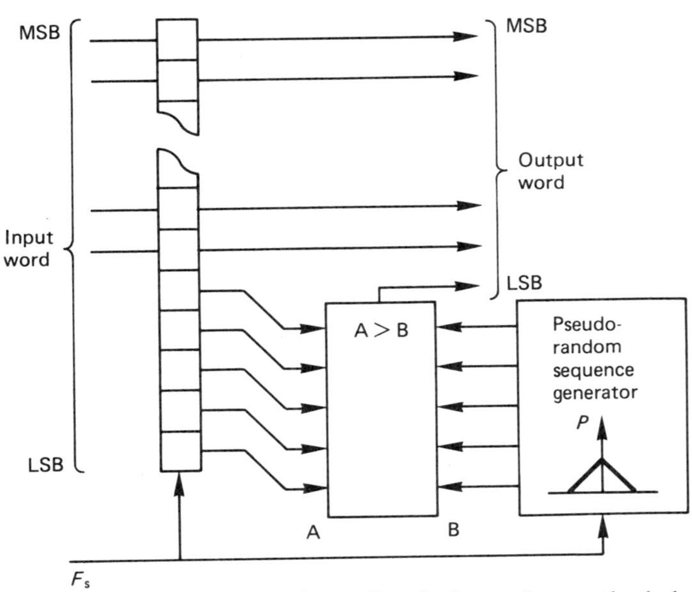

Figure 3.14 In order to shorten the wordlength of a sample correctly, the least significant bit must contain a random element to prevent quantizing distortion. By comparing the triangular probability pseudo-random sequence with the low-order bits, the resulting LSB contains information from those bits in the form of duty cycle modulation

linearizes the system. If several such sample streams are added together, the random element in the low-order bits may now be some way below the least significant bit in the shortened word. If the word is simply truncated by ignoring the low-order bits below the desired wordlength, or rounded, the result will be quantizing distortion, because the dither component has been removed. The wordlength of samples must be shortened so as to replace the lost dither, a process called digital dithering. A pseudo-random sequence generator is necessary, which is compared with the last bit to be retained and those below it. The comparison, as shown in Figure 3.14, rounds up or down to yield the least significant bit of the shortened word, which has a linearizing random component. The probability density of the pseudo-random sequence is important. Vanderkooy and Lipshitz[1] found that uniform probability density produced noise modulation, in which the amplitude of the random component varies as a function of the amplitude of the samples. A triangular probability density function obtained by adding together two pseudo-random sequences eliminated the noise modulation to yield a signal-independent white-noise component in the least significant bit. It is vital that such steps are taken when sample wordlength is to be reduced.

3.7 Introduction to digital interfacing

Of all the advantages of digital recording applied to video the most important of these for production work is the ability to record through multiple generations without quality loss. For the highest-quality post-production work, digital interconnection between such items as switchers and recorders is highly desirable to avoid the degradation due to repeated conversion and filtering stages.

Video converters universally use parallel connection, where all bits of the pixel value are applied simultaneously to separate pins. Although rotary-head digital recorders lay data on the tape serially, within the circuitry of the recorder parallel presentation is in the majority, because it allows slower, and hence cheaper, memory chips to be used for interleaving and timebase correction. The Reed–Solomon error correction depends upon symbols assembled from 8 bits at once. Digital colour processors universally operate upon pixel values in parallel.

Bearing all of this in mind, there is a strong argument for parallel connection, because the video naturally appears in the parallel format in typical machines. All that is necessary is a set of suitable driver chips, running at an appropriate sampling rate, to send video down cables having separate conductors for each bit of the sample, and clocks to tell the receiver when to sample the bit values. The cost in electronic components is very small, and for short distances this approach represents the optimum solution. An example would be where it is desired to dub a digital recording from one D-2 machine to another standing beside it.

Parallel connection has drawbacks too; these come into play when longer distances are contemplated. A multicore cable is expensive, and the connectors are physically large. More seriously, there are electronic problems with multicore cables. The propagation speeds of pulses down all of the cores in the cable will not be exactly the same, and so, at the end of a long cable, some bits may still be in transition from the previous state when the clock arrives, and others may have begun to change to the value in the next pixel. In the presence of crosstalk between the conductors, and reflections due to suboptimal termination, the data integrity eventually becomes marginal.

Where it is proposed to interconnect a large number of units with a router, that device will be extremely complex because of the number of parallel signals to be handled.

The answer to these problems is the serial connection. All of the digital samples are multiplexed into a serial bit stream, and this is encoded to form a self-clocking channel code which can be sent down a single channel. Skew caused by differences in propagation speed cannot then occur. The bit rate necessary is in the order of 150 Mbits/s, but this is easily accommodated by coaxial cable. Provided suitable equalization and termination is used, it may be possible to send such a signal over cables intended for analog video. A similar approach was pioneered in digital audio by the AES/EBU serial interface, which allowed digital audio to be transmitted down existing twisted-pair analog cabling.

The cabling savings implicit in such a system are obvious, but the electronic complexity of a serial interconnect is greater, as high-speed multiplexers or shift registers are necessary at the transmitting end, and a phase-locked loop, data separator and deserializer are needed at the receiver to regenerate the parallel signal needed with the equipment. The increased complexity will not raise cost excessively if specialized chips are available, and there will be a saving in cabling.

A distinct advantage of serial transmission is that a matrix distribution unit or router is more easily realized. Where numerous pieces of video equipment need to be interconnected in various ways for different purposes, a crosspoint matrix is an obvious solution. With serial signals, only one switching element per signal is needed, whereas in a parallel system, a matrix is unwieldly.

The third interconnect to be considered uses fibreoptics. The advantages of this technology are numerous: the bandwidth of an optical fibre is staggering, as it is determined primarily by the speed of the light source and sensor, and it is possible to multiplex many digital TV signals down one cable, although the cost of the multiplexing circuitry would preclude this except for long distances. The optical transmission is immune to electromagnetic interference from other sources, nor does it contribute any. The cable can be made completely from insulating materials, so that ground loops cannot occur.

Drawbacks of fibreoptics are few. They do not like too many connectors in a given channel, as the losses at a connection are much greater than with an electrical plug and socket. It is preferable for the only breaks in the fibre to be at the transmitting and receiving points. For similar reasons, fibreoptics are less suitable for distribution, where one source feeds many destinations. The familiar loop-through connection of analog video is just not possible. The bidirectional open collector or tri-state buses of electronic systems cannot be implemented with fibreoptics, nor is it easy to build a crosspoint matrix.

Fibreoptics are currently costly, but wider adoption would reduce prices.

3.8 D-2 composite digital interface

Composite digital samples at four times subcarrier frequency, and so there will be major differences between the standards.

The PAL and NTSC versions of the composite digital interface will be described separately. The electrical interface is the same for both, and is described in Section 3.11.

3.9 PAL D-2 interface

The quantizing range of digital PAL was shown in Figure 3.3, and this section should be read in conjunction with that diagram.

In PAL, the composite digital interface samples at $4 \times F_{sc}$ with sample phase aligned with burst phase. PAL burst swing results in burst phases of ±35°, and samples are taken at these phases and at ±45°, precisely half-way between the *U* and *V* axes. It will be seen later in this chapter that this is useful for colour processing. It is most important that samples are taken exactly at the points specified, since any residual phase error in the sampling clock will cause the equivalent of a chroma phase error when the samples are added to other samples in a switcher.

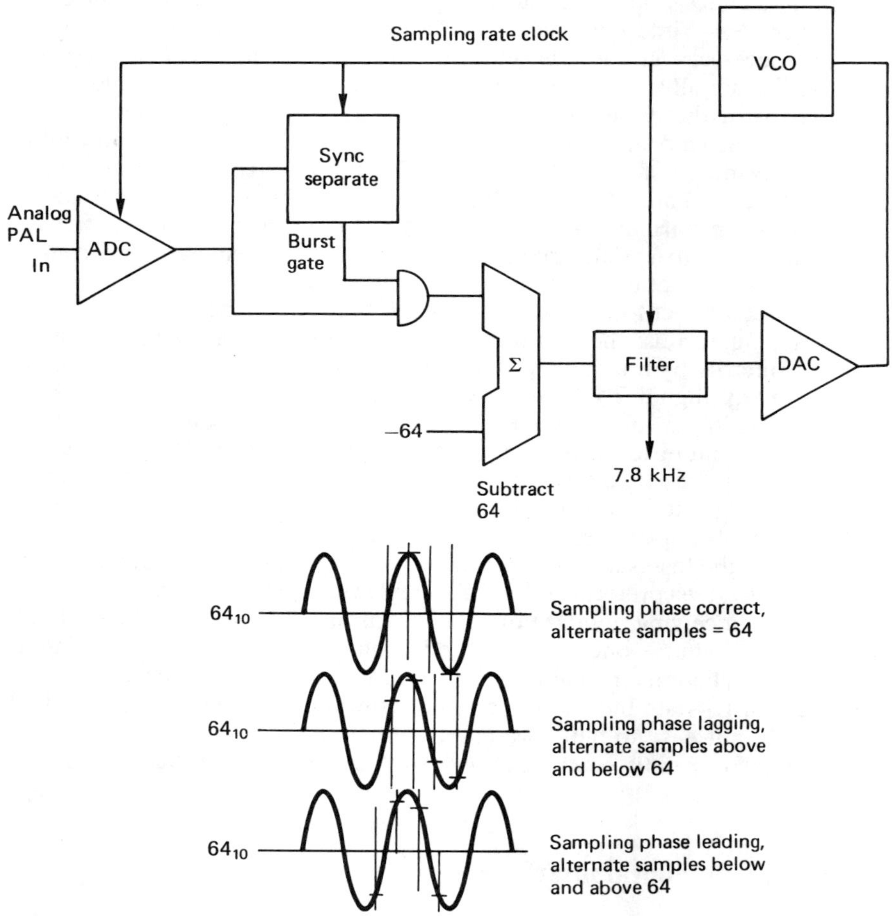

Figure 3.15 Obtaining the sampling clock in PAL D-2. The values obtained by sampling burst are analysed. When phase is correct, burst will be sampled at zero crossing and sample value will be 64_{10} or blanking level. If phase is wrong, sample will be above or below blanking. Filter must ignore alternate samples at burst peaks and shift one sample every line to allow for burst swing. It also averages over several burst crossings to reduce jitter. Filter output drives DAC and thus controls sampling clock VCO

Figure 3.15 shows how the sampling clock may be derived. The incoming sync is used to derive a burst gate, during which the samples of burst are analysed. If the clock is correctly phased, the sampled burst will give values of 95_{10}, 64_{10}, 32_{10}, 64_{10}, repeated, whereas if a phase error exists, the values at the burst crossings will be above or below 64_{10}. The difference between the sample values and blanking level can be used to drive a DAC which controls the sampling VCO. In this way any phase errors in the ADC are eliminated, because the sampling clock will automatically servo its phase to be identical to digital burst. Burst swing causes the burst-peak and burst-crossing samples to change places, so a phase comparison is always possible during burst. The sample numbering is such that sample zero is always at the start of active line.

In PAL, the subcarrier frequency contains a 25 Hz offset, and so $4 \times F_{sc}$ will contain a 100 Hz offset. The sampling rate is not h-coherent, and the sampling structure is not quite orthogonal. As subcarrier is given by:

$$F_{sc} = 283\tfrac{3}{4} \times F_h + F_v/2$$

the sampling rate will be given by:

$$F_s = 1135F_h + 2F_v$$

This results in 709 379 samples per frame, and there will not be a whole number of samples in a line. In practice, 1135 sample periods are considered to be one line, with an additional four sample periods which are skipped over during the frame blanking.

Since sampling is not h-coherent, the position of sync pulses will change relative to the sampling points from line to line. The relationship can also be changed by the ScH phase of the analog input. Zero ScH is defined as coincidence between sync and zero degrees of subcarrier phase at line one of field one. Since composite digital samples on burst phase, not on subcarrier phase, the definition of zero ScH will be as shown in Figure 3.15, where it will be seen that two samples occur at exactly equal distances either side of the 50% sync point. If the input is not zero ScH, the samples conveying sync will have different values. Measurement of these values will allow ScH phase to be computed. Non-standard ScH is only a problem on the initial conversion from analog because the D-2 DVTR does not record sync, and will regenerate zero ScH syncs on replay.

3.10 NTSC D-2 interface

Although they have some similarities, PAL and NTSC are quite different when analysed at the digital sample level.

Figure 3.16 shows how the NTSC waveform fits into the quantizing structure. Blanking is at 60_{10} in an 8 bit system, and peak white is at 200_{10}, so that 1 IRE unit is the equivalent of 1.4 Q.

Subcarrier in NTSC has an exact half-line offset, so there will be an integer number of cycles of subcarrier in two lines. F_{sc} is simply $227.5 \times F_h$, and as sampling is at $4 \times F_{sc}$, there will be $227.5 \times 4 = 910$ samples per line period, and the sampling will be orthogonal.

As with PAL, the sampling phase is chosen to facilitate processing in the digital domain. In NTSC the I axis is phase shifted 123° from subcarrier and 57° from

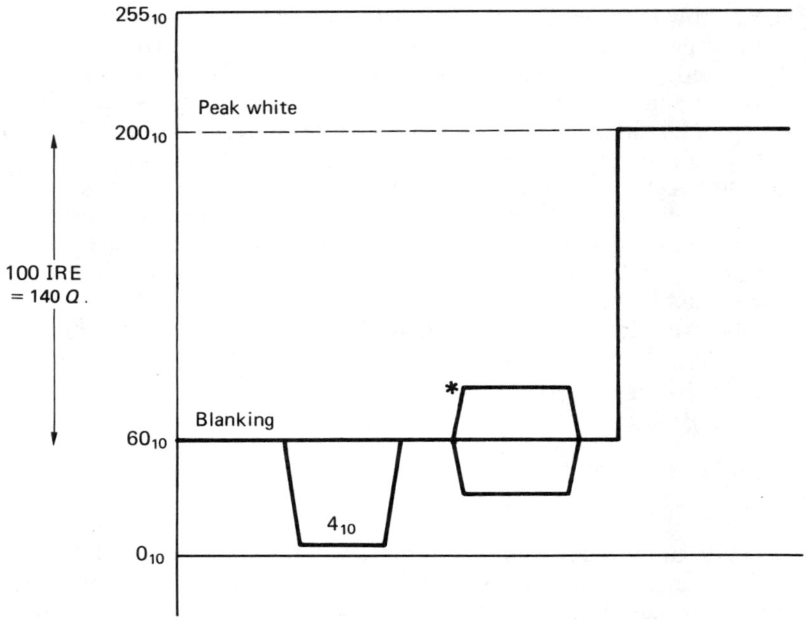

Figure 3.16 The composite NTSC signal fits into the quantizing range as shown here. Note that there is sufficient range to allow the instantaneous voltage to exceed peak white in the presence of saturated, bright colours. Values shown are decimal equivalents in a 10 or 8 bit system. In a 10 bit system the additional 2 bits increase resolution, not magnitude, so they are below the radix point and the decimal equivalent is unchanged. *Note that, unlike PAL, NTSC D-2 does not sample on burst phase and so values during burst are not shown here. See Figure 3.17 for burst sample details

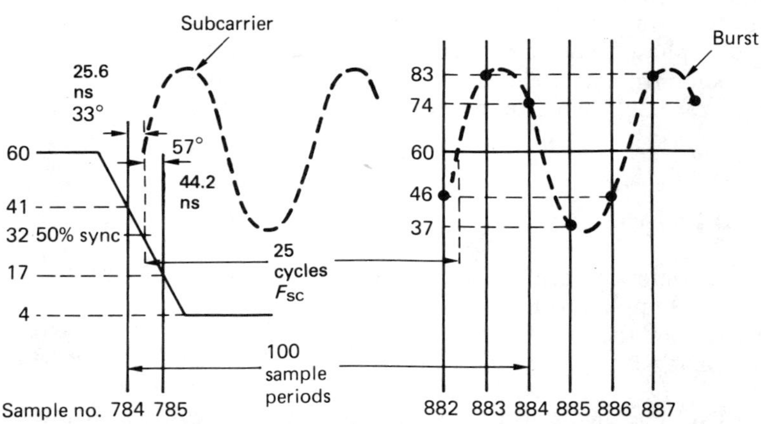

Figure 3.17 NTSC D-2 ScH phase. Sampling is not performed in phase with burst as in PAL, but on the *I* and *Q* axes. Since in NTSC there is a phase angle of 57° between burst and *I*, this will also be the phase at which burst samples should be taken. If ScH phase is zero, then phase of subcarrier taken at 50% sync will be zero, and the samples will be taken 33° before and 57° after sync. 25 cycles of subcarrier or 100 samples later, during burst, the sample values will be obtained. Note that in NTSC burst is inverted subcarrier, so sample 785 is positive, but sample 885 is negative

burst. Composite digital NTSC samples on the *I* and *Q* axes, at 33°, 123°, 213° and 303° of subcarrier, exactly as an NTSC decoder would.

Figure 3.17 shows how this approach works in relation to sync and burst. Zero ScH is defined as zero degrees of subcarrier at the 50% point on sync, but the 57° sampling phase means that the sync edge is actually sampled 25.6 ns ahead of, and 44.2 ns after, the 50% point. Similarly, when the burst is reached, the phase shift means that burst-sample values will be 46_{10}, 83_{10}, 74_{10} and 37_{10}, repeating. The phase-locked loop which produces the sampling clock will digitally compare the samples of burst with the values given here. Since it is not sampling burst at a zero crossing, the slope will be slightly less, so the gain of the phase error detector will also be less, and more prone to burst noise than in the PAL process. The phase error can, however, be averaged over several burst samples to overcome this problem.

The diagram shows sample numbers which, as for PAL, start at zero at the beginning of the active line. Again, if the analog input does not have zero ScH phase, the sync pulse values will change, but the burst values will not. The D-2 recorder does not record sync and will regenerate zero ScH syncs on replay.

3.11 The parallel electrical interface

D-2 uses the same electrical and mechanical interface as is used for 4:2:2 component wording,[2–4] although clearly it is not possible to connect one to the

Pin no.	Function		Pin no.	Function	
1	Clock +		14	Clock −	
2	System ground		15	System ground	
3	Data 7 (MSB) +		16	Data 7 (MSB) −	
4	Data 6 +		17	Data 6 −	
5	Data 5 +		18	Data 5 −	
6	Data 4 +		19	Data 4 −	
7	Data 3 +		20	Data 3 −	
8	Data 2 +		21	Data 2 −	
9	Data 1 +		22	Data 1 −	
10	Data 0 +		23	Data 0 −	
11	Data −1 +	10 bit systems only	24	Data −1 −	10 bit systems only
12	Data −2 +		25	Data −2 −	
13	Cable shield				

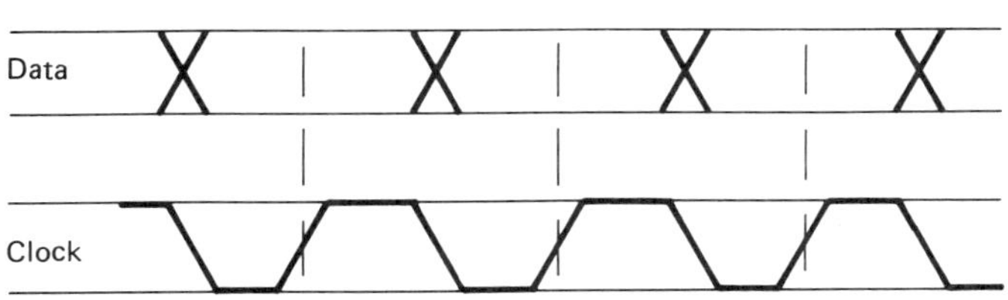

Figure 3.18 In the parallel interface there is capacity for 10 bit words plus a clock where each signal is differential. In 8 bit systems DATA −1 and DATA −2 are not used. Clock transition takes place in the centre of a data bit cell as shown in the lower diagram

other directly. Each signal in the interface is carried by a balanced pair using ECL drive levels. As Figure 3.18 shows, there are eight signal pairs and two optional pairs, so that a 10 bit word can be accommodated. The optional signals are used to add bits at the least significant end of the word. A separate clock and a number of grounding and shielding pins complete the connection. A 25 pin D-type connector to ISO 2110-1980 is specified. Equipment always has female connectors, cables always have male connectors. Connector latching is by screws, with suitable posts provided on the female connector.

Figure 3.18 also shows the relationship between the clock and the data. A positive-going clock edge is used to sample the signal lines after the level has settled between transitions. The clock will be four times the frequency of PAL or NTSC subcarrier.

3.12 Scrambled serial interface

The interface described here has been developed to allow up to 10 bit samples of component or composite digital video to be communicated serially.[5]

Scrambling, or pseudo-random coding, uses concepts which will be introduced in Chapter 4. The serial interface uses convolutional coding, which is simpler to implement in a cable installation because no separate synchronizing of the randomizing is needed.

The components necessary for a composite serial link are shown in Figure 3.19. Parallel D-2 data having a wordlength of up to 10 bits form the input. These are fed to a 10 bit shift register which is clocked at ten times the input rate, or $40 \times F_{sc}$. If there are less than 10 bits in the input words, the missing bits are forced to zero for transmission. The serial data from the shift register are then passed through the scrambler, which produces the exclusive–OR of a given bit and 2 bits which are five and nine clocks ahead. This is followed by another stage, which converts channel ones into transitions. The resulting signal can be fed down 75 W coaxial cable using BNC connectors.

The receiver must generate a bit clock at $40 \times F_{sc}$ from the input signal, and this clock drives a circuit which simply reverses the scrambling at the transmitter. The first stage returns transitions to ones, and the second stage is a mirror image of the encoder which reverses the exclusive–OR calculation to output the original data. Since transmission is serial, it is necessary to obtain word synchronization, so that correct deserialization can take place. A Timing Reference and Identification Signal (TRS-ID) is added during blanking, and the receiver can detect the patterns which it contains. TRS-ID consists of five words which are inserted just after the leading edge of video sync. The first four are for synchronizing and consist of a single word of all ones, followed by three words of all zeros. The fifth word is for identification and carries line and field numbering information shown in Figure 3.20. In PAL the field numbering will go from zero to seven, whereas in NTSC it will only reach three.

On detection of the synchronizing symbols, a divide-by-ten circuit is reset, and the output of this will clock words out of the shift register at the correct times. This output will also become the output word clock.

It is also possible to merge four channels of digital audio with the scrambled serial interface, and this is described in Chapter 8.

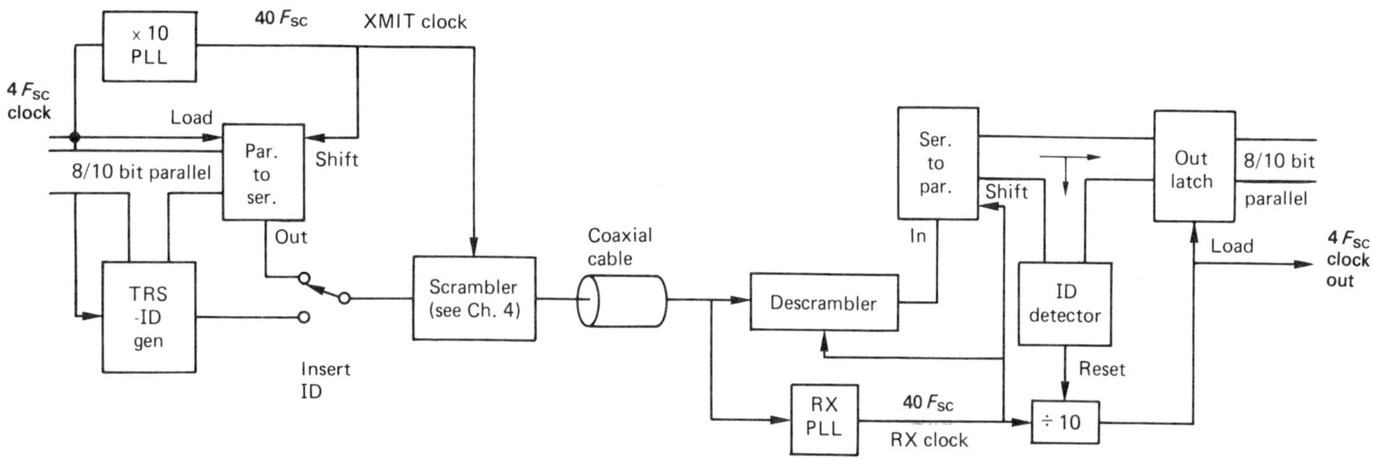

Figure 3.19 In the scrambled serial link, parallel data are loaded into a shift register and clocked out ten times as fast. The phase-locked loop in the receiver regenerates the bit clock from the scrambled transmission and uses it to clock the serial to parallel converter. Every tenth bit, the parallel data must be loaded to the output latch. The phase of this loading is controlled by detection of the ID pattern which is inserted at the transmitter

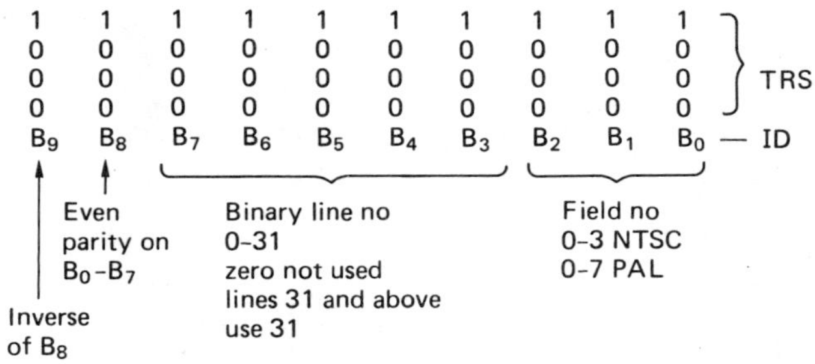

Figure 3.20 The contents of the TRS-ID pattern which is added to the transmission during the horizontal sync pulse just after the leading edge. The field number conveys the composite colour framing field count, and the line number carries a restricted line count intended to give vertical positioning information during the vertical interval. This count saturates at 31 for lines of that number and above

3.13 Timebase correction

In Chapter 1 it was stated that a strength of digital technology is the ease with which delay can be provided. Accurate control of delay is the essence of timebase correction, necessary whenever the instantaneous time of arrival or rate from a data source does not match the destination.

In magnetic tape recording, the destination will almost always have perfectly regular timing, namely the reference sampling-rate generator, and one aspect of timebase correction consists of aligning jittery offtape signals with the stable reference. Rotary-head recorders suffer from impulsive jitter due to the heads striking the tape.

All digital video recorders assemble data into blocks to facilitate editing and error correction as well as to permit head switching between clocks. Owing to the spaces between blocks, data arrive in bursts on replay, but must be fed to the output converters in an unbroken stream at the sampling rate. In this way, timing errors are rendered unmeasurable. An extension of the timebase correction concept is *time compression*, which is used in rotary-head machines to give more freedom in the mechanical design of the scanner. This will be treated in detail in Chapter 6.

Although delay is easily implemented, it is not possible to advance a data stream. Most real machines cause instabilities balanced about the correct timing: the output jitters between too early and too late. Since the information cannot be advanced in the corrector, only delayed, the solution is to run the machine in advance of real time. In this case, correctly timed output signals will need a nominal delay to align them with reference timing. Early output signals will receive more delay, and late output signals will receive less delay.

3.14 RAM timebase correction

There are three basic ways of obtaining delay in the digital domain: shift registers, random access memories (RAMs) and first-in–first-out devices (FIFOs). Where the delay is large, the RAM is the only practical device.

The basic memory element in logic circuits is the latch, which is constructed from two gates and can be set or reset. A more useful variant is the D-type latch which remembers the state of the input at the time a separate clock either changes state for an edge-triggered device, or after it goes false for a level-triggered device. D-type latches are commonly available with four or eight latches to the chip. A shift register can be made by connecting them one after the other, so that data are delayed by the number of stages in the register.

Where large numbers of bits are to be stored, cross-coupled latches are less suitable because they are more complicated to fabricate inside integrated circuits than dynamic memory, and consume current.

In large random access memories (RAMs), the data bits are stored as the presence or absence of charge in a tiny capacitor. The charge will suffer leakage, and the value would become indeterminate after a few milliseconds. Where the delay needed is less than this, decay is of no consequence, as data will be read out

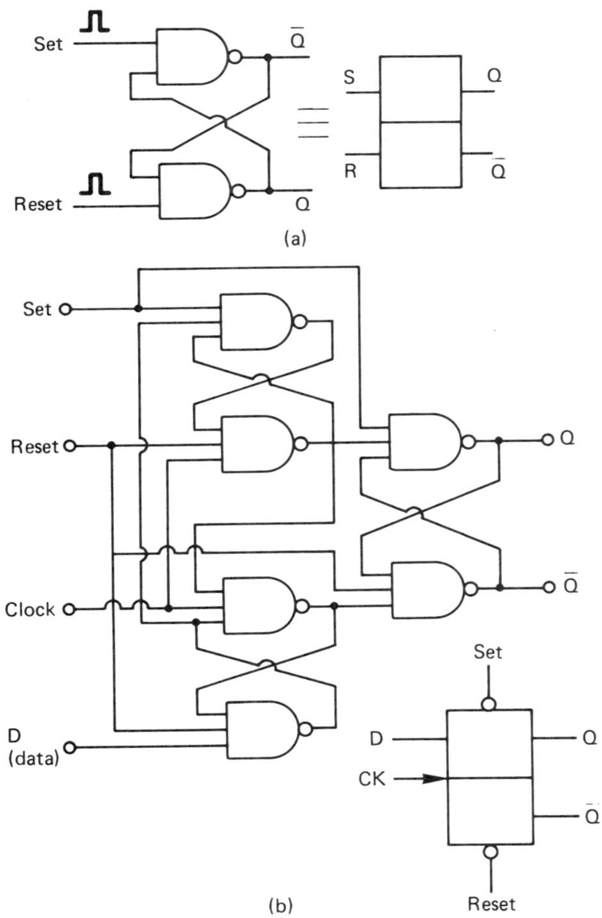

Figure 3.21 Digital semiconductor memory types. At (a), one data bit can be stored in a simple set–reset latch, which has little application because the D-type latch at (b) can store the state of the single data input when the clock occurs. These devices can be implemented with bipolar transistors of FETs, and are called static memories because they can store indefinitely. They consume a lot of power

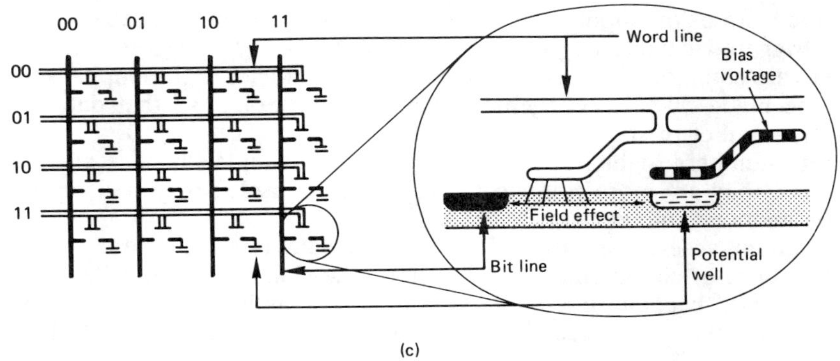

Figure 3.21 (*Continued*). At (c), a bit is stored as the charge in a potential well in the substrate of a chip. It is accessed by connecting the bit line with the field effect from the word line. The single well where the two lines cross can then be written or read. These devices are called dynamic RAMs because the charge decays, and they must be read and rewritten (refreshed) periodically

before they have had a chance to decay. Where longer delays are necessary, such memories must be refreshed periodically by reading the bit value and writing it back to the same place. Most modern RAM chips have suitable circuitry built in. Large RAMs store thousands of bits, and it is clearly impractical to have a connection to each one. Instead, the desired bit has to be addressed before it can be read or written. The size of the chip package restricts the number of pins available, so that large memories use the same address pins more than once. The bits are arranged internally as rows and columns, and the row address and the column address are specified sequentially on the same pins. Figure 3.21 shows some examples of different memory devices.

The shift-register approach and the memory approach to delay are very similar, as a shift register can be thought of as a memory whose address increases automatically when clocked. The data rate and the maximum delay determine the capacity of the memory required. Figure 3.22 shows that the addressing of the memory is by a counter that overflows endlessly from the end of the memory back to the beginning, giving the memory a ring-like structure. The write address is

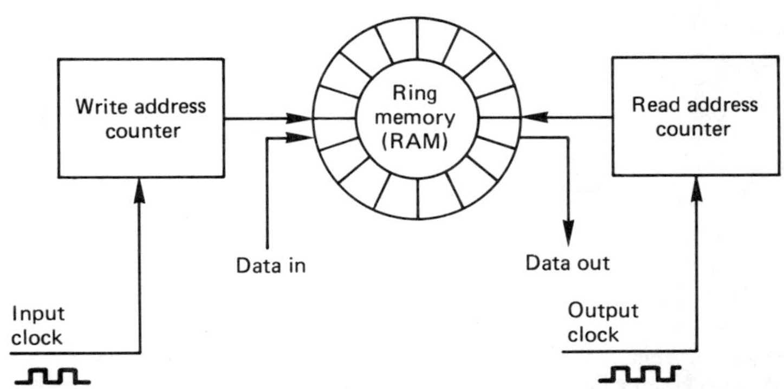

Figure 3.22 TBC memory is addressed by a counter which periodically overflows to give a ring structure. Memory allows read side to be non-synchronous with write side

determined by the incoming data, and the read address is determined by the outgoing data. This means that the memory has to be able to read and write at the same time. The switching between read and write involves not only a data multiplexer but also an address multiplexer, as can be seen in Figure 3.23. In general the arbitration between read and write will be done by signals from the stable side of the TBC, which in the case of digital recorder replay will be the read side. The stable side of the memory will read a sample when it demands, and the writing will be locked out for that period. The input data cannot be interrupted in many applications, however, so a small buffer silo is installed before the memory, which fills up as the writing is locked out and empties again as writing is permitted. Alternatively, the memory will be split into blocks, such that when one block is reading a different block will be writing and the problem does not arise. In most digital video applications, the sampling rate exceeds the rate at which economically available RAM chips can operate. The solution is to arrange several video samples into one longer word, usually known as a superword, and to construct the memory so that it stores superwords in parallel.

3.15 Digital filtering

Earlier sections of this chapter have described how delay and gain control are achieved in the digital domain. These are the vital components required for digital filtering. One of the strengths of digital filtering is that it can be performed with

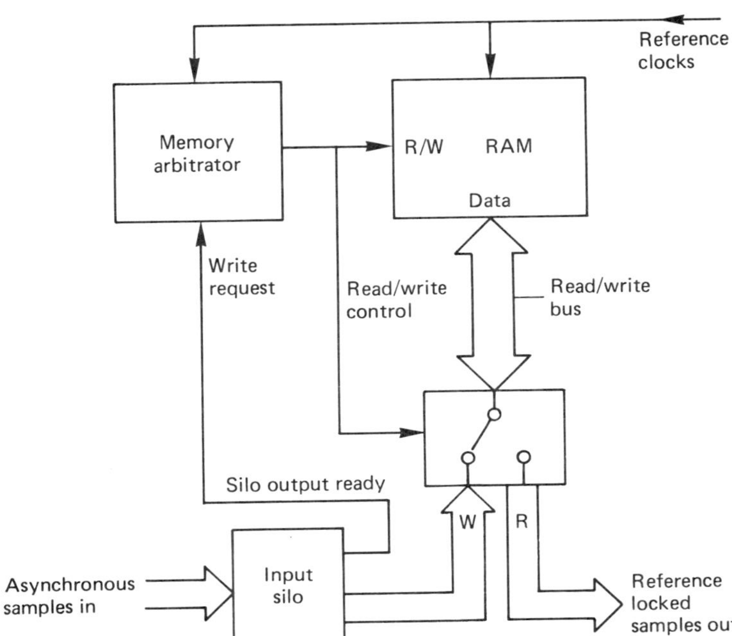

Figure 3.23 In a RAM-based TBC, the RAM is reference synchronous, and an arbitrator decides when it will write. During reading, asynchronous input data backs up in the input silo, asserting a write request to the arbitrator. Arbitrator will then cause a write cycle between read cycles

stable binary logic instead of the inductors and capacitors needed for analog filters. In analog filtering, the frequency response is usually the most quoted parameter, followed by the phase response and the impulse response. These last two are the most difficult to get right in an analog filter.

Figure 3.24 shows that impulse response testing tells a great deal about a filter. In a perfect filter, all frequencies should experience the same time delay; this is the group delay. If not, there is a group-delay error. As an impulse contains an infinite spectrum, a filter fed with an impulse will separate the different frequencies in time if it suffers from group-delay error. Group-delay error is particularly unacceptable in video, because it causes the different frequencies in a sharp edge to appear at different places across the screen.

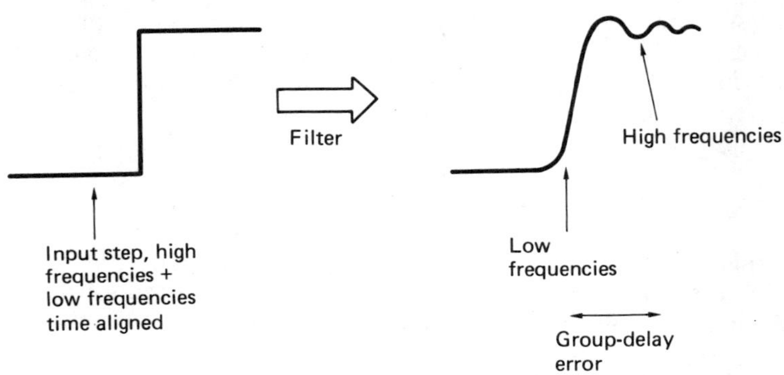

Figure 3.24 Group-delay error time displaces signals as a function of frequency

A pure delay will cause a phase shift proportional to frequency, and a filter with this characteristic is said to be phase linear. The impulse response of a phase-linear filter is symmetrical. If a filter suffers from group-delay error it cannot be phase linear. It is almost impossible to make a phase-linear analog filter, and many filters have a group-delay equalization stage following them which is often as complex as the filter itself. In the digital domain it is reasonably straightforward to make a phase-linear filter, and phase equalization becomes unnecessary. Because of the sampled nature of the signal, whatever the response at low frequencies may be, all digital channels act as low-pass filters cutting off at the Nyquist limit, or half the sampling frequency.

3.16 FIR and IIR filters compared

Filters can be described in two main classes, as shown in Figure 3.25, according to the nature of the impulse response. Finite-impulse response (FIR) filters are always stable and, as their name suggests, respond to an impulse once, as they have only a forward path. In the temporal domain, the time for which the filter responds to an input is finite, fixed and readily established. The same is therefore true about the distance over which an FIR filter responds in the spatial domain.

Most filters intended for video use fall into this category. Infinite-impulse response (IIR) filters respond to an impulse indefinitely and are not necessarily

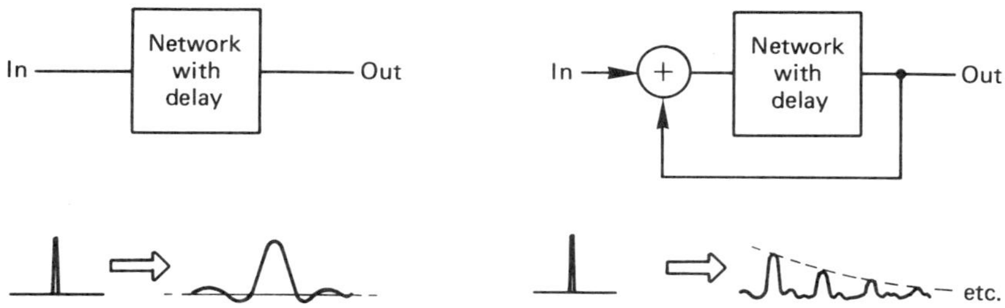

(a) Finite impulse response (FIR)

(b) Infinite impulse response (IIR)

Figure 3.25 An FIR filter (a) responds only to an input, whereas the output of an IIR filter (b) continues indefinitely rather like a decaying echo

stable, as they have a return path from the output to the input. For this reason they are also called recursive filters. Digital noise reducers and effects units which generate trails employ recursive filters, but since these processes are not necessary in D-2 recorders, the subject of IIR filters will not be explored here.

3.17 FIR filters

An FIR filter works by graphically constructing the impulse response for every input sample. It is first necessary to establish the correct impulse response. Figure 3.26(a) shows an example of a low-pass filter which cuts off at one-quarter of the sampling rate. The impulse response of a perfect low-pass filter is a $\sin x/x$ curve, where the time between the two central zero crossings is the reciprocal of the cut-off frequency. According to the mathematics, the waveform has always existed and carries on for ever. The peak value of the output coincides with the input impulse. This means that the filter is not causal, because the output has changed before the input is known. Thus in all practical applications it is necessary to truncate the extreme ends of the impulse response, which causes an aperture effect, and to introduce a time delay in the filter equal to half the duration of the truncated impulse in order to make the filter causal. As an input impulse is shifted through the series of registers in Figure 3.26(b), the impulse response is created, because at each point it is multiplied by a coefficient as in Figure 3.26(c). These coefficients are simply the result of sampling and quantizing the desired impulse response. Clearly the sampling rate used to sample the impulse must be the same as the sampling rate for which the filter is being designed. In practice the coefficients are calculated, rather than attempting to sample an actual impulse response. The coefficient wordlength will be a compromise between cost and performance. Because the input sample shifts across the system to create the shape of the impulse response, the configuration is also known as a transversal filter. In operation with real sample streams, there will be several consecutive sample values in the filter registers at any time in order to convolve the input with the impulse response.

Simply truncating the impulse response causes an abrupt transition from input samples which matter and those which do not. This aperture effect results in a tendency for the response to peak just before the cut-off frequency. This peak is

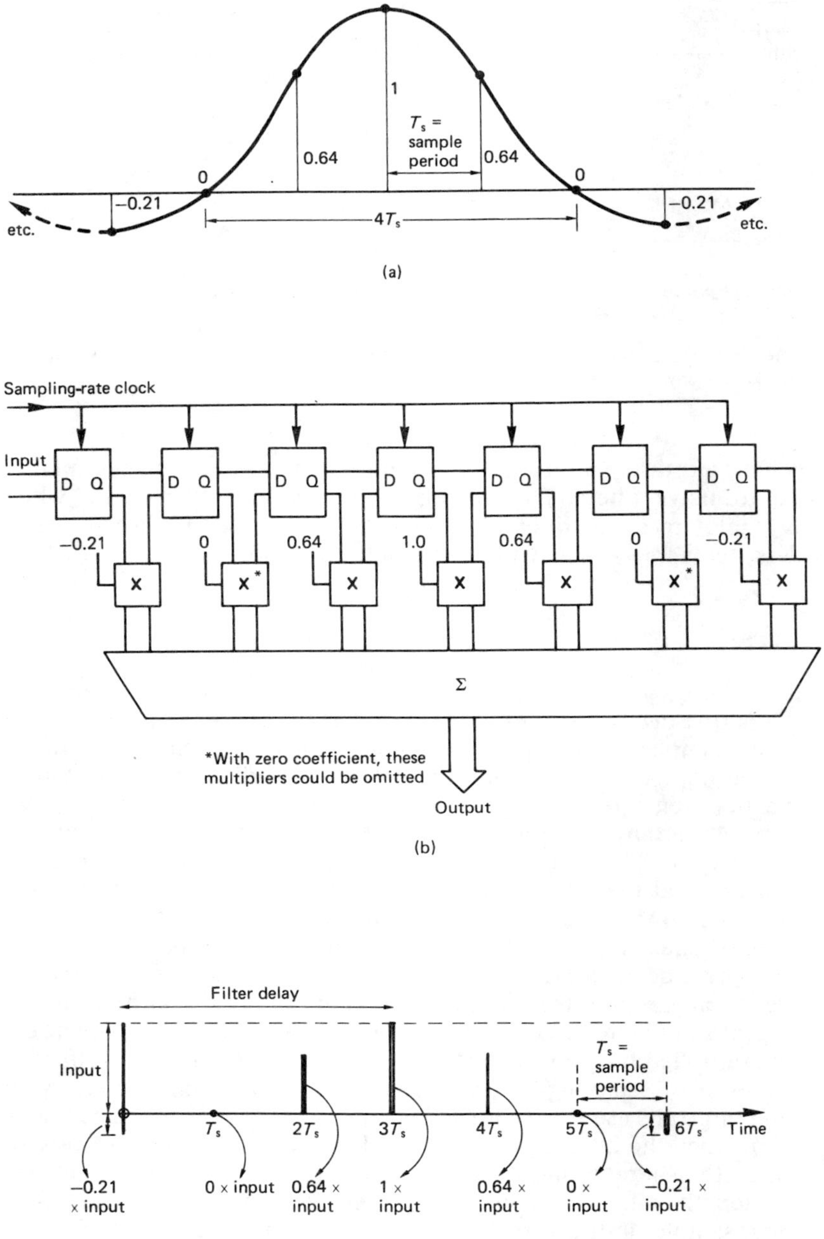

Figure 3.26 (a) The impulse response of an LPF is a sin x/x curve which stretches from $-\infty$ to $+\infty$ in time. The ends of the response must be neglected, and a delay introduced to make the filter causal. (b) The structure of an FIR LPF. Input samples shift across the register and at each point are multiplied by different coefficients. (c) When a single unit sample shifts across the circuit of Figure 3.26(b), the impulse response is created at the output as the impulse is multiplied by each coefficient in turn

known as Gibb's phenomenon; it causes ripples in both passband and stopband.[6,7] As a result, the length of the impulse which must be considered will depend not only on the frequency response, but also on the amount of ripple which can be tolerated. If the relevant period of the impulse is measured in sample periods, the result will be the number of points or multiplications needed in the filter.

Figure 3.27 compares the performance of filters with different numbers of points. A typical digital video FIR filter may need eight points.

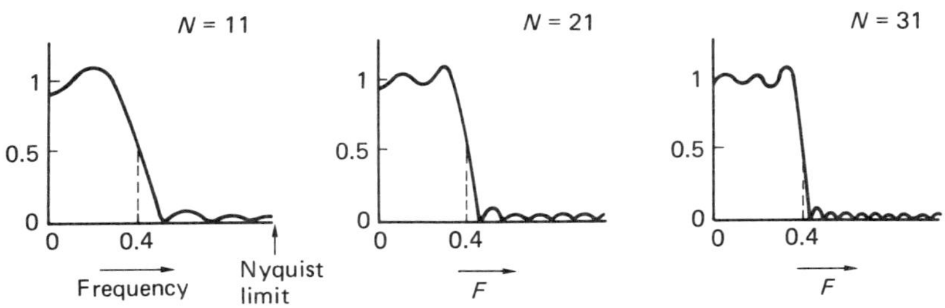

Figure 3.27 The truncation of the impulse in an FIR filter caused by the use of a finite number of points (N) results in ripple in the response. Shown here are three different numbers of points for the same impulse response. The filter is an LPF which rolls off at 0.4 of the fundamental interval (Courtesy *Philips Technical Review*)

Rather than simply truncate the impulse response time, it is better to make a smooth transition from samples which do not count to those that do. This can be done by multiplying the coefficients in the filter by a window function which peaks in the centre of the impulse. Figure 3.28 shows some different window functions and their responses. Clearly the rectangular window is the same as truncation, and the response is shown at I. A linear reduction in weight from the centre of the window to the edges characterizes the Bartlett window II, which trades ripple for an increase in transition region width. At III is shown the Hanning window, which is essentially a raised cosine shape. Not shown is the similar Hamming window, which offers a slightly different trade-off between ripple and the width of the main lobe. The Blackman window introduces an extra cosine term into the Hamming window at half the period of the main cosine period, reducing Gibb's phenomenon and ripple level, but increasing the width of the transition region. The Kaiser window is a family of windows based on the Bessel function, allowing various trade-offs between ripple ratio and main lobe width. Two of these are shown in IV and V. The drawback of the Kaiser windows is that they are complex to implement.

Filter coefficients can be optimized by computer simulation. One of the best-known techniques used is the Remez exchange algorithm, which converges on the optimum coefficients after a number of iterations.

In the example of Figure 3.29, the low-pass filter of Figure 3.26 is shown with a Bartlett window. Acceptable ripple determines the number of significant sample periods embraced by the impulse. This determines in turn both the number of points in the filter and the filter delay. As the impulse is symmetrical, the delay will be half the impulse period. The impulse response is a $\sin x/x$ function, and this has been calculated in the figure. The $\sin x/x$ response is next multiplied by the window function to give the windowed impulse response. If the coefficients are not quantized finely enough, it will be as if they had been calculated inaccurately, and

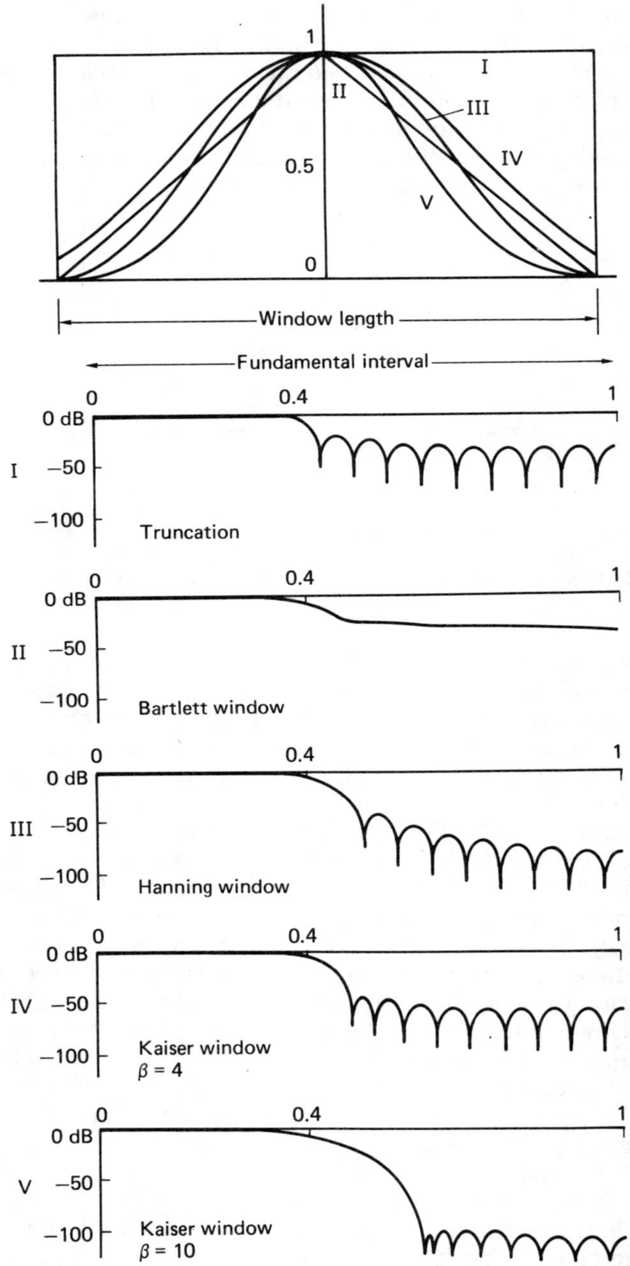

Figure 3.28 The effect of window functions. At top, various window functions are shown in continuous form. Once the number of samples in the window is established, the continuous functions shown here are sampled at the appropriate spacing to obtain window coefficients. These are multiplied by the truncated impulse response coefficients to obtain the actual coefficients used by the filter. The amplitude responses I–V correspond to the window functions illustrated (Responses courtesy *Philips Technical Review*)

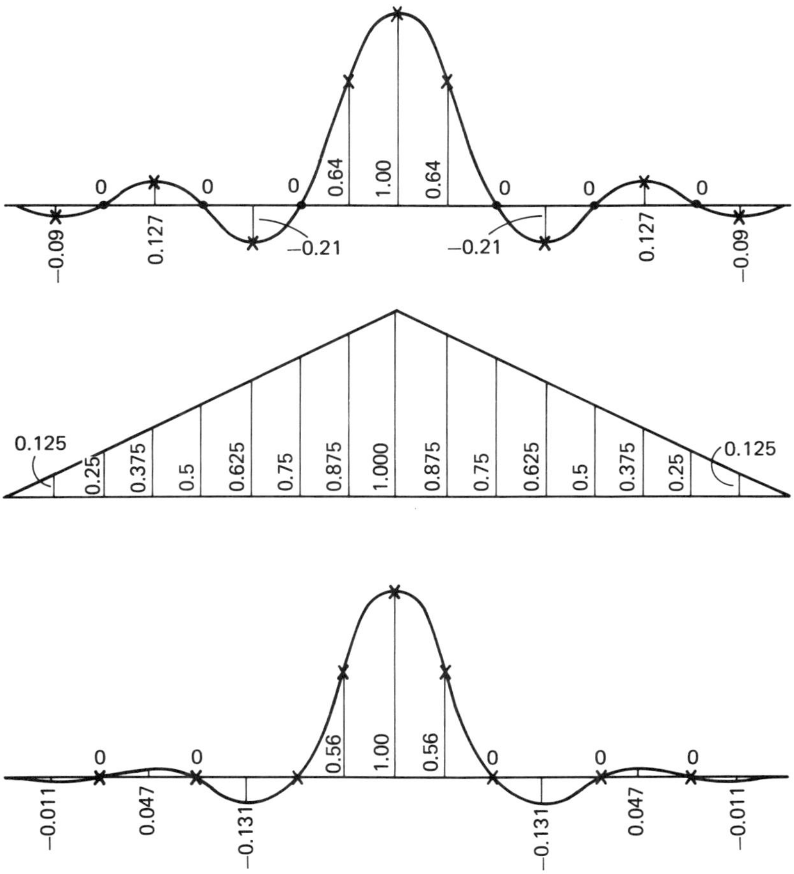

Figure 3.29 A truncated sin *x*/*x* impulse (top) is multiplied by a Bartlett window function (centre) to produce the actual coefficients used (bottom)

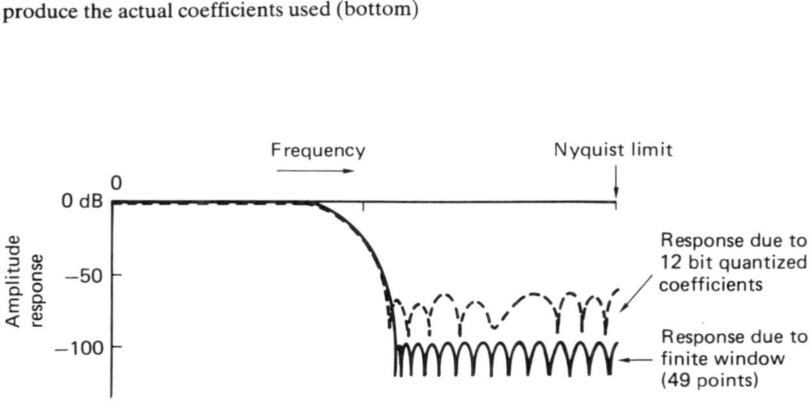

Figure 3.30 Frequency response of a 49 point transversal filter with infinite precision (solid line) shows ripple due to finite window size. Quantizing coefficients to 12 bits reduces attenuation in the stopband (Responses courtesy *Philips Technical Review*)

the performance of the filter will be less than expected. Figure 3.30 shows an example of quantizing coefficients. Conversely, raising the wordlength of the coefficients increases cost.

The FIR structure is inherently phase linear because there is rigid time control as a result of the way samples are shifted through the filter. The individual samples in a digital system do not know in isolation what frequency they represent, and they can only pass the filter at a rate determined by the clock. Because of this inherent phase linearity, an FIR filter can be designed for a specific impulse response, and the frequency response will follow.

The frequency response of the filter can be changed at will by changing the coefficients. A programmable filter only requires a series of PROMs to supply the coefficients; the address supplied to the PROMs will select the response. The frequency response of a digital filter will also change if the clock rate is changed, so it is often less ambiguous to specify a frequency of interest in a digital filter in terms of a fraction of the fundamental interval rather than in absolute terms. This approach also helps in spatial filters where the fundamental interval is a spatial frequency of one-half the number of pixels per unit distance. The configuration shown in Figure 3.26 serves to illustrate the principle. The units used on the diagrams are sample periods, but these could be replaced by pixel spacings for a spatial filter. The response is proportional to these periods or spacings, and so it is not necessary to use actual figures.

Where the impulse response is symmetrical, it is often possible to reduce the number of multiplications, because the same product can be used twice, at equal distances before and after the centre of the window. This is known as folding the filter. A folded filter is shown in Figure 3.31.

3.18 Applications of digital filters in D-2

Digital filters are found in numerous places in the signal system of a D-2 recorder. In order to take advantage of oversampling converters, an increase in sampling rate is necessary for DACs and a reduction in sampling rate is necessary for ADCs. In oversampling the factors by which the rates are changed are simple; either two or four. The timing of the system is thus simplified because all samples (input and output) are present on edges of the higher-rate sampling clock.

When a composite VTR plays at other than normal speed, the subcarrier sequence will be broken by omitting or repeating fields, and it will be necessary to filter chroma from luminance so that the chroma phase can be corrected. Additionally, the VTR may play an odd field when an even field is required, and a digital interpolator working in the vertical dimension is required to produce the required lines from those which are available.

3.19 Oversampling filters

As was seen in Chapter 2, oversampling allows the use of simpler analog anti-aliasing and reconstruction filters, but sampling-rate conversion will then be necessary. Sampling-rate reduction will be needed after the ADC and before the

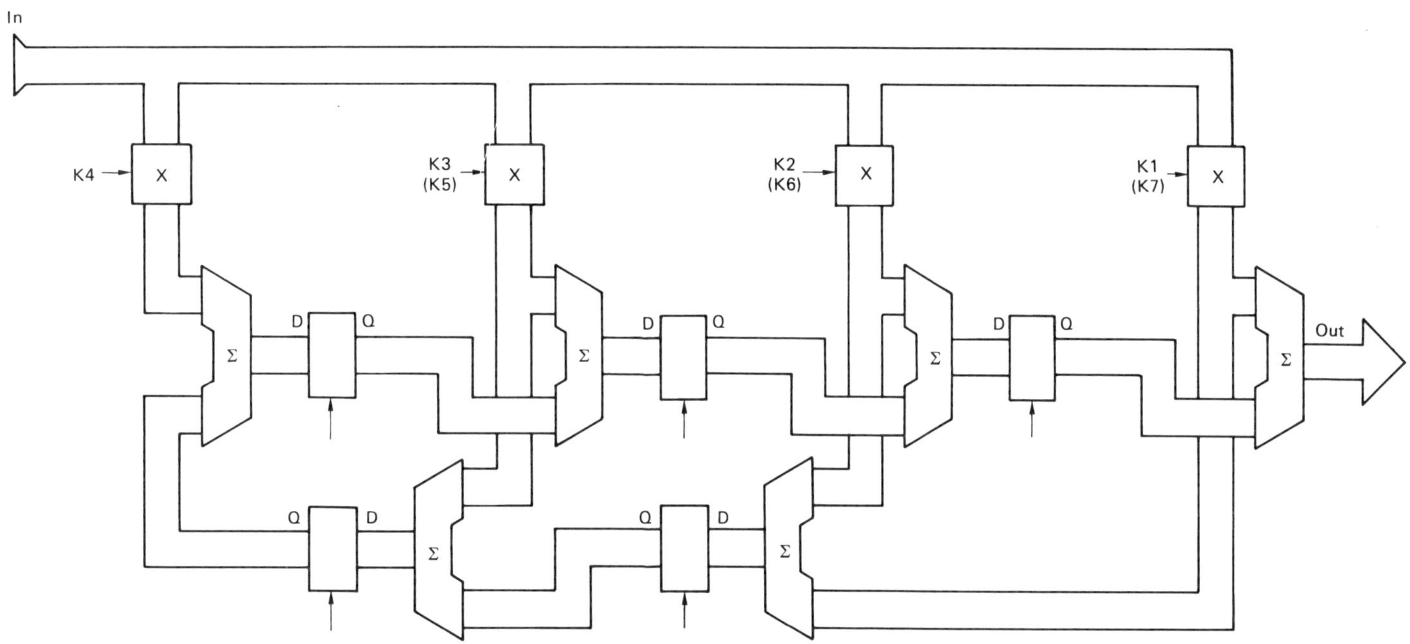

Figure 3.31 A seven-point folded filter for a symmetrical impulse response. In this case K1 and K7 will be identical, and so the input sample can be multiplied by either in one multiplier, and the product fed into the output shift system in two different places. The centre coefficient K4 appears once. In an even-numbered filter the centre coefficient would also be used twice

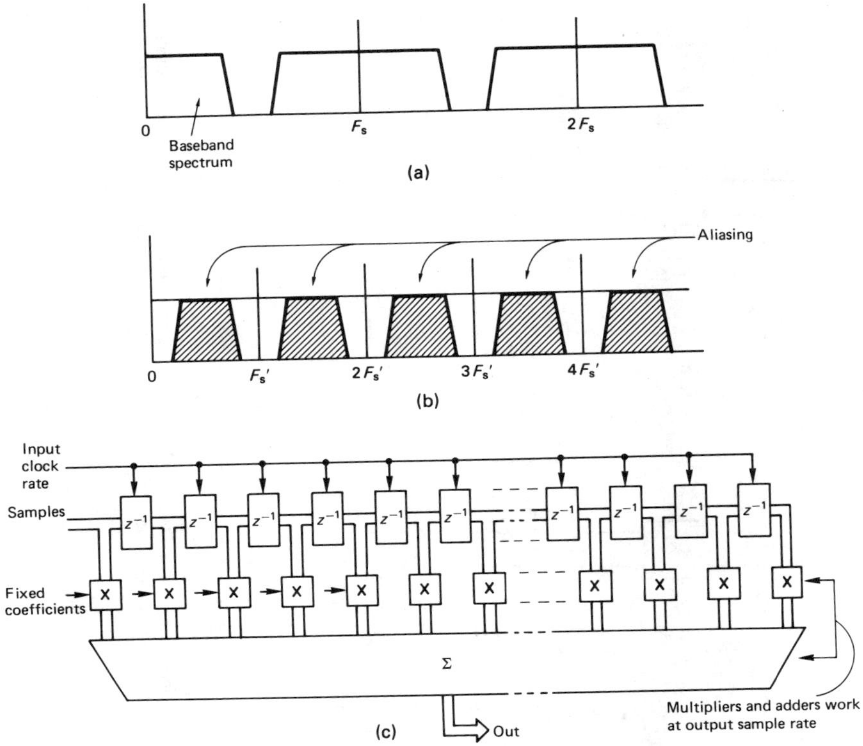

Figure 3.32 The spectrum of a typical digital sample stream at (a) will be subject to aliasing as in (b) if the baseband width is not reduced by an LPF. At (c) an FIR low-pass filter prevents aliasing. Samples are clocked transversely across the filter at the input rate, but the filter only computes at the output sample rate. Clearly this will only work if the two are related by an integer factor

DAC. The filter is simplified enormously if the oversampling factor is an integer. Rate reduction following an ADC is dealt with first.

Figure 3.32(a) shows the spectrum of a typical sampled system where the sampling rate is a little more than twice the analog bandwidth. Attempts to reduce the sampling rate by simply omitting samples, a process known as decimation, will result in aliasing, as shown in Figure 3.32(b). Intuitively it is obvious that omitting samples is the same as if the original sampling rate was lower. In order to prevent aliasing, it is necessary to incorporate low-pass filtering into the system where the cut-off frequency reflects the new, lower, sampling rate. An FIR-type low-pass filter could be installed immediately prior to the stage where samples are omitted, but this would be wasteful, because for much of its time the FIR filter would be calculating sample values which are to be discarded. A more effective method is to combine the low-pass filter with the decimator so that the filter only calculates values to be retained in the output sample stream. Figure 3.32(c) shows how this is done. The filter makes one accumulation for every output sample, but that accumulation is the result of multiplying all relevant input samples in the filter window by an appropriate coefficient. The number of points in the filter is determined by the number of *input* samples in the period of the filter window, but the number of multiplications per second is obtained by multiplying that figure by the *output* rate. If the filter is not integrated with the decimator, the number of points has to be multiplied by the input rate. The larger the rate-reduction factor the more advantageous the decimating filter ought to be, but this is not quite the case, as the greater the reduction in rate, the longer the filter window will need to be to accommodate the broader impulse response.

When the sampling rate is to be increased by an integer factor, additional samples must be created at even spacing between the existing ones. There is no need for the bandwidth of the input samples to be reduced since, if the original sampling rate was adequate, a higher one must also be adequate.

Figure 3.33 shows that the process of sampling-rate increase can be thought of in two stages. First, the correct rate is achieved by inserting samples of zero value at the correct instant, and then the additional samples are given meaningful values by passing the sample stream through a low-pass filter which cuts off at the Nyquist frequency of the original sampling rate. This filter is known as an interpolator, and one of its tasks is to prevent images of the lower input-sampling spectrum from appearing in the extended baseband of the higher-rate output spectrum.

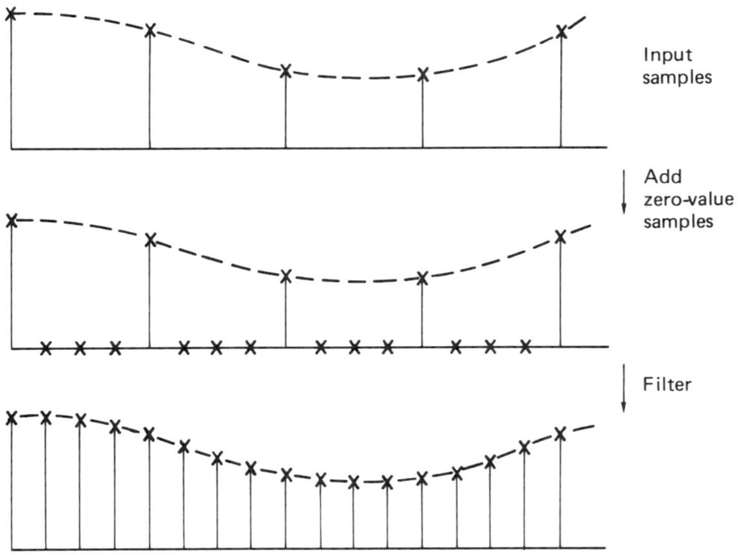

Figure 3.33 In integer ratio sampling, rate increase can be obtained in two stages. First zero-value samples are inserted to increase the rate, and then filtering is used to give the extra samples real values. The filter necessary will be an LPF with a response which cuts off at the Nyquist frequency of the input samples

How do interpolators work? Remember that, according to Nyquist, all sampled systems have finite bandwidth. An individual digital sample value is obtained by sampling the instantaneous voltage of the original analog waveform, and because it has zero duration, it must contain an infinite spectrum. However, such a sample can never be seen in that form because of the reconstruction process, which limits the spectrum of the impulse to the Nyquist limit. After reconstruction, one infinitely short digital sample actually represents a $\sin x/x$ pulse whose central peak width is determined by the response of the reconstruction filter, and whose amplitude is proportional to the sample value. This implies that, in reality, one sample value has meaning over a considerable timespan, rather than just at the sample instant. In the spatial domain, one can consider that a pixel has meaning over a considerable distance, in two dimensions in a still frame, in three in moving video. If this were not true, it would be impossible to build an interpolator.

As in rate reduction, performing the steps separately is inefficient. The bandwidth of the information is unchanged when the sampling rate is increased; therefore the original input samples will pass through the filter unchanged, and it is superfluous to compute them. The combination of the two processes into an interpolating filter minimizes the amount of computation.

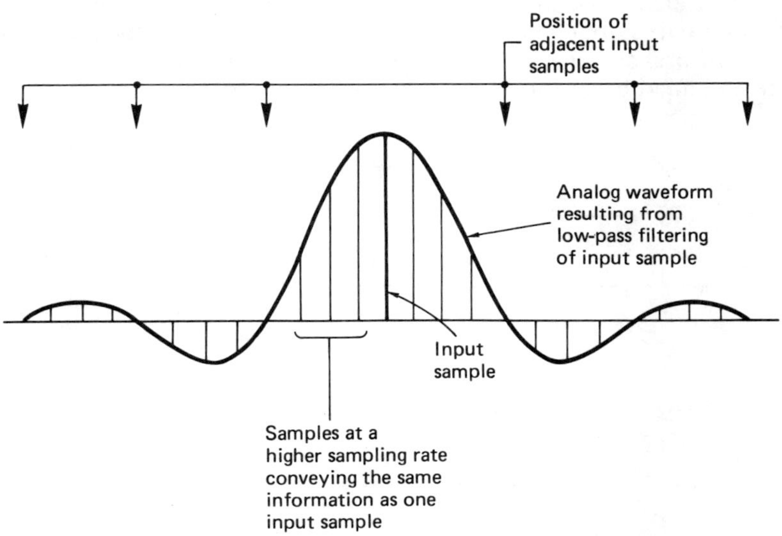

Figure 3.34 A single sample results in a sin *x*/*x* waveform after filtering in the analog domain. At a new, higher, sampling rate, the same waveform after filtering will be obtained if the numerous samples of differing size shown here are used. It follows that the values of these new samples can be calculated from the input samples in the digital domain in an FIR filter

As the purpose of the system is purely to increase the sampling rate, the filter must be as transparent as possible, and this implies that a linear phase configuration is mandatory, suggesting the use of an FIR structure. Figure 3.34 shows that the theoretical impulse response of such a filter is a $\sin x/x$ curve which has zero value at the position of adjacent input samples. In practice this impulse cannot be implemented because it is infinite. The impulse response used will be truncated and windowed as described earlier. To simplify this discussion, assume that a $\sin x/x$ impulse is to be used. To see how the process of interpolation works, recall the principle of the reconstruction filter described in Chapter 2. The analog voltage is returned to the time-continuous state by summing the analog impulses due to each sample. In a digital interpolating filter, this process is duplicated.[8]

If the sampling rate is to be doubled, new samples must be interpolated exactly half-way between existing samples. The necessary impulse response is shown in Figure 3.35; it can be sampled at the *output* sample period and quantized to form coefficients. If a single input sample is multiplied by each of these coefficients in turn, the impulse response of that sample at the new sampling rate will be obtained. Note that every other coefficient is zero, which confirms that no computation is

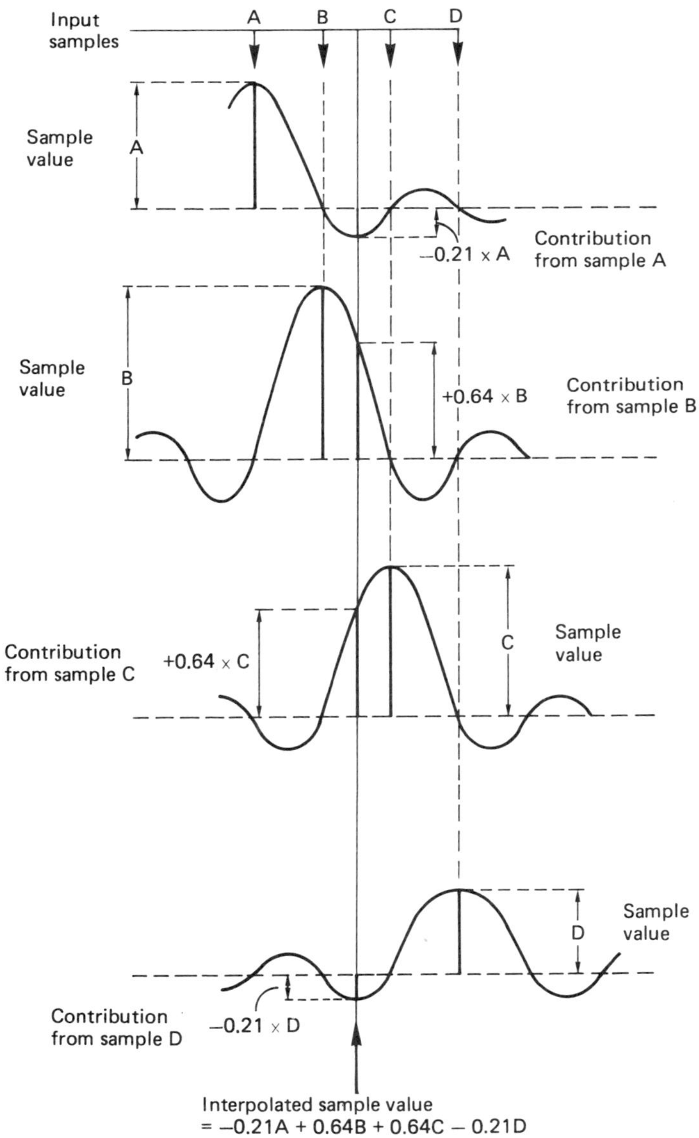

Figure 3.35 A 2× oversampling interpolator. To compute an intermediate sample, the input samples are imagined to be sin *x*/*x* impulses, and the contributions from each at the point of interest can be calculated. In practice, rather more samples on either side need to be taken into account

necessary on the existing samples; they are just transferred to the output. The intermediate sample is computed by adding together the impulse responses of every input sample in the window. The figure shows how this mechanism operates. If the sampling rate is to be increased by a factor of four, three sample values must be interpolated between existing input samples. Figure 3.36 shows that it is only

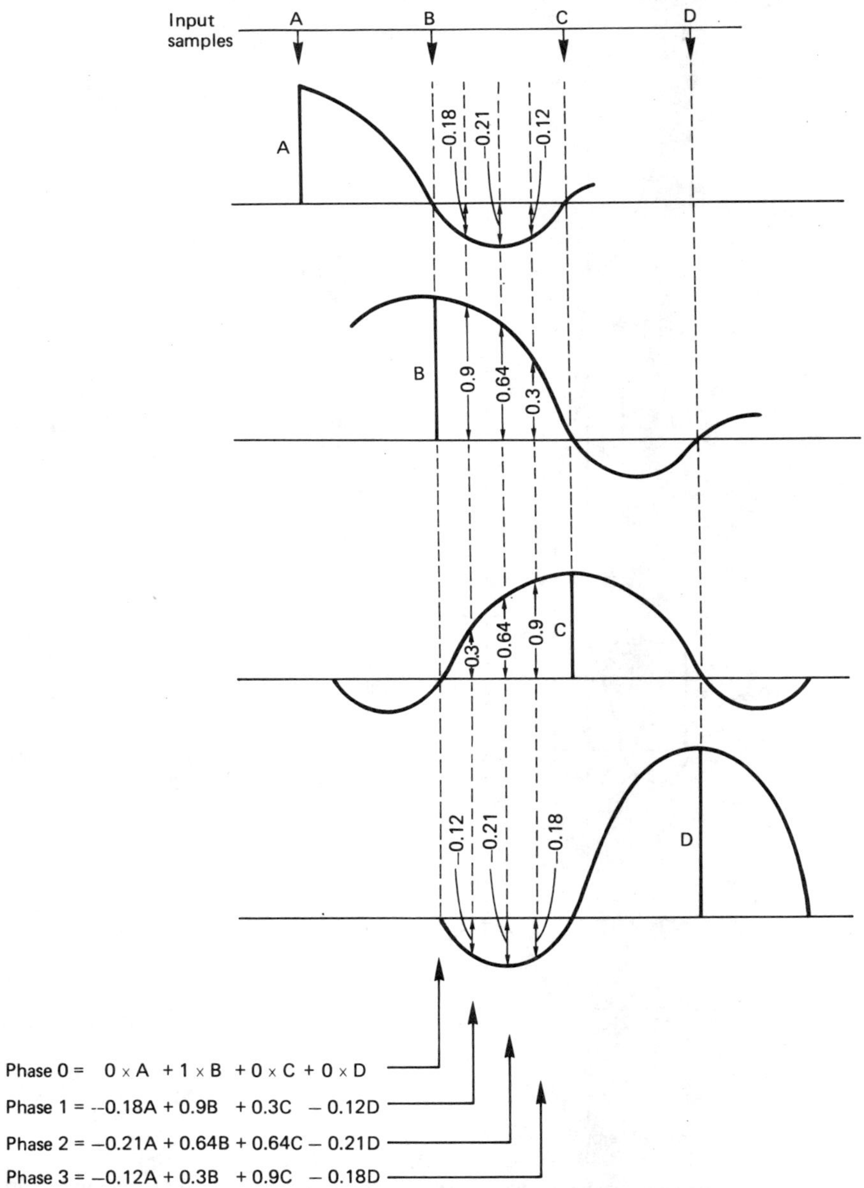

Figure 3.36 In 4× oversampling, for each set of input samples, four phases of coefficients are necessary, each of which produces one of the oversampled values

necessary to sample the impulse response at one-quarter the period of input samples to obtain three sets of coefficients which will be used in turn. In hardware-implemented filters, the input sample which is passed straight to the output is transferred by using a fourth filter phase where all coefficients are zero except the central one which is unity.

3.20 Field interpolators

Owing to the use of interlace, the head jumps necessary in variable-speed operation may result in an odd field being played when an even field is needed, and vice versa. The solution is to use vertical interpolation.

If interpolation is only used when the interlace is incorrect, the picture focus will be seen to change as the interpolator is brought in and out of operation. To keep the resolution constant, interpolation should be used continuously. The output raster and the input raster are shifted slightly so that neither input field aligns with either output field. By using memory to produce three delays of one line each, samples representing four points in a vertical line on the input field can be made simultaneously available to a four-point FIR filter. If the output field is to be the same as the input field, the phase of the interpolator will be such that the sample value for a line one-quarter of the way down from an input line will be computed. If the opposite type of field is required, the order of the coefficients to the FIR filter will be reversed such that a line three-quarters of the way down from an input line will be computed. Figure 3.37 illustrates how this process results in the image staying at the same vertical position on the screen irrespective of the relation of input and output fields.[9]

The luminance conveys the subjective resolution, and so the vertical chroma interpolation can be done with a simpler filter having only one line of delay to give two points.

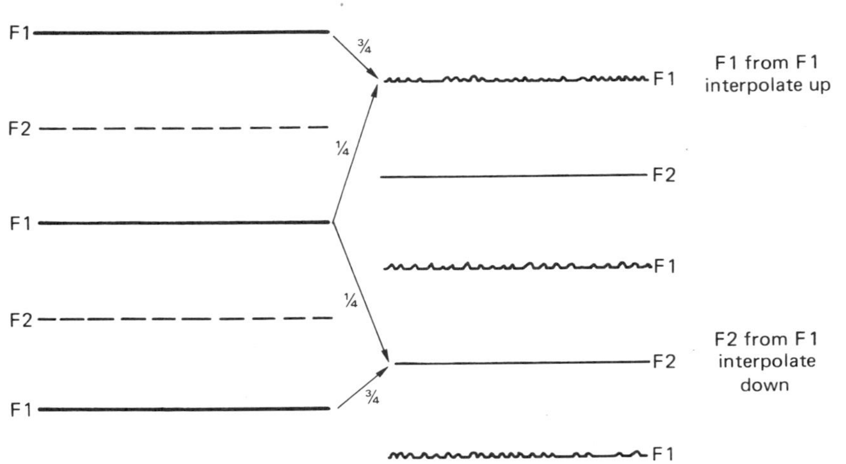

Figure 3.37 Vertical interpolation of either output field from either input field using a half line shift means that output resolution is not changed whether VTR is interlaced or not, and vertical movement is eliminated

The impulse response of the luminance filter is determined by the vertical spatial sampling rate, which is the same as the field line spacing. The interpolator has the same structure as the oversampling filter in Section 3.17, but in this case the input and output (spatial) sampling rates are the same, since it is the sample phase which is being changed.

3.21 Luminance/chrominance separation

The use of composite video recording results in a need for luminance/chrominance separation in colour processors.

Composite video interleaves the basically similar spectra of chroma and luminance by shifting the chroma spectrum by half a line with respect to luminance. They can only be fully separated by using a comb filter. There are various requirements for luminance/chrominance separation at different levels of cost and complexity, each appropriate to a given application.

3.22 Simple bandpass filters

This type of filter will be used for minimum cost and complexity and can be readily described because operation is directly parallel to that of an analog filter.

Figure 3.38(a) shows that an input signal enters an operational amplifier both directly and through a delay. The phase difference between the ends of the delay is a function of the input frequency, and will be given by:

$$\frac{\text{delay period}}{\text{signal period}} \times 360°$$

The op-amp output is the vector sum of the two inputs:

$$\tfrac{1}{2}V_{\text{in}} + \tfrac{1}{2}V_{\text{in}}\sin T$$

where T is the phase angle between the inputs. When the delay period exceeds the signal period, the phase shift exceeds 360°, and clearly subtracting 360° or multiples thereof does not change the vector sum. There will thus be a series of signal periods all of which will have the same gain. The frequency response becomes repetitive, hence the term comb filter. Figure 3.38(b) shows several frequencies, all of which suffer complete cancellation owing to effective inversion by the delay, and Figure 3.38(c) shows frequencies which give unity gain because the delay is an integer multiple of the signal period. Figure 3.38(d) shows the overall frequency response, which has a rectified cosine shape. Note that the peak spacing is at the reciprocal of the delay period.

A sharper response peak can be obtained by using two delays, as in Figure 3.39(a). The op-amp now has three inputs. Figure 3.39(b) shows the situation where a frequency which suffers a 180° shift in each delay is applied. Since the twice-delayed signal and the input signal will add, they must be given half as much gain as the signal from the centre tap. The op-amp thus has weighted gains of $+\tfrac{1}{4}$, $-\tfrac{1}{2}$ and $+\tfrac{1}{4}$. The frequency response shown in Figure 3.39(d) is a cosinusoid, again with the peaks spaced at the reciprocal of the delay.

Transferring to the digital domain, signal voltage is represented by a binary number, delay is achieved using latches or RAM and the operational amplifier is replaced by an adder. Figure 3.40 shows the digital equivalent of Figure 3.39. The constraint in this kind of filter is that the delay periods can only be integer multiples of the sampling period.

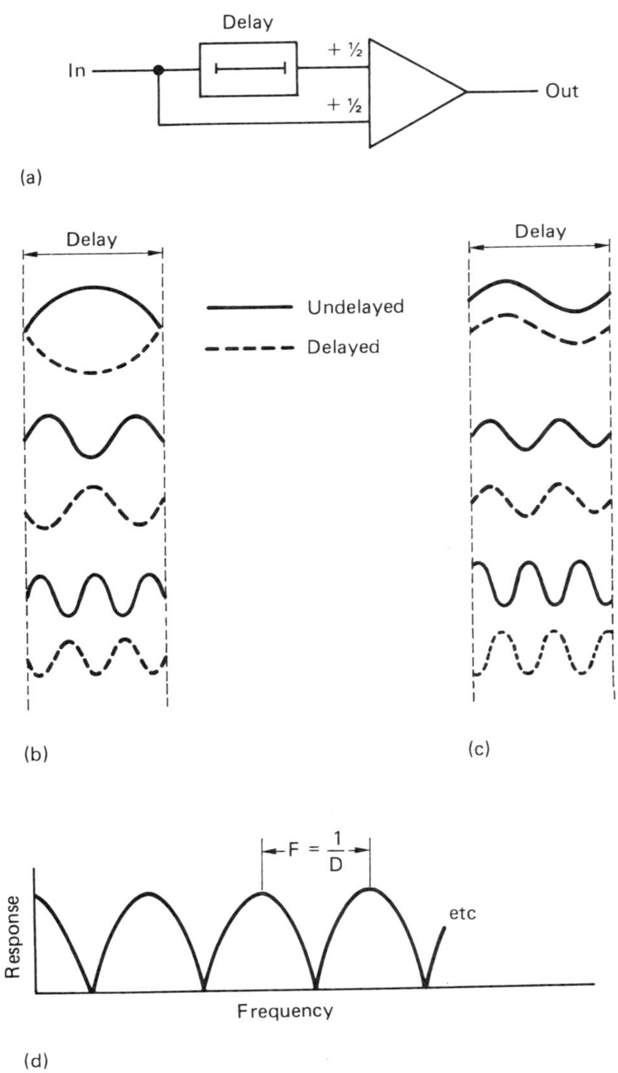

Figure 3.38 Simple comb filter at (a) has delay in one input of adder. At some frequencies, (b) delay causes cancellation, whereas at others (c) delayed and undelayed signals are in phase. Frequency response shown at (d)

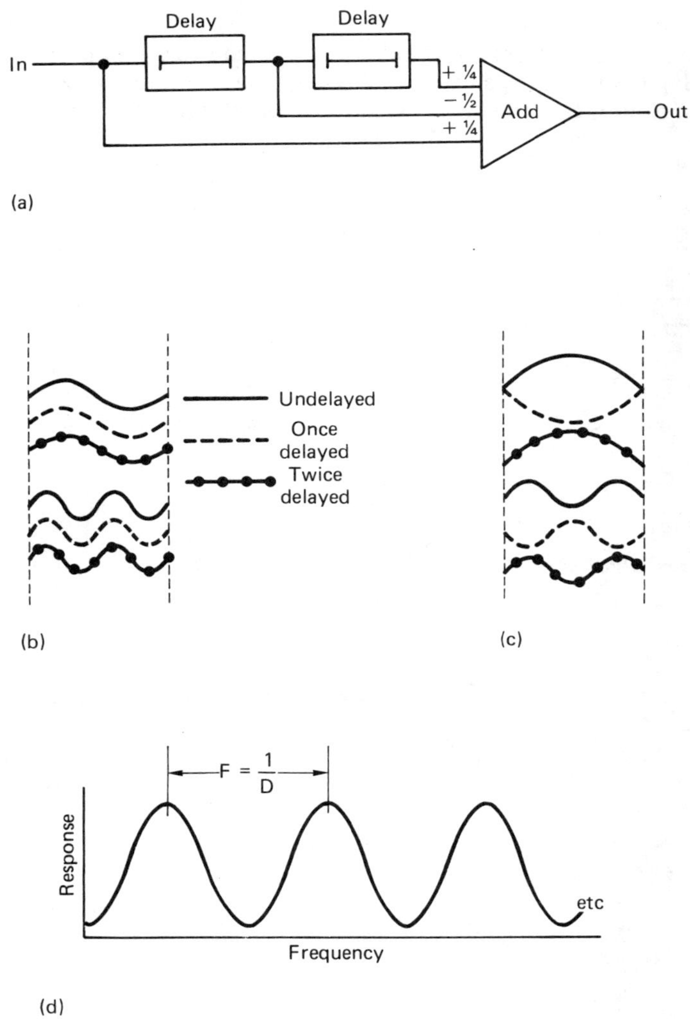

Figure 3.39 Two-delay comb filter at (a). In (b) certain frequencies suffer 360° shift in one delay, but since once-delayed signal has twice the weighting there is cancellation. At (c) is 180° resultant phase shift in one delay, removed by inverting input, permitting unity gain. Response at (d) is cosinusoid

Figure 3.39 and Figure 3.40 show filters where a phase shift of 180° is necessary at the null frequency. This phase shift is readily achieved in a $4 \times F_{sc}$ system by using two sample periods of delay.

The $4 \times F_{sc}$ filter of Figure 3.40 has a gain peak at subcarrier frequency, and the output will be chroma. To obtain luminance, output values are subtracted from input values, with a compensating delay.

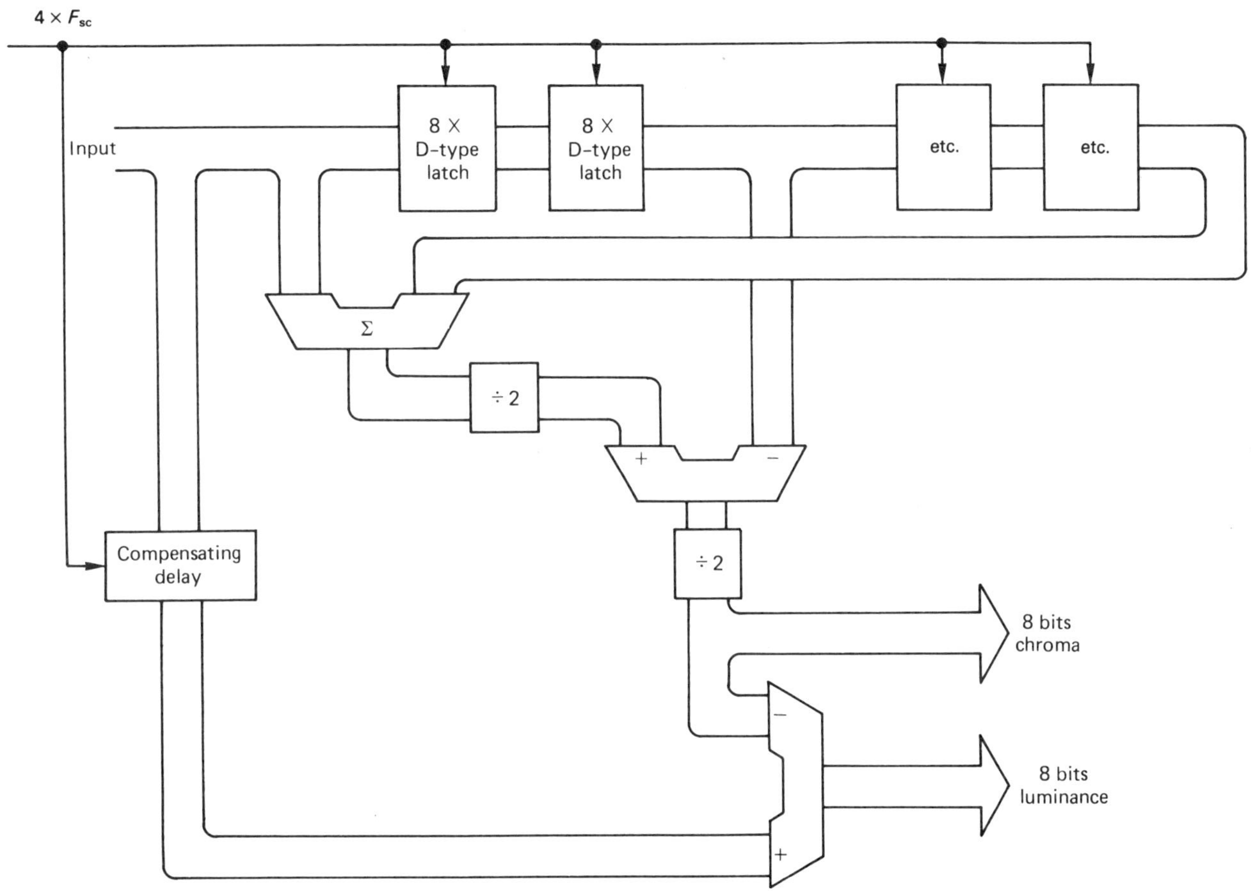

Figure 3.40 Digital *Y*/*C* separation at $4 \times F_{sc}$ uses pairs of latches to give 180° phase shift at subcarrier frequency. Binary adders and dividers give weighting of +¼, −½, +¼ to three samples. Chroma output is subtracted from the input to give luminance. There is only one response peak at F_{sc}, since Nyquist limit for $4 \times F_{sc}$ sampling is reached at $2 \times F_{sc}$

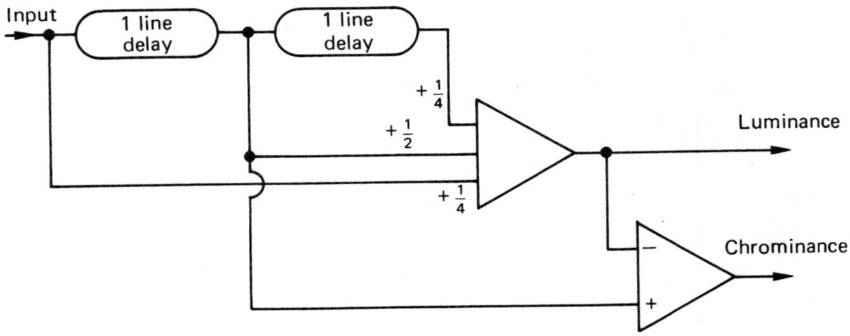

Figure 3.41 Simple line comb filter. Frequency response is ideal, but there are practical shortcomings explained in the text

3.23 Advanced filters

In a D-2 colour processor, the *Y*/*C* separation is in the main video path, and it must be of high quality. Since the spectral interleave of a composite signal is primarily on multiples of line rate, it follows that delays of one line are necessary in the comb filter, so that the frequency response peaks will be at multiples of F_h. Figure 3.41 shows a simple line comb filter. Although this device has a good spectral response, promising to give minimum cross-colour and cross-luminance, in fact there are some snags. First, the summing junction which is designed to reject chroma is adding together adjacent points on different lines and is actually reducing the vertical resolution of the picture. Secondly, the filter is working with real chroma, whose phase can change according to picture content, whereas perfect comb response is only achieved when the chroma phase remains consistent over the three points feeding the summing junction. Vertical changes in chroma phase will cause the comb to fail to cancel or *mesh*, and residual chroma will appear in the luminance as a series of white dots at horizontal boundaries between colours.

Most of the vertical resolution can be restored by using a bandpass filter in conjunction with the line comb. Figure 3.42 shows that incoming video passes through two filters in parallel. One is a low-pass filter which cuts just below chroma frequencies and the other is a band limiting filter. The outputs of the two filters are subtracted, to give a bandpass response, which contains chroma and high-frequency luminance. This signal is applied to the line comb, which rejects chroma without impairing vertical resolution at frequencies below the bandpass. The high-frequency luminance can then be added back to the low-pass filtered signal to give full bandwidth luminance. Comb mesh failure can be detected by analysing the chroma signals at the ends of the comb, and if chroma will not be cancelled, the high-frequency luminance is not added back to the main channel, and a low-pass response results.

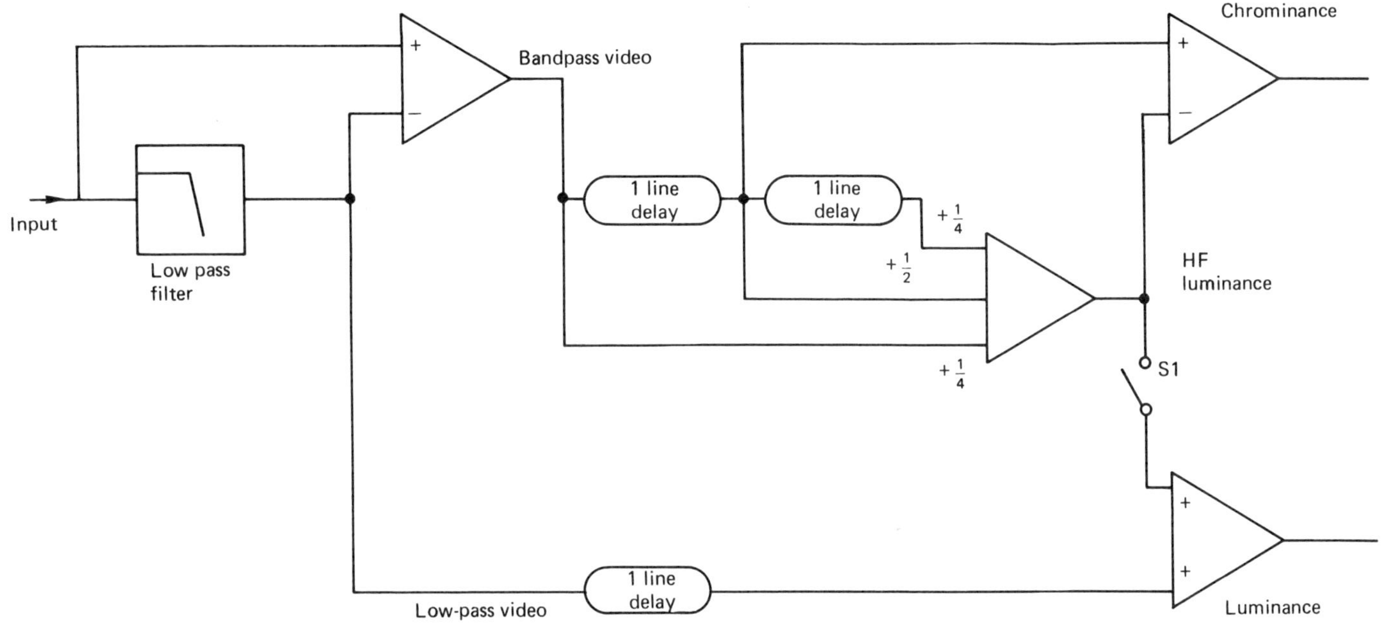

Figure 3.42(a) A more complex line comb filter where the line comb works on bandpass video to reduce the loss of vertical resolution. HF luminance is added back to low-pass signal to restore luminance bandwidth. If comb fails to mesh due to vertical chroma phase change, switch S1 opens to prevent chroma breaking through into luminance channel. With S1 open luminance bandwidth is reduced

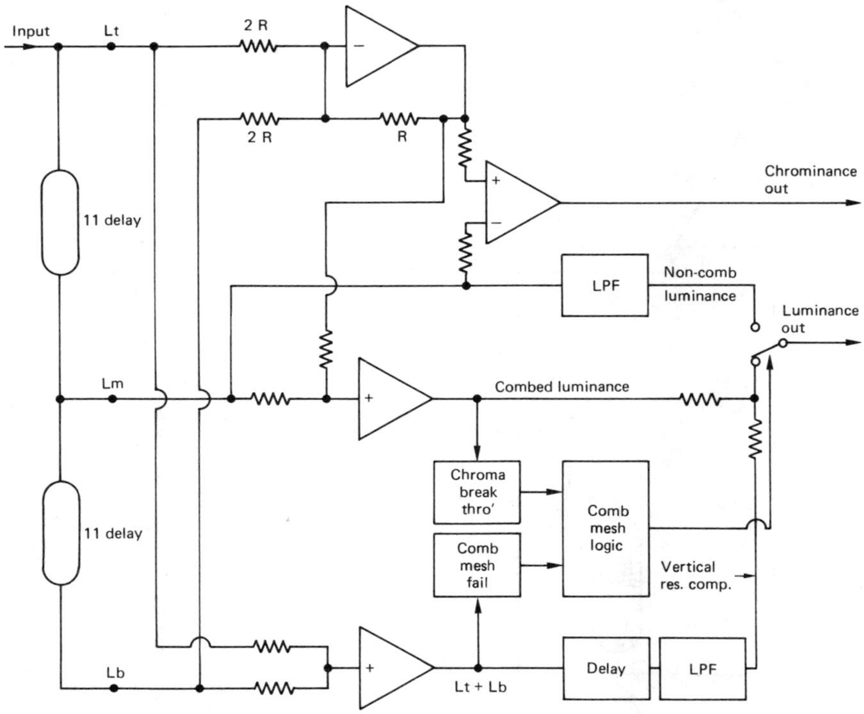

Figure 3.42(b) Rearranging the filter of Figure 3.42(a) in this way eliminates a delay line by restoring vertical resolution a different way. Vertical resolution compensation is achieved by taking signals from top and bottom of the delays and adding them back to luminance via a low-pass filter shown bottom right. In the digital domain, replace operational amplifiers with adders, delays with RAM. Low-pass filters will be FIR types as described

References

1. VANDERKOOY, J. and LIPSHITZ, S. P. Digital dither. Presented at the 81st Audio Engineering Society Convention (Los Angeles, 1986), preprint 2412(c-8)
2. SMPTE RP 125
3. Ampex Corporation, Ampex digital format for video and audio tape recording of composite video signals using 19 mm type D-2 cassette. Redwood City, CA (1988)
4. CCIR Recommendation 656
5. Proposed American National Standard – Serial digital interface for 10 bit 4:2:2 component and NTSC composite digital signals
6. VAN DEN ENDEN, A. W. M. and VERHOECKX, N. A. M. Digital signal processing: theoretical background. *Philips Tech. Rev.*, **42**, 110–144 (1985)
7. McCLELLAN, J. H., PARKS, T. W. and RABINER, L. R. A computer program for designing optimum FIR linear-phase filters. *IEEE Trans. Audio Electroacoust.*, **AU-21**, 506–526 (1973)
8. CROCHIERE, R. E. and RABINER, L. R. Interpolation and decimation of digital signals – a tutorial review. *Proc. IEEE*, **69**, 300–331 (1981)
9. CROLL, M. G. The general design of a digital still storage system for use with an electronic rostrum colour camera. *BBC Res. Dep. Rep.*, no. 1980/3

Chapter 4

Digital magnetic recording

Although the physics of the record/replay process are unaffected by the meaning attributed to signals, the techniques used in digital recording are rather different from those found in analog recording, although often the same phenomenon shows up in a different guise. In this chapter the fundamentals of digital magnetic recording are treated along with the necessary coding methods.

4.1 Magnetic recording

In analog recording, the characteristics of the medium affect the signal recorded directly, whereas by expressing a signal in binary numerical form by sampling and quantizing, the quality becomes independent of the medium. The dynamic range required no longer directly decides the track width needed. In digital circuitry there is a great deal of noise immunity because the signal can only have two states, which are widely separated compared with the amplitude of noise. In digital magnetic recording there are also only two states of the medium, N–S and S–N, but paradoxically the noise immunity is much reduced. As noise immunity is a function of track width, reduction of the working SNR of a digital track allows the same information to be carried in a smaller area of the medium, improving economy of operation. It also increases the random error rate, but as an error-correction system is already necessary to deal with dropouts, it is simply made to work harder.

It is interesting to compare tape consumption between analog and digital machines. In the C-format, 1 inch tape is used, moving at 230 mm/s, whereas the D-2 format uses ¾ inch tape at only 130 mm/s. In digital recording narrower tracks are to be expected, and steps must be taken to register the heads accurately with the tracks by appropriate mechanical design and improved slitting accuracy of the tape.

4.2 Head noise and head-to-tape speed

There are several important sources of replay noise in a magnetic recorder, which will be examined later in this chapter. One of these is the noise from the head. All components with resistance generate noise according to their temperature, and the replay head is no exception. If a given recording exists on a tape, a better SNR will be obtained by moving the head relative to the tape at a higher speed, since the head noise is constant and the signal induced is proportional to speed. This is one reason why rotary-head recorders offer better packing density than stationary-head recorders. The other reason is much more obvious. A rotary-head machine determines track spacing by linear tape speed, whereas stationary heads are difficult to fabricate with narrow spacing between tracks. In digitizing a video waveform, there has been an exchange in the importance of SNR and bandwidth. The bandwidth of a digital channel always exceeds the bandwidth of the original analog signal, but the extra bandwidth is only required with poor SNR. This explains the paradox that greater bandwidth is needed but less tape is used. As in analog recording, the rotating head can be used to obtain high bandwidth without excessively short tape wavelengths and at moderate linear tape speed. A further advantage of rotary-head machines is that, by changing the scanner geometry, the best compromise can be reached between bandwidth and SNR. Figure 4.1 shows

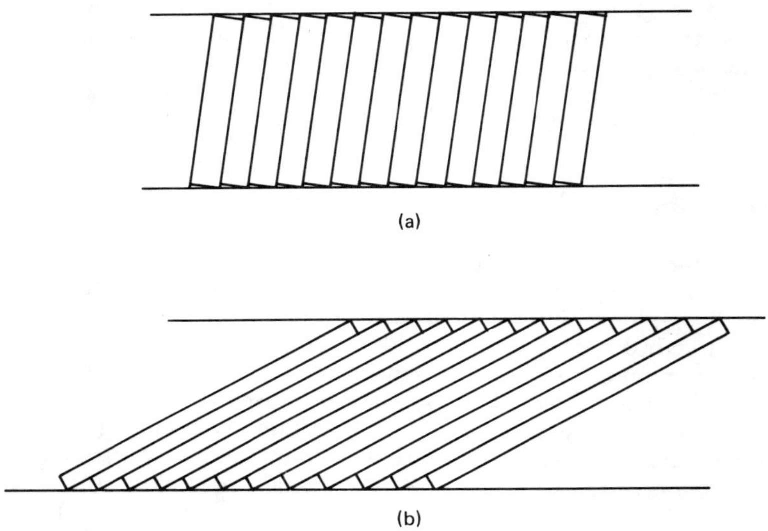

Figure 4.1 Two different rotary-head formats having the same tape consumption but different characteristics

that, in the same area of tape, two different recordings can be made. The first has a lower head-to-tape speed because the tracks are shorter, but better SNR because they are wider. The second has higher bandwidth because of the longer tracks, but these tracks are narrower. In the absence of head noise, the information capacity of both formats would be the same. Where head noise is a factor, the second format will be superior. The lower limit to track width is generally set by the ability to register a head with adequate precision. Another technique is to use azimuth recording which gives some protection from residual tracking error. The comparison of tape consumption between D-2 and the C-format is even more dramatic because of the adoption of azimuth recording.

Whether the machine is a stationary- or rotary-head recorder, the recorded wavelengths must be kept short to conserve tape in the direction of the track. Very short wavelengths can only be replayed with consistent intimate contact between the head and the medium, so that the surface finish of the medium must be of the highest order. The roughness of the tape backcoat must be limited to prevent the back of one layer embossing an adjacent magnetic layer when the tape is wound on a spool. Digital tape has a thin coating, because thickness loss prevents flux from a thick coat being of much use at short wavelengths and because a thin coating is less prone to self-demagnetization. The thin coat needs high-energy particles to allow useful replay signals with reduced magnetic volume. The backing material, or substrate, is relatively thin to allow the tape to accommodate head irregularities without losing contact, but print-through is not an issue in digital recording. A happy consequence of the use of thin tape is that more can be accommodated in a given cassette, but it does require careful transport design to avoid damage. Transports are considered in Chapter 6.

4.3 Basic digital magnetic recording

The basic principle of digital magnetic recording is remarkably simple. Since it is intended that the medium should have only two states, the record waveform will typically be a current whose direction reverses but whose magnitude remains constant, as in Figure 4.2. To provide the best SNR on replay, the current required

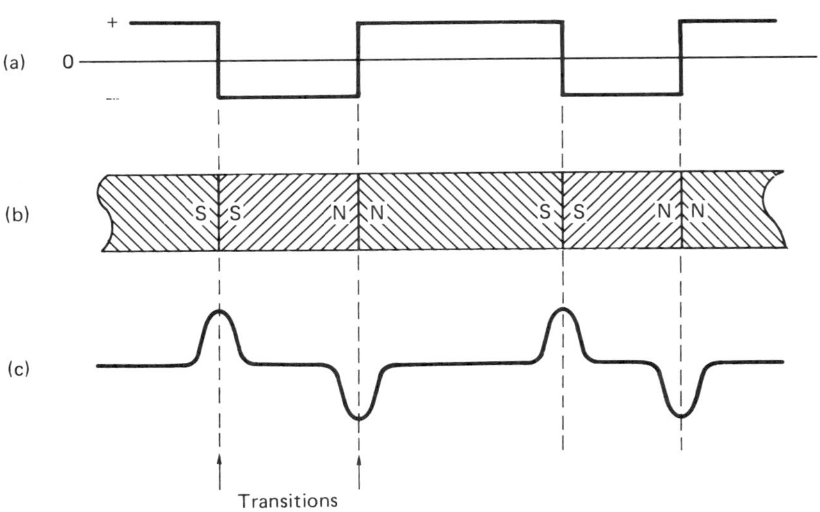

Figure 4.2 Basic digital recording. At (a) the write current in the head is reversed from time to time, leaving a binary magnetization pattern shown at (b). When replayed, the waveform at (c) results because an ouput is only produced when flux in the head changes. Changes are referred to as transitions

is a little less than that needed to saturate the tape, as saturation causes fringing fields around the head and crosstalk in adjacent tracks. In recent machines, the record current may be produced in an analog amplifier which has a response which corrects for losses in the record-head material at high frequencies. The tape encounters a strength of flux which increases and then decreases as it passes the head. The recording is actually made near the trailing pole of the head, as shown in Figure 4.3, where the flux from the head falls below the coercive force needed to change the state of the particles. The steeper the flux gradient on the trailing pole, the shorter the wavelength which can be recorded. This is generally obtained with a relatively wide gap. Bias is unnecessary in digital recording because linearity is not a goal.

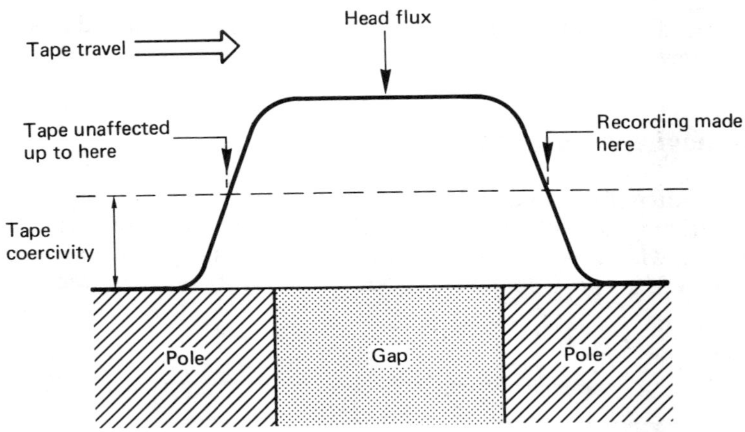

Figure 4.3 The recording is actually made near the trailing pole of the head where the head flux falls below the coercivity of the tape

4.4 Head construction

Design of a head for D-2 machines is not easy, since there are a number of conflicting requirements. The high coercivity of the tape used requires that the head should be able to pass high flux densities into the tape without saturating. The bit rate of digital video is such that eddy current losses in the head are significant.

Ferrite is attractive as a head material because it is a non-conductor and this limits losses. Unfortunately currently available ferrites saturate before 1500 Oe tape can be fully modulated. Metals are available which will carry the required flux density, but they suffer from eddy current losses. One solution to the problem is to make a composite head, where the bulk of the head is made from ferrite, but metal poles are fitted which concentrate the flux from a large cross-section of ferrite into a small track width. The construction of such a head is shown in Figure 4.4.

Another approach to limiting eddy current losses is lamination. Laminated metal heads have been used in stationary-head recorders for many years, but it is difficult to fabricate laminated heads for digital recorders because of the small dimensions involved. Developments in plating technology mean that it is now possible to fabricate laminated heads on an extremely small scale by plating layers of magnetic material alternately with thinly deposited insulating layers.

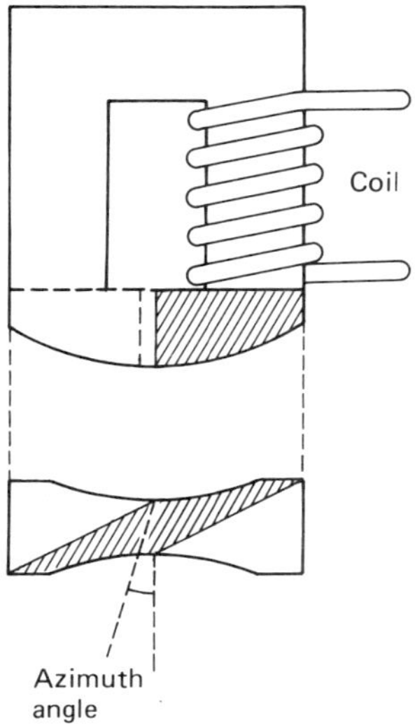

Figure 4.4 As ferrite will saturate before high coercivity tape, it cannot be used for head poles, but as shown here, it can be used for the bulk of the head, with metal tips to focus the flux to the tip. Eddy current losses in the metal are minimized. Alternatives are laminated or amorphous metal heads

4.5 Playback

When a digital recording is replayed, the output of the head will be a differentiated version of the record waveform, because the head only responds to the rate of change of flux.

The initial task of the replay circuits is to reconstruct the record waveform. The amplitude of the signal is of no consequence; what matters is the time at which the write current, and hence the flux stored on the medium, reverses. This can be

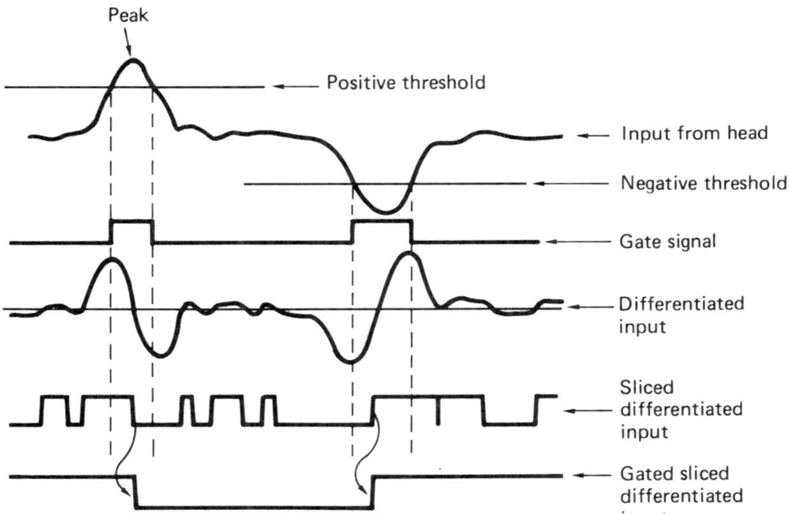

Figure 4.5 Gated peak detection rejects noise by disabling the differentiated output between transitions

determined by locating the peaks of the replay impulses. At high data rates this can conveniently be done by differentiating the signal and looking for zero crossings. Figure 4.5 shows that this results in noise between the peaks. This problem is overcome by the gated peak detector, where only zero crossings from a pulse which exceeds the threshold will be counted. This method is almost universally used in digital video and in disk drives whereas at the relatively low data rates of floppy disks and in digital audio the record waveform can also be restored by integration, which opposes the differentiation of the head as in Figure 4.6.[1]

There are a number of details which must be added to this simplistic picture in order to appreciate the real position. Figure 4.7 shows that the differentiating effect of the replay process causes the head output initially to rise at 6 dB per octave from a DC response of zero. Although a high-frequency recording can be made throughout the thickness of the medium, the flux deep within the medium cannot

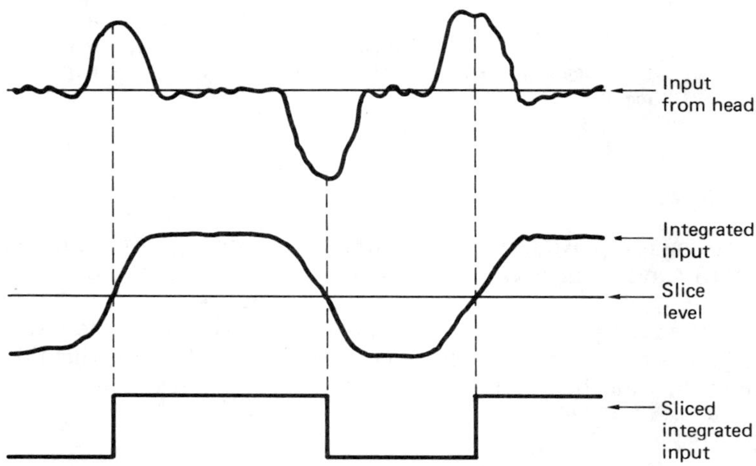

Figure 4.6 Integration method for recreating write-current waveform

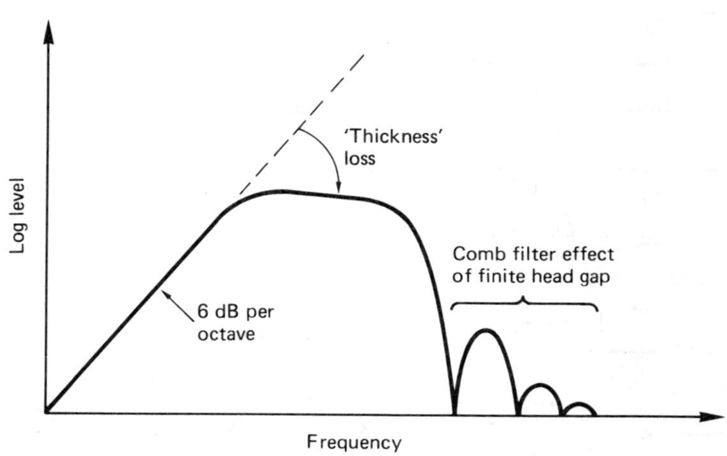

Figure 4.7 The major mechanisms defining magnetic channel bandwidth

couple with the replay head at short wavelengths, and so as recorded wavelength falls, a thinner and thinner layer near the surface remains responsible for the replay flux.[2] This is called thickness loss, although it is a form of separation loss, and it causes a loss of 6 dB per octave, which cancels the differentiating effect to give a region of constant frequency response. The construction of the head results in the same action as that of a two-point transversal filter, as the two poles of the head see the tape with a small delay interposed due to the finite gap. As expected, the head response is like a comb filter with the well-known nulls where flux cancellation takes place across the gap. Clearly the smaller the gap the shorter the wavelength of the first null. This contradicts the requirement of the record head to have a large gap. In quality analog audio recorders, it is the norm to have different record and replay heads for this reason, and the same is sometimes true in digital recording.

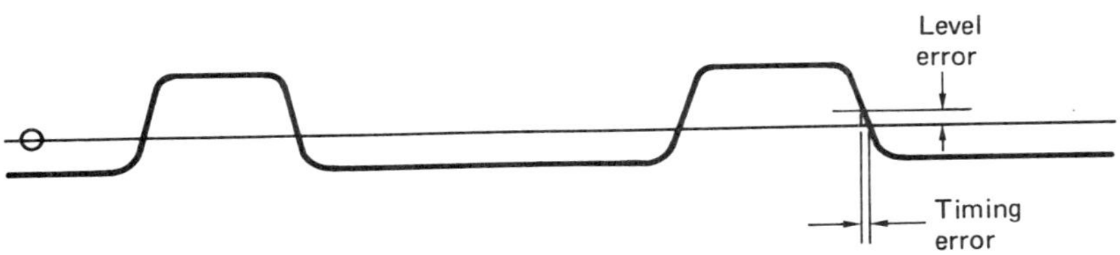

Figure 4.8 A DC offset can cause timing errors

Figure 4.8 shows that when an uneven duty cycle is recorded, there are a number of problems. The lack of DC response causes a level shift. Combined with the finite rate of change of voltage, the shift can cause timing errors unless care is taken to slice the signal about its own centre. The finite gap in the replay head causes closely spaced flux reversals to interfere with one another, which causes peak shift distortion (also known as intersymbol interference or pulse crowding) and tends to reduce the asymmetry of the waveform, again causing timing errors. The mechanism responsible for peak shift is shown in Figure 4.9(a). The results of two independent and opposite transitions passing the head are shown, and summing these gives the result of replaying two close together. Interaction between the two transitions reduces the amplitude of the signal and moves the peaks apart. Avoidance of peak shift requires equalization of the channel,[3] and this can be done by a network after the replay head, termed an equalizer or pulse sharpener,[4] as in Figure 4.9(b), or before the record head, where it is called precompensation, as in Figure 4.9(c). Both of these techniques use transversal filtering to oppose the inherent transversal effect of the head.

4.6 Azimuth recording and rotary heads

Conventional magnetic recorders record the transitions on the tape track at right angles to the edge of the track, and Figure 4.10 shows that it is necessary to leave so-called guard bands between tracks to allow some tracking error without causing crosstalk from adjacent tracks. These guard bands represent wasted tape.

Figure 4.11(a) shows that, in azimuth recording, the transitions are laid down at an angle to the track by using a head which is tilted. Machines using azimuth recording must always have an even number of heads, so that adjacent tracks can

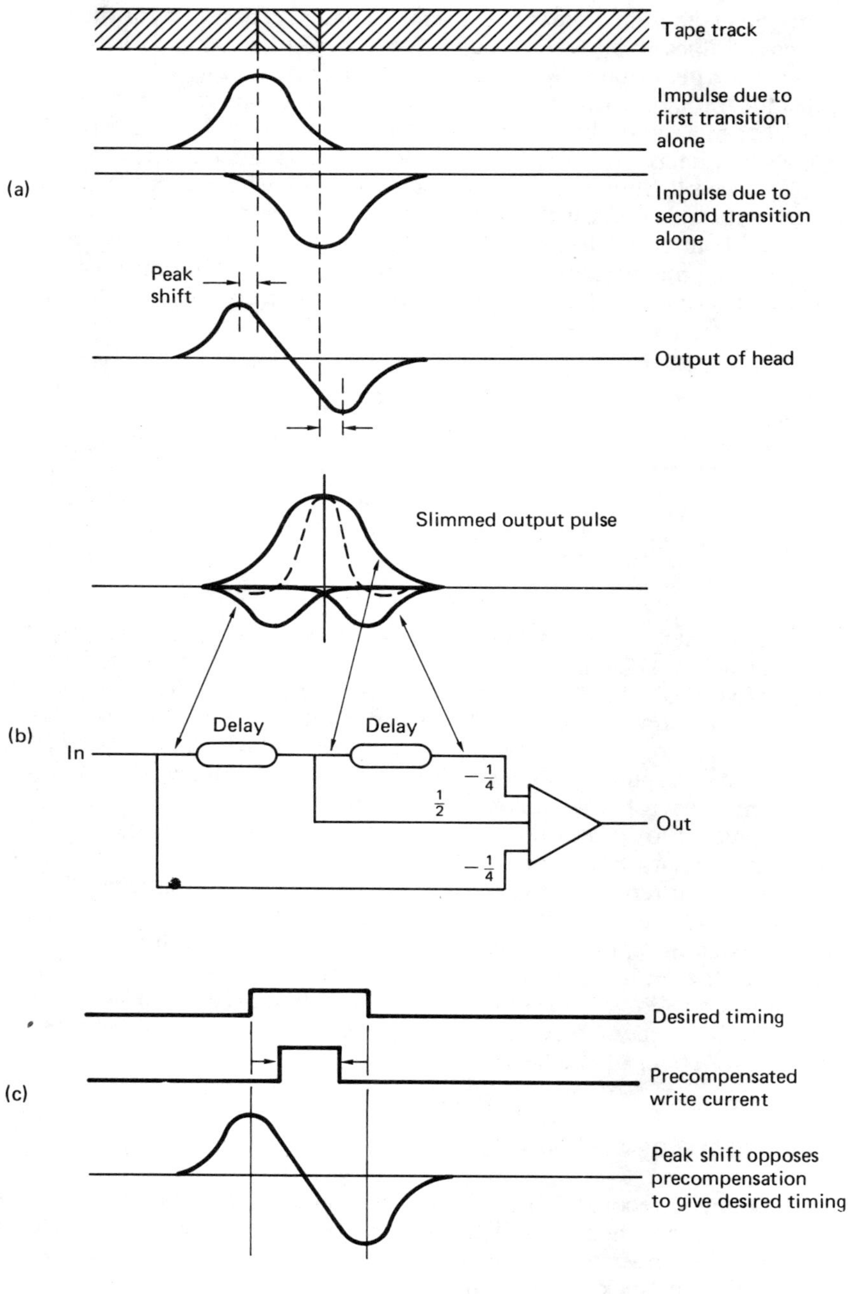

Figure 4.9 (a) Peak shift distortion can be reduced by (b) equalization in replay, or (c) precompensation

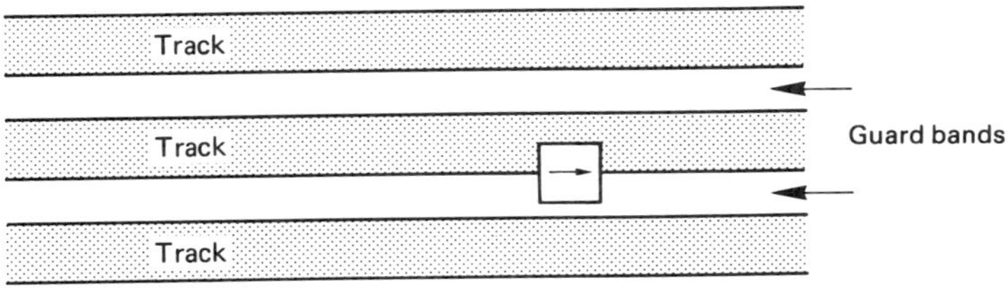

Figure 4.10 In conventional recording, a space or guard band must be left between tracks so that if a head is misaligned, the output signal simply reduces instead of becoming a composite signal from two tracks. The guard bands represent unused tape

be recorded with opposite azimuth angle. The two track types are usually referred to as A and B. Figure 4.11(b) shows the effect of playing a track with the wrong type of head. The playback process suffers from an enormous azimuth loss. The effect of azimuth error can be understood by imagining the tape track to be made from many identical parallel strips. In the presence of azimuth error, the strips at one edge of the track are played back with a phase shift relative to strips at the other side. At some wavelengths, the phase shift will be 180°, and there will be no output, while at other wavelengths, especially long wavelengths, some output will reappear. The effect is rather like that of a comb filter and serves to attenuate crosstalk due to adjacent tracks. Since no tape is wasted between the tracks, efficient use is made of the tape. The term guard-band-less recording is often used instead of, or in addition to, the term azimuth recording. The failure of the azimuth effect at long wavelengths is a characteristic of azimuth recording, and it is necessary to ensure that the spectrum of the signal to be recorded has a small low-frequency content.

In digital recording there is often no separate erase process, and erasure is achieved by overwriting with a new waveform. When a rotary-head machine uses overwriting in conjunction with azimuth recording, the recorded tracks can be made rather narrower than the head pole simply by reducing the linear speed of the

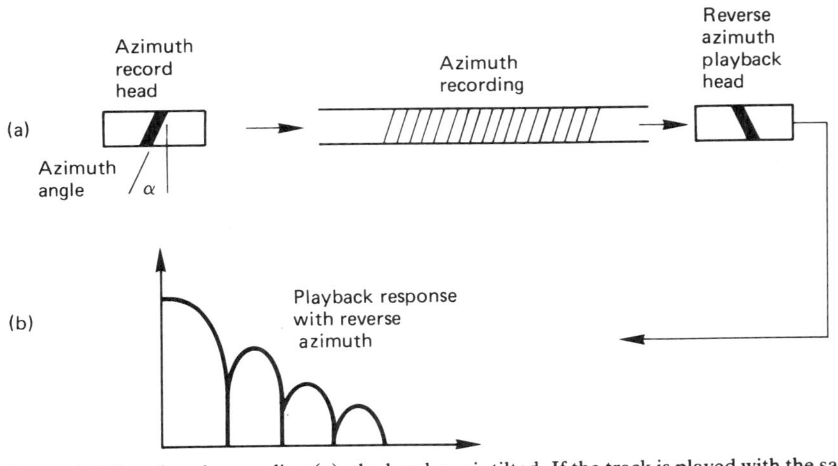

Figure 4.11 In azimuth recording (a), the head gap is tilted. If the track is played with the same head, playback is normal, but the response of the reverse azimuth head is attenuated (b)

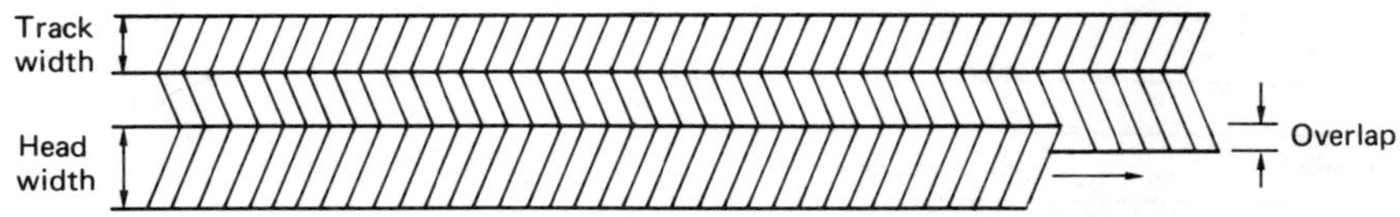

Figure 4.12 In azimuth recording, the tracks can be made narrower than the head pole by overwriting the previous track

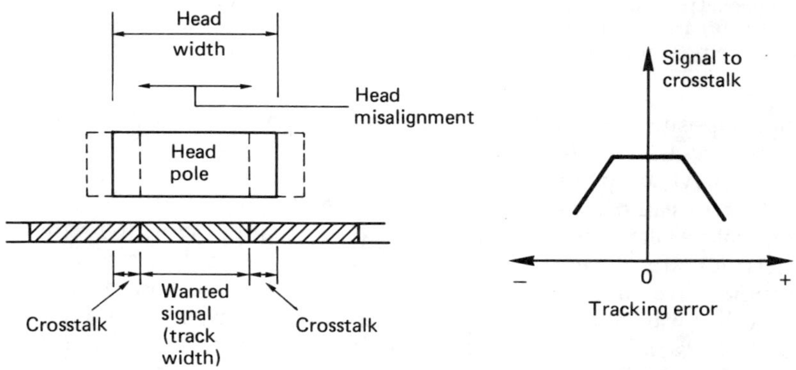

Figure 4.13 When the head pole is wider than the track, the wanted signal is picked up along with crosstalk from the adjacent tracks. If the head is misaligned, the signal-to-crosstalk ratio remains the same until the head fails to register with the whole of the wanted track

tape so that it does not advance so far between sweeps of the rotary heads. This can be seen in Figure 4.12. The same head can be used for replay, even though it is wider than the tracks. It can be seen from Figure 4.13 that there will be crosstalk from tracks at both sides of the home track, but this crosstalk is attenuated by the azimuth effect. The amount by which the head overlaps the adjacent track determines the spectrum of the crosstalk, since it changes the delay in the azimuth comb-filtering effect. More importantly, the signal-to-crosstalk ratio becomes independent of tracking error over a small range, because as the head moves to one side the loss of crosstalk from one adjacent track is balanced by the increase of crosstalk from the track on the opposite side. This phenomenon allows for some loss of track straightness and for the residual error which is present in all track-following servo systems.

The azimuth angle used has to be chosen with some care. The greater the azimuth angle, the less will be the crosstalk, but the effective writing speed is the head-to-tape speed multiplied by the cosine of the azimuth angle. A further smaller effect is that the tape is anisotropic because of particle orientation. Noise due to the tape, head or amplifier is virtually unaffected by the azimuth angle, and there is no point in reducing crosstalk below the noise. The value of $\pm 15°$ used in D-2 causes a loss of less than 1 dB due to the apparent reduction in writing speed.

4.7 Equalization

In practice there are difficulties in providing correct equalization at all times. Tape surface asperities and substrate irregularities cause variations in the intimacy of head contact, which changes the response at high frequencies much more than at low frequencies, thereby undermining any fixed equalization. Equalization can be used on recording in the form of precompensation, which moves recorded transitions in such a way as to oppose the effects of peak shift. Optimum equalization is difficult under dynamic conditions, although an adaptive equalizer can be made which uses the timing errors caused by poor equalization to change the response. This approach is used in the Ampex D-2 machines.

In most of the above, a clearer picture has been obtained by studying the impulse response of devices than from the frequency response, and this follows from the impulsive nature of digital techniques.

4.8 Shortcomings of recording channels

This chapter has outlined the basic physics of the recording process and introduced numerous shortcomings and restrictions from which real recording channels will suffer. Now these shortcomings will be categorized, and coding methods to overcome them will be described.

Noise in the channel causes uncertainty about the voltage of the reproduced signal. Jitter causes uncertainty about the time at which a particular event occurred. The frequency response of the channel then places an overall limit on the spacing of events in the channel. Particular emphasis must be placed on the interplay of bandwidth, jitter and noise, which will be shown here to be the key to the design of a successful channel code.

4.9 Jitter windows

Figure 4.14 shows several possibilities for a complete digital recording channel. The reconstituted waveform at the output of this channel will now be a replica of the timing of the record signal, with the addition of time uncertainty in the position of the edges due to noise, jitter and dubious equalization. In the same way that binary circuits reject noise by using two voltage levels which are spaced further apart than the uncertainty due to noise, digital recording combats time uncertainty by using events, known as transitions, at multiples of some basic time period, which is larger than the typical time uncertainty. Figure 4.15 shows how this jitter rejection mechanism works.

As digital transitions occur at multiples of a basic period, an oscilloscope, which is triggered on random data, will show an eye pattern if connected to the output of the equalizer. Study of the eye pattern reveals how well the coding used suits the channel.[5] Noise closes the eyes in a vertical direction, and jitter closes the eyes in a horizontal direction, as in Figure 4.16. In the centre of the eyes, at regular intervals, the receiver must make binary decisions about the state of the signal, high or low. If the eyes remain sensibly open, this will be possible. Clearly more jitter can be tolerated if there is less noise, and vice versa. Information theory usually only takes account of SNR and bandwidth when assessing channel capacity. Magnetic and optical recorder channels will never achieve these capacities because of jitter.

It is not possible to record data directly on to the medium, because in real data continuous ones and continuous zeros can occur and, as shown in Figure 4.17, this

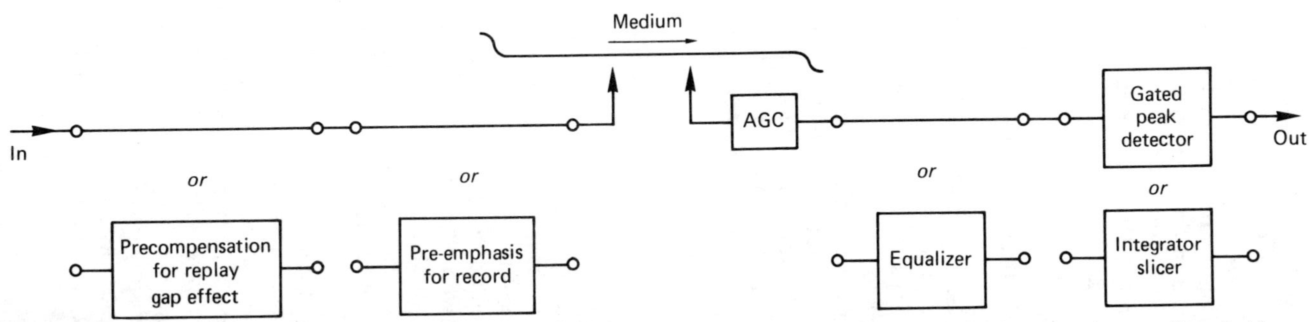

Figure 4.14 A general recording channel showing the various processes described in the text. The system shown here permits recording of a binary waveform

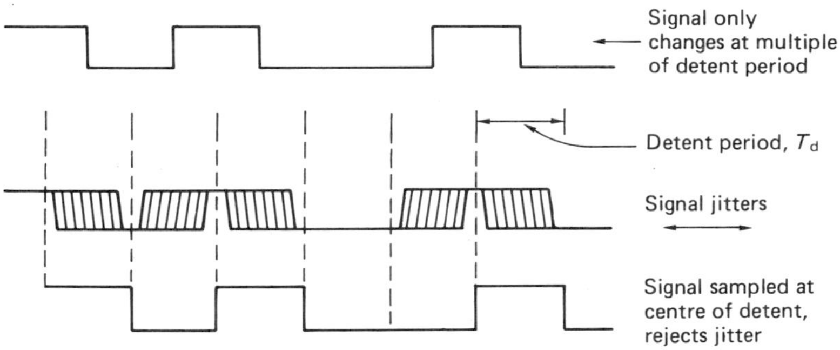

Figure 4.15 A certain amount of jitter can be rejected by changing the signal at multiples of the basic detent period T_d

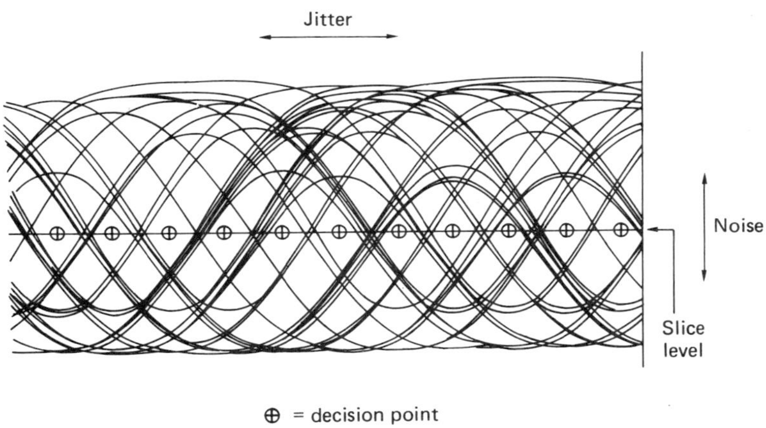

Figure 4.16 At the decision points, the receiver must make binary decisions about the voltage of the signal, whether it is above or below the slicing level. If the eyes remain open, this will be possible in the presence of noise and jitter

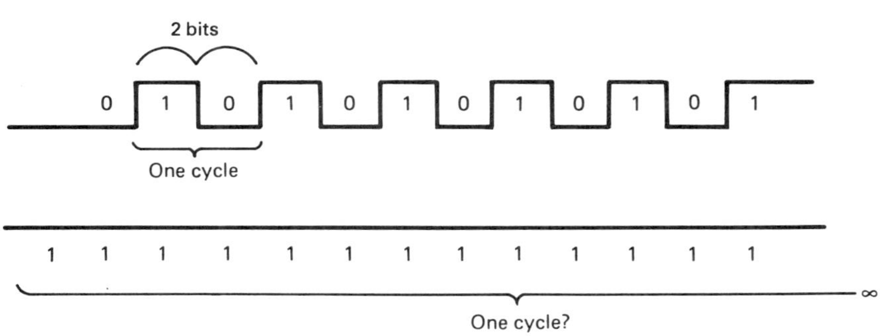

Figure 4.17 The extreme cases of real data. Alternate ones and zeros gives the highest Nyquist rate (= half bit rate). Continuous ones (or zeros) gives DC. Real data fill the spectrum from DC to Nyquist rate

is effectively a DC component of the source data. Alternate ones and zeros represent the other extreme, a frequency of half the bit rate, which is known as the Nyquist rate. Magnetic recorders will not respond to DC, nor is it possible to discriminate between successive identical bits in a channel subject to time instability.

Both of these problems can be solved with a suitable channel code, which will combine a clock with the data prior to recording, in a way which reduces the DC content and permits separation of adjacent symbols on replay. Figure 4.18 shows that a channel coder is necessary prior to the record stage, and that a decoder, known as a data separator, is necessary after the replay stage.

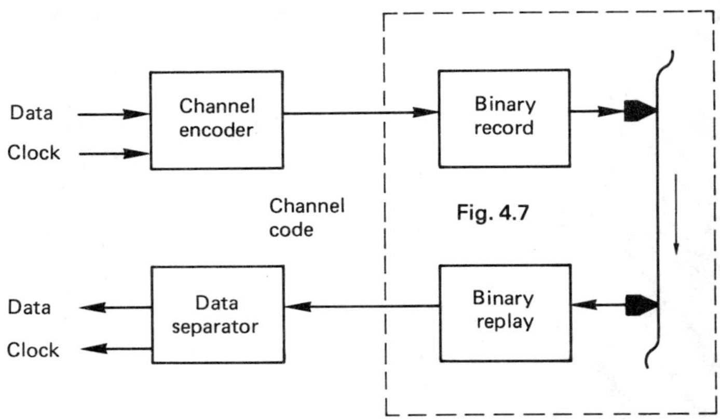

Figure 4.18 In channel coding, the input data and clock are combined into a single waveform called the channel code. On replay the channel code is restored to the original data stream by the data separator

Some codes eliminate DC content entirely, which is advantageous for rotary-head recording. Some codes can reduce the channel bandwidth needed by lowering the upper spectrum limit. This permits higher linear density, but usually at the expense of jitter rejection. A code with a narrow spectrum has a number of advantages. The reduction in asymmetry will reduce peak shift, and data separators can lock more readily where the possible frequencies are fewer. In theory, the narrower the spectrum, the less noise, but excessive noise filtering can ruin the equalization, nullifying any gain.

A convenient definition of a channel code (for there are certainly others) is: 'A method of modulating real data such that they can be reliably received despite shortcomings of a real channel, while making maximum economic use of the channel capacity'.

One of the fundamental parameters of a channel code is the density ratio (DR). One definition of density ratio is that it is the worst-case ratio of the number of data bits recorded to the number of transitions in the channel. It can also be thought of as the ratio between the Nyquist rate of the data and the frequency response of the channel. When better hardware is available to increase the capacity of a channel, the use of a higher density ratio code multiplies the capacity further.

The basic time periods of the recorded signal are called positions or detents, in which the recorded flux will be reversed or stay the same according to the state of a channel bit which describes the detent. The symbol used for the units of channel time is T_d. Channel coding is the art of converting real data into channel bits. It is important to appreciate that the convention in coding is that a channel bit one represents a flux change, whereas a zero represents no change. This is confusing because the input data only change where successive bits differ. The differentiating action of magnetic playback has a lot to do with these conventions.

As jitter is such an important issue in digital recording, a parameter has been introduced to quantify the ability of a channel code to reject time instability. This parameter, the jitter margin, also known as the window margin or phase margin (T_w), is defined as the permitted range of time over which a transition can still be received correctly, divided by the data bit-cell period (T).

Since equalization is often difficult in practice, a code which has a large jitter margin will sometimes be used with short recorded wavelengths because it resists the effects of peak shift distortion. Such a code may achieve a working density better than a code with a higher density ratio but poor jitter performance.

A more realistic comparison of code performance will be obtained by taking into account both density ratio and jitter margin. This is the purpose of the figure of merit (FoM), which is defined as $DR \times T_w$.

4.10 Simple codes

The essence of channel coding is to convert real data into channel bits. Some actual codes will now be examined.

The adoption of FM in analog recorders permitted the recording of DC levels for instrumentation and video. When a binary signal is fed to a frequency modulator, the result is frequency shift keying (FSK) shown in Figure 4.19(a). This is

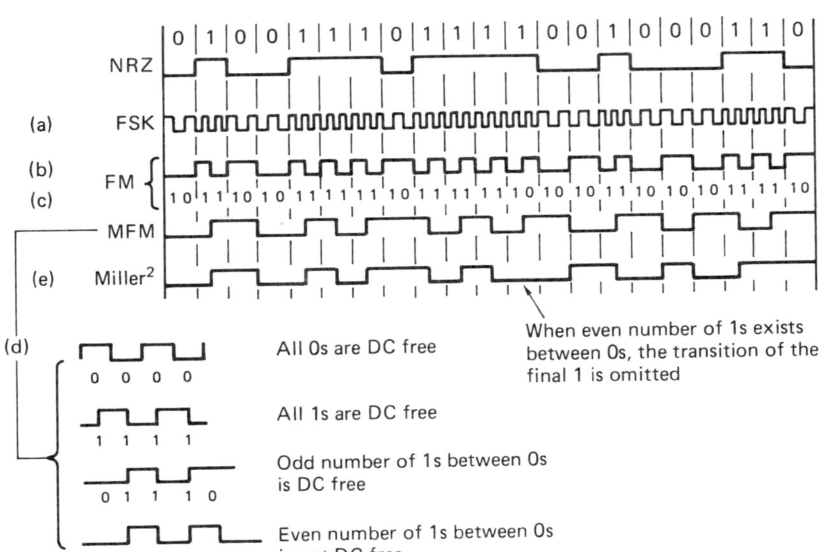

Figure 4.19 Evolution from FSK to Miller². Note that although Miller² is DC free T_{max} and L_c are worse than MFM

inherently DC free and suits radio transmission and rotary-head recorders. FSK has a poor density ratio and has been superseded by more recent codes.

The limiting case of FSK is binary FM (also known as Manchester code) shown in Figure 4.19(b). This was the first practical self-clocking binary code. It is DC free and very easy to encode and decode. It remains in use today where recording density is not of prime importance, for example in single-density floppy disks and in SMPTE/EBU timecode for VTRs. It is also specified for the AES/EBU digital audio interconnect standard which transmits audio serially down cables.

In FM there is always a transition at the bit-cell boundary which acts as a clock. For a data one, there is an additional transition at the bit-cell centre. Figure 4.19(c) shows that each data bit is represented by two channel bits. For a data zero, they will be 10, and for a data one they will be 11. Since the first bit is always one, it conveys no information and is responsible for the density ratio of only one-half. Since there can be two transitions for each data bit, the jitter margin can only be half a bit, and the FoM resulting is only 0.25. The high clock content of FM does, however, mean that data recovery is possible over a wide range of speeds; hence the use for timecode.

In MFM[6] the highly redundant clock content of FM was reduced by the use of a phase-locked loop in the receiver which could flywheel over missing clock transitions. This technique is implicit in all the more advanced codes. The bit-cell centre transition on a data one was retained, but the bit-cell boundary transition is now only required between successive zeros. There are still two channel bits for every data bit, but adjacent channel bits will never be one, doubling the minimum time between transitions, and giving a DR of 1. Clearly the coding of the current bit is now influenced by the preceding bit. The maximum number of prior bits which affect the current bit is known as the constraint length L_c, measured in data-bit periods. For MFM $L_c = T$. Another way of considering the constraint length is that it assesses the number of data bits which may be corrupted if the receiver misplaces one transition. If L_c is long, all errors will be burst errors.

MFM doubled the density ratio compared with FM without changing the jitter performance; thus the FoM also doubles. It was adopted for many rigid disks at the time of its development and remains in use on double-density floppy disks. It is not, however, DC free. Figure 4.19(d) shows how MFM can have DC content under certain conditions.

4.11 Miller^2 code

The Miller^2 code used in D-2 is derived from MFM, and Figure 4.19(e) shows that the DC content is eliminated by a slight increase in complexity.[7] Wherever an even number of ones occurs between zeros, the transition at the last one is omitted. This creates two additional, longer run lengths and increases the T_{max} of the code. The decoder can detect these longer run lengths in order to reinsert the suppressed ones.

4.12 Synchronizing

Although the Miller^2 code of D-2 is DC free, this is achieved with the penalty of a reduction of clock content under certain conditions, so that the number of channel bits which have to be measured between transitions becomes quite large. The only way in which the channel code can be decoded is to use a phase-locked loop to regenerate the channel-bit clock. A fixed-frequency clock would be of no use, since even if the head-to-tape speed could be controlled accurately enough for the channel-bit rate to match the clock rate, the instantaneous errors due to jitter would be insuperable. In phase-locked loops, the voltage-controlled oscillator is driven by a phase error measured between the output and some reference, such that the oscillator eventually runs at the same frequency as the reference. If a divider is placed between the VCO and the phase comparator, as in Figure 4.20,

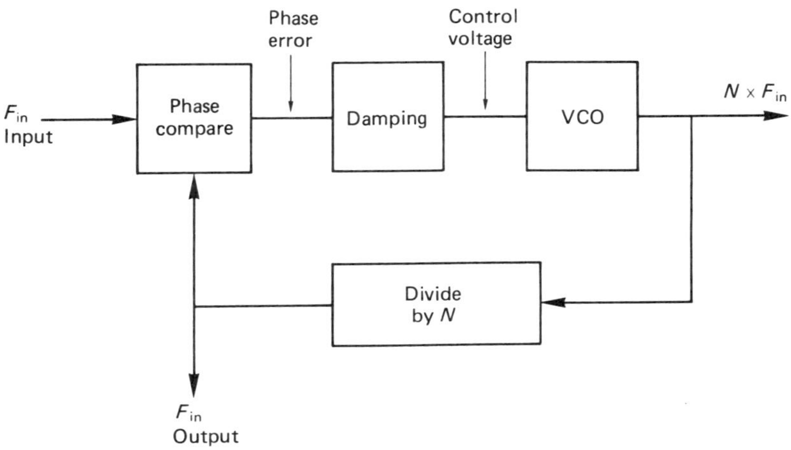

Figure 4.20 A typical phase-locked loop where the VCO is forced to run at a multiple of the input frequency. If the input ceases, the output will continue for a time at the same frequency until it drifts

the VCO frequency can be made to be a multiple of the reference. This also has the effect of making the loop more heavily damped. If a channel code is used as a reference to a PLL, the loop will be able to make a phase comparison whenever a transition arrives, but when there are channel zeros between transitions, the loop will flywheel at the last known frequency and phase until it can rephase at a subsequent transition. In this way, cycles of the VCO can be counted to measure the number of channel zeros between transitions and hence to decode the information. Figure 4.21 illustrates this mechanism.

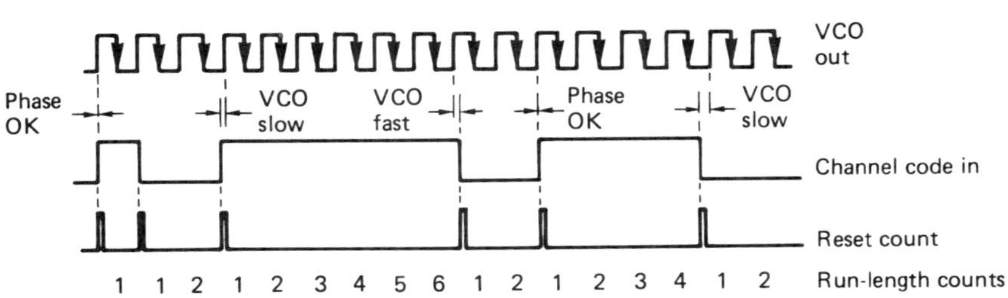

Figure 4.21 In order to reconstruct the channel patterns, a phase-locked loop is fed with the channel code and freewheels between transitions, correcting its phase at each one. Counting the VCO edges (↓) between transitions reconstructs the channel bits. If the medium changes speed, the VCO will track. If the maximum run length is too long, the VCO will not be able to phase-correct often enough and may miscount channel bits in the presence of jitter

Clearly data cannot be separated if the PLL is not locked, but it cannot be locked until it has seen transitions for a reasonable period. The solution is to precede each data block with a pattern of transitions whose sole purpose is to provide a timing reference for synchronizing the phase-locked loop. This pattern is known as a preamble. In D-2, the preamble will be the result of encoding alternate ones and zeros, which is a square wave at half the bit rate to avoid the attenuation of the highest frequencies when the PLL is attempting to lock. In magnetic recording there is almost always a postamble at the end of the data block. Again a few bits are recorded after the real data. When magnetic heads are turned off at the end of a write, there is often a transient which corrupts the last bits written. The postamble can be damaged in this way without consequence.

Once the PLL has locked to the preamble, a data stream and a clock will emerge from the data separator. It is then vital to know at what point in the data stream the preamble finishes and the actual data commence. In serial recording, words are recorded one after the other, one bit at a time, with no spaces in between, so that although the designer knows that a block contains, say, 24 words of 8 bits each, the medium simply holds 192 bits in a row. If the exact position of the first bit is not known, then it is not possible to put all the bits in the right places in the right words. The effect of sync slippage is devastating, because a 1 bit disparity between the data-bit count and the bit stream will corrupt every word in the block, which is just as bad as a massive dropout.[8]

The synchronization of the data separator and the synchronization to the block format are two distinct problems, and they are often solved separately. At the end of the preamble, a so-called sync pattern may be inserted. This is a pattern which is identical for every block; it will be recognized by the replay circuitry and used to reset the bit count through the block. By counting bits from the sync pattern and dividing by the wordlength, the replay circuitry will be able to determine the position of the boundaries between words. The sync pattern must be chosen with care, so that a bit or bits in error does not cause sync to be recognized in the wrong place. Such a pattern is designed to be as different as possible to itself however many places it is shifted, which is the same as saying it has a low autocorrelation.

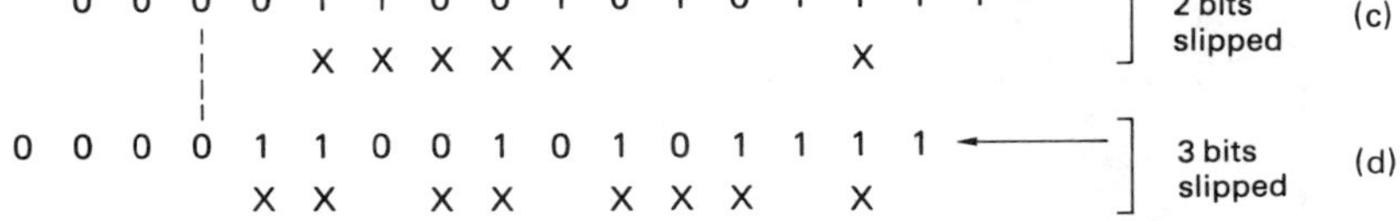

Figure 4.22 A synchronizing pattern having a low autocorrection is shown here at (a). In (b), the pattern is 1 bit from synchronization, and fails to match in seven places (shown as X). In (c) the pattern is 2 bits away from synchronization, and fails to match in six places. At (d) the pattern is 3 bits away from synchronization, and fails in eight places.

Owing to the large number of differences when the pattern shifts, there is a much reduced probability of an error reading the pattern causing false synchronization. The pattern shown is used in the D-2 format DVTR for both audio and video data blocks

The solution adopted in D-2 is a good example of a low-autocorrelation sync pattern. The pattern used is 30F5 hex. (As the data are recorded a byte at a time, LSB first, the bit pattern will be 0000110010101111.) Figure 4.22 shows how many differences are caused for different false sync states, which helps to ensure that only the correct timing is recognized. A smaller number of differences, due to bit errors on playback, will not prevent the recognition of sync.

In run-length-limited codes, it is possible to combine the function of preamble and sync pattern by producing a transition pattern at the start of the block which contains timing to phase the channel-bit rate PLL, but which contains run lengths which violate the limits. There is no way that these run lengths will be interpreted as data, but they can be detected by the replay circuitry. Such a technique is used in the AES/EBU digital audio interconnect (see Chapter 8) and is illustrated in Figure 4.23.

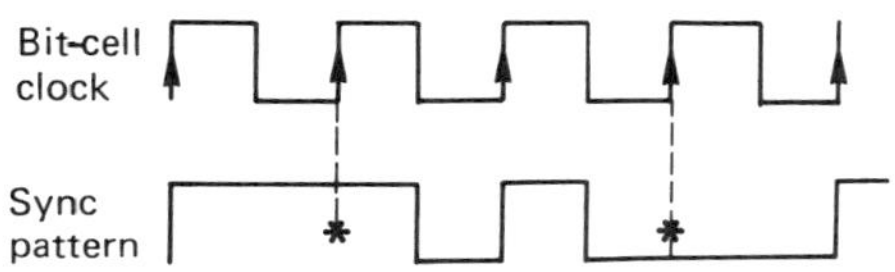

Figure 4.23 Synchronization by run-length violation. The AES/EBU audio interface uses FM code which *must* have a transition on every bit-cell clock. One of the possible sync patterns is shown here to violate the code at points *

Where reading in both directions is required, a reverse sync pattern will be placed between the data and the postamble. In SMPTE/EBU timecode, the sync pattern is asymmetrical, so that the read circuitry can tell which way the tape is moving without any other source of information.

References

1. DEELEY, E. M. Integrating and differentiating channels in digital tape recording. *Radio Electron. Eng.*, **56**, 169–173 (1986)
2. MEE, C. D. *The Physics of Magnetic Recording*. Amsterdam: Elsevier–North-Holland (1978)
3. JACOBY, G. V. Signal equalization in digital magnetic recording. *IEEE Trans. Magn.*, **MAG-11**, 302–305 (1975)
4. SCHNEIDER, R. C. An improved pulse-slimming method for magnetic recording. *IEEE Trans. Magn.*, **MAG-11**, 1240–1241 (1975)
5. NAKAJIMA, H. *et al. Digital Audio Technology*, pp.150–152. Blue Ridge Summit, PA: TAB Books (1983)
6. MALLINSON, J. C. and MILLER, J. W. Optimum codes for digital magnetic recording. *Radio Electron. Eng.*, **47**, 172–176 (1977)
7. MILLER, J. W. DC-free encoding for data transmission system. *US Patent* 4 027 335 (1977)
8. GRIFFITHS, F. A. A digital audio recording system. Presented at 65th Audio Engineering Society Convention (London, 1980), preprint 1580(C1)

Chapter 5

Error correction

The subject of error correction is almost always described in mathematical terms by specialists for the benefit of other specialists. Such mathematical approaches are quite inappropriate for a proper understanding of the concepts of error correction and only become necessary to analyse the quantitative behaviour of a system. The description below will use the minimum possible amount of mathematics, and it will then be seen that error correction is, in fact, quite straightforward.

5.1 Sensitivity of message to error

Before attempting to specify any piece of equipment, it is necessary to quantify the problems to be overcome and how effectively they need to be overcome. For a digital recording system the causes of errors must be studied to quantify the problem, and the sensitivity of the destination to errors must be assessed. In audio and video, the sensitivity to errors must be subjective. In both, the effect of a single bit in error depends upon the significance of the bit. If the least significant bit of a sample is wrong, the chances are that the effect will be lost in the noise. Conversely, if a high-order bit is in error, a massive transient will be added to the sound or picture waveform. The effect of uncorrected high-order errors in PCM video is to produce dots in the picture, which will be visible as bright dots in dark areas and vice versa.

If the error rate demanded by the destination cannot be met by the unaided channel, some form of error handling will be necessary.

5.2 Error mechanisms

In magnetic recording, there are many mechanisms available to corrupt data, from mechanical problems such as media dropout and poor head contact, to Gaussian thermal noise in replay circuits and heads.

The results of all of these mechanisms are twofold. There are large isolated corruptions, called error bursts, where numerous bits are corrupted all together in an area which is otherwise error free, and there are random errors affecting single bits or symbols. Whatever the mechanism, the result will be that the received data will not be exactly the same as those sent.

5.3 Error handling

Figure 5.1 shows the broad subdivisions of error handling. The data pass through the channel, which causes whatever corruptions it feels like. On receipt of the data the occurrence of errors is first detected, and this process must be extremely reliable, as it does not matter how effective the correction or how good the concealment algorithm, if it is not known that they are necessary! The detection of an error then results in a course of action being decided.

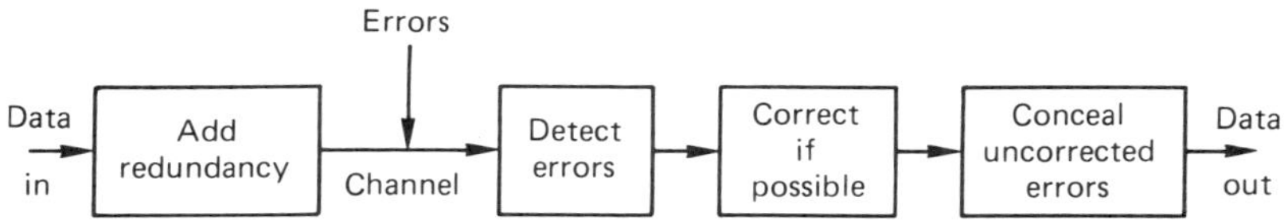

Figure 5.1 The basic stages of an error-correction system. Of these the most critical is the detection stage, since this controls the subsequent actions

5.4 Interpolation

Although audio and video recorders are at a disadvantage with respect to computer recorders in that they cannot preformat and verify the medium, and there is no time for retries, they have the advantage that there is a certain amount of redundancy in the information conveyed. If an error cannot be corrected, then it can be concealed. In audio systems, if a sample is lost, it is possible to obtain an approximation to it by interpolating between the samples before and after the missing one. In video, there is a two-dimensional array of samples making up a field, and so concealment in more than one dimension is possible. Video concealment is a complex subject which will be dealt with later in this chapter.

If there is too much corruption for concealment, the only course in audio is to mute, whereas a video system will typically freeze and display the last available field.

In general, if use is to be made of concealment on replay, the data must generally be reordered or shuffled prior to recording. To take a simple example, odd-numbered samples are recorded in a different area of tape from even-numbered samples. On playback, if a gross error occurs on the tape, depending on its position, the result will be either corrupted odd samples or corrupted even samples, but it is most unlikely that both will be lost. Interpolation is then possible if the power of the correction system is exceeded.

5.5 Noise and probability

Whilst it is generally stated that inside a piece of logic circuitry the binary signal is always greater than the noise, so that there will be no errors, the same is not true in high-density recording. To see why this is true, it is necessary to look at the characteristics of noise.

Noise is random, and although under given conditions the noise power in a system may be constant, this value only determines the heat that would be developed in a resistive load. In digital recording, it is the instantaneous voltage of noise which is of interest, since it is a form of interference which could alter the state of a binary signal if it were large enough. Unfortunately it cannot be predicted; indeed if it could the interference could not be called noise.

Noise can only be quantified statistically, by measuring or predicting the likelihood of a given noise amplitude.

Figure 5.2 shows a graph relating the probability of occurrence to the amplitude of noise. The noise amplitude increases away from the origin along the horizontal

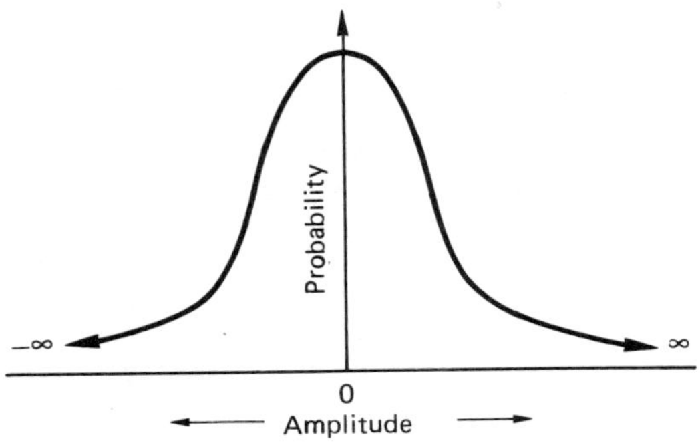

Figure 5.2 The probability of occurrence of noise of a given amplitude falls as that amplitude increases, but never reaches zero; as a consequence, however good signal-to-noise ratio may be, errors will still occur

axis, and for any amplitude of interest the probability of that noise amplitude occurring can be read from the curve. The shape of the curve is known as a Gaussian distribution, which crops up whenever the overall effect of a large number of independent phenomena is considered. Magnetic recording depends on superimposing some average magnetism on vast numbers of magnetic particles.

If it were possible to isolate an individual noise-generating microcosm of a tape or a head on the molecular scale, the noise it could generate would have physical limits owing to the finite energy present. The noise distribution might then be rectangular as shown in Figure 5.3(a), where all amplitudes below the physical limit are equally likely. If the combined effect of two of these microcosms is considered, clearly the maximum amplitude is now doubled, because the two effects can add, but since the two effects are uncorrelated, they can also subtract, so the probability is no longer rectangular, but becomes triangular as in Figure 5.3(b). Probability falls to zero at peak amplitude because the chances of two independent mechanisms reaching their peak value with the same polarity at the same time are understandably small.

If the number of mechanisms summed together is now allowed to increase without limit, the result is the Gaussian curve shown in Figure 5.3(c), where it will be seen that the curve has no limit, because it is just possible that all mechanisms will simultaneously reach their peak value together, although the chances of this happening are incredibly remote.

Some conclusions can be drawn from the Gaussian distribution of noise.

First, it is not possible to make error-free digital recordings, because however high the signal-to-noise ratio of the recording, there is still a small but finite chance that the noise can exceed the signal. Measuring the signal-to-noise ratio of a

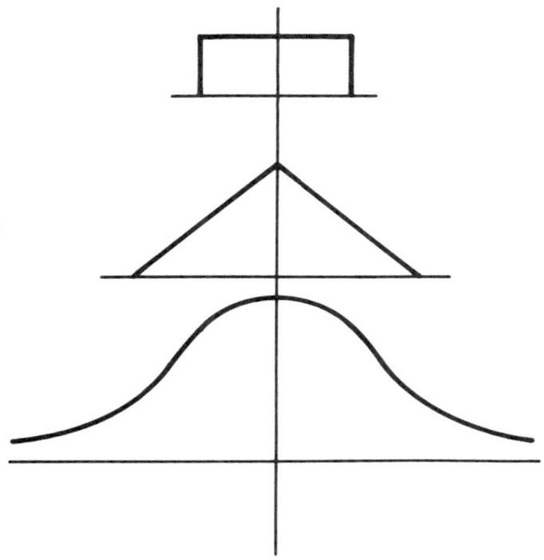

Figure 5.3 At (a) is a rectangular probability; all values are equally likely, but between physical limits. At (b) is the sum of two rectangular probabilities, which is triangular, and at (c) is the Gaussian curve which is the sum of an infinite number of rectangular probabilities

channel establishes the noise power, which determines the width of the noise distribution curve relative to the signal amplitude. When in a binary system the noise amplitude exceeds the signal amplitude, a bit error will occur. Knowledge of the shape of the Gaussian curve allows the conversion of signal-to-noise ratio into bit error rate (BER). It can be predicted how many bits will fail because of noise in a given recording, but it is not possible to say which bits will be affected.

The logical reaction to this statement is to decide that error correction is necessary if progress is to be made with recording density.

Error correction works by adding some bits to the data which are calculated from the data. This creates an entity called a codeword which spans a greater length of time than one bit alone. The statistics of noise means that whilst one bit may be lost in a codeword, the loss of the rest of the codeword due to noise is highly improbable. As will be described later in this chapter, codewords are designed to be able to correct totally a finite number of corrupted bits. The greater the timespan over which the coding is performed, the greater will be the reliability achieved, although this does mean that an encoding delay will be experienced on recording, and a similar delay on reproduction. Shannon[1] discovered that a message can be sent to any desired degree of accuracy provided that it is spread over a sufficient timespan. Engineers have to compromise, because an infinite delay in the recovery of an error-free signal is not acceptable.

Having decided that error correction is necessary, it is then only a small step to put it to maximum use. All error correction depends on adding bits to the original message, and this of course increases the number of bits to be recorded, although it does not increase the information recorded. It might be imagined that error correction is going to reduce storage capacity, because space has to be found for all the extra bits. Nothing could be further from the truth. Once an error-correction system is used, the signal-to-noise ratio of the channel can be reduced, because the raised BER of the channel will be overcome by the error-correction system. Reduction of SNR by 3 dB in a magnetic tape track can be achieved by halving the track width, provided that the system is not dominated by head noise. This doubles the recording density, making the storage of the additional bits needed for error correction a trivial matter. In short error correction is not a nuisance to be tolerated, but a vital tool needed to maximize the efficiency of recorders. High-density recording would not be possible without it.

Although the foregoing is true, error-correction systems must be able to handle burst errors due to dropouts as well as random errors due to noise.

5.6 Parity

The error detection and correction processes are closely related and will be dealt with together here. The actual correction of an error is simplified tremendously by the adoption of binary. As there are only two symbols, 0 and 1, it is enough to know that a symbol is wrong, and the correct value is obvious. Figure 5.4 shows a minimal circuit required for correction once the bit in error has been identified. The exclusive–OR gate shows up extensively in error correction and the figure also shows the truth table. One way of remembering the characteristics of this useful device is that there will be an output when the inputs are different. Inspection of the truth table will show that there is an even number of ones in each row (zero is an even number) and so the device could also be called an even parity gate.

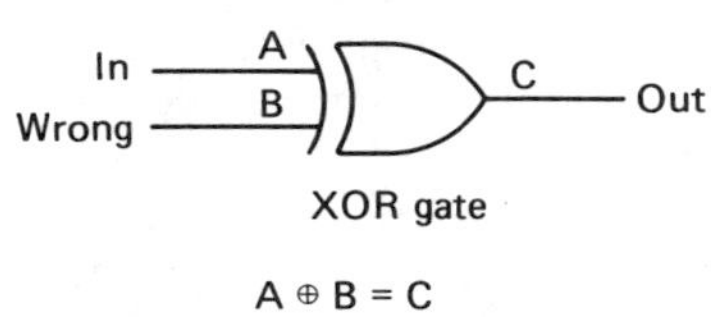

Truth table of XOR gate

A	B	C
0	0	0
0	1	1
1	0	1
1	1	0

Figure 5.4 Once the position of the error is identified, the correction process in binary is easy

Parity is a fundamental concept in error detection. In Figure 5.5, the example is given of a 4 bit data word which is to be protected. If an extra bit is added to the word which is calculated in such a way that the total number of ones in the 5 bit word is even, this property can be tested on receipt. The generation of the parity bit in Figure 5.5 can be performed by a number of the ubiquitous exclusive–OR (XOR) gates configured into what is known as a parity tree. In the figure, if a bit is corrupted, the received message will be seen no longer to have an even number of ones. If 2 bits are corrupted, the failure will be undetected. This example can be used to introduce much of the terminology of error correction. The extra bit added to the message carries no information of its own, since it is calculated from the other bits. It is therefore called a redundant bit. The addition of the redundant bit gives the message a special property, i.e. the number of ones is even. A message having some special property irrespective of the actual data content is called a codeword. All error correction relies on adding redundancy to real data to form codewords for transmission. If any corruption occurs, the intention is that the received message will not have the special property; in other words if the received message is not a codeword there has definitely been an error. If the received message is a codeword, there probably has not been an error. The word 'probably'

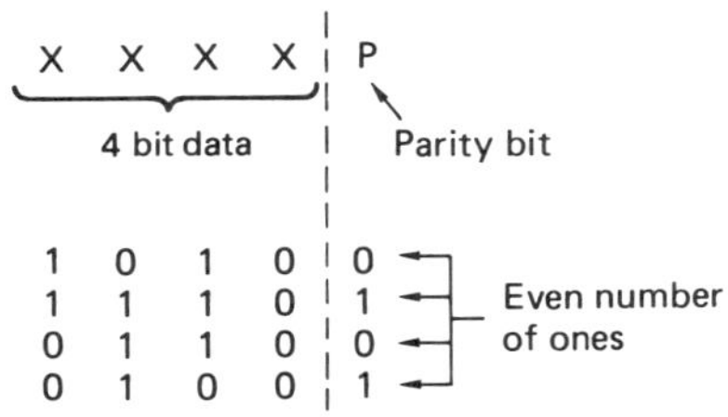

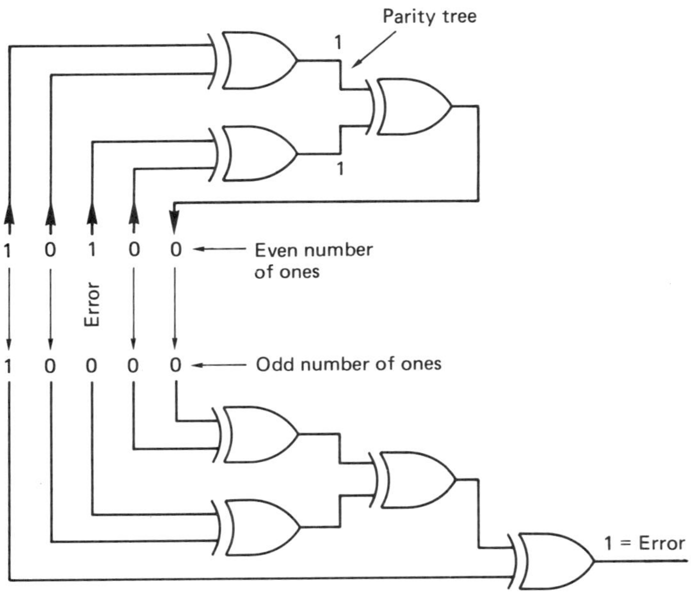

Figure 5.5 Parity checking adds up the number of ones in a word using, in this example, parity trees. One error bit and odd numbers of errors are detected. Even numbers of errors cannot be detected

must be used because the figure shows that 2 bits in error will cause the received message to be a codeword, which cannot be discerned from an error-free message. If it is known that generally the only failure mechanism in the channel in question is loss of a single bit, it is *assumed* that receipt of a codeword means that there has been no error. If there is a probability of two error bits, that becomes very nearly the probability of failing to detect an error, since all odd numbers of errors will be detected, and a 4 bit error is much less likely. It is paramount in all error-correction systems that the protection used should be appropriate for the probability of errors to be encountered. An inadequate error-correction system is actually worse than not having any correction. Error correction works by trading probabilities. Error-free performance with a certain error rate is achieved at the expense of performance at higher error rates. Figure 5.6 shows the effect of an error-correction system on the output error rate for a given raw, or input, error

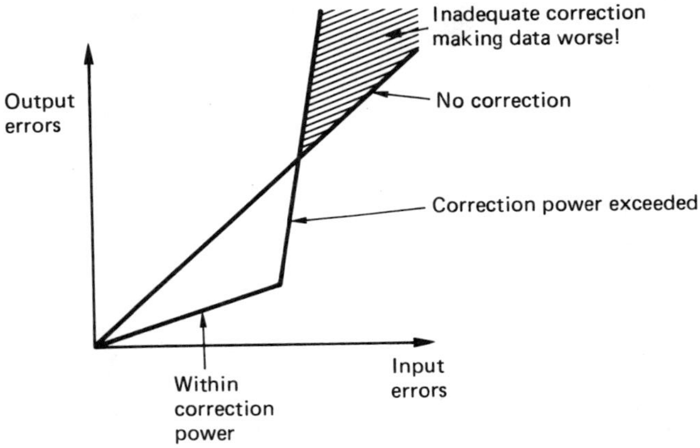

Figure 5.6 An error-correction system can only reduce errors at normal error rates at the expense of increasing errors at higher rates. It is most important to keep a system working to the left of the knee in the graph

rate. It will be seen that there is a characteristic knee in the graph. If the expected raw error rate has been misjudged, the consequences can be disastrous. Another result demonstrated by the example is that we can only guarantee to detect the same number of bits in error as there are redundant bits.

In the example of Figure 5.5, the error was detected but it was not possible to say which bit was in error. Even though the code used can only detect errors, correction is still possible if a suitable strategy is used. Figure 5.7 shows the use of a crossword code, also known as a product code. The data are formed into a two-dimensional array, with parity taken on rows and columns. If a single bit fails, one row check and one column check will fail, and the failing bit can be located at the intersection of the two failing checks. Although 2 bits in error confuse this simple scheme, using more complex coding in a two-dimensional structure is very powerful, and further examples will be given throughout this chapter.

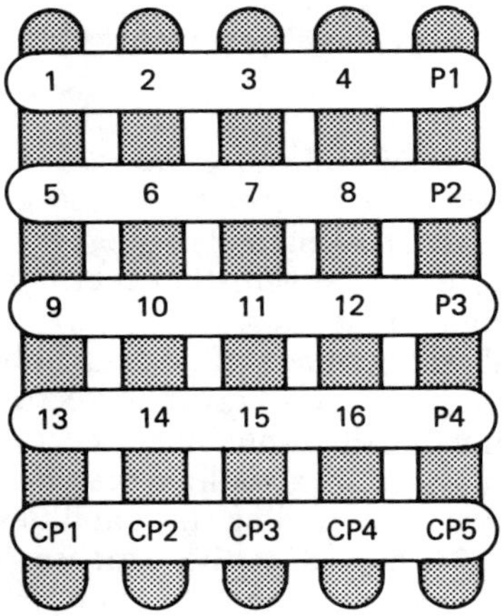

Figure 5.7 A crossword parity-check system. Horizontal checks are made by P1, P2, etc., and cross parity checks on columns are made by CP1, CP2, etc. If, for example, bit 10 were in error, it would be located by CP2 and P3 intersecting

5.7 Hamming code

In a one-dimensional code, the position of the failing bit can be determined by using more parity checks. In Figure 5.8, the four data bits have been used to compute three redundancy bits, making a 7 bit codeword. The four data bits are examined in turn, and each bit which is a one will cause the corresponding row of a generator matrix to be added to an exclusive–OR sum. For example, if the data were 1001, the top and bottom rows of the matrix would be XORed. The matrix used is known as an identity matrix, because the data bits in the codeword are identical to the data bits to be conveyed. This is useful because the original data can be stored unmodified, and the check bits are simply attached to the end to make a so-called systematic codeword. Almost all digital recording equipment uses systematic codes. The way in which the redundancy bits are calculated is simply that they do not all use every data bit. If a data bit has not been included in a parity check, it can fail without affecting the outcome of that check. The position of the error is deduced from the pattern of successful and unsuccessful checks in the check matrix. This pattern is known as a syndrome.

In the figure the example of a failing bit is given. Bit 3 fails, and because this bit is included in only two of the checks, there are two ones in the failure pattern, 011. As some care was taken in designing the matrix pattern for the generation of the check bits, the syndrome, 011, is the address of the failing bit. This is the fundamental feature of the Hamming codes due to Richard Hamming.[2] The performance of this 7 bit codeword can be assessed. In 7 bits there can be 128 combinations, but in four data bits there are only 16 combinations. Thus out of 128 possible received messages, only 16 will be codewords, so if the message is completely trashed by a gross corruption, it will still be possible to detect that this has happened 112 times out of 127, as in these cases the syndrome will be non-zero (the 128th case is the correct data). There is thus only a probability of detecting that all of the message is corrupt. In an idle moment it is possible to work out, in a similar way, the number of false codewords which can result from different numbers of bits being assumed to have failed. For less than 3 bits, the failure will always be detected, because there are three check bits. Returning to the example, if 2 bits fail, there will be a non-zero syndrome, but if this is used to point to a bit in error, a miscorrection will result. From these results can be deduced another important feature of error codes. The power of detection is always greater than the power of correction, which is also fortunate, since if the correcting power is exceeded by an error it will at least be a known problem, and steps can be taken to prevent any undesirable consequences.

5.8 Hamming distance

It is useful at this point to introduce the concept of Hamming distance. This is the minimum number of bits that must be changed in any codeword without turning it into another codeword. Clearly if errors corrupt a codeword so that it is no longer a codeword, it will definitely be detectable and possibly correctable. If errors convert one codeword into another, it will not even be detected.

Figure 5.9 shows Hamming distance diagrammatically. A 3 bit codeword is used with two data bits and one parity bit. With 3 bits, a received code could have eight combinations, but only four of these will be codewords. The valid codewords are shown in the centre of each of the disks, and these will be seen to be identical to the

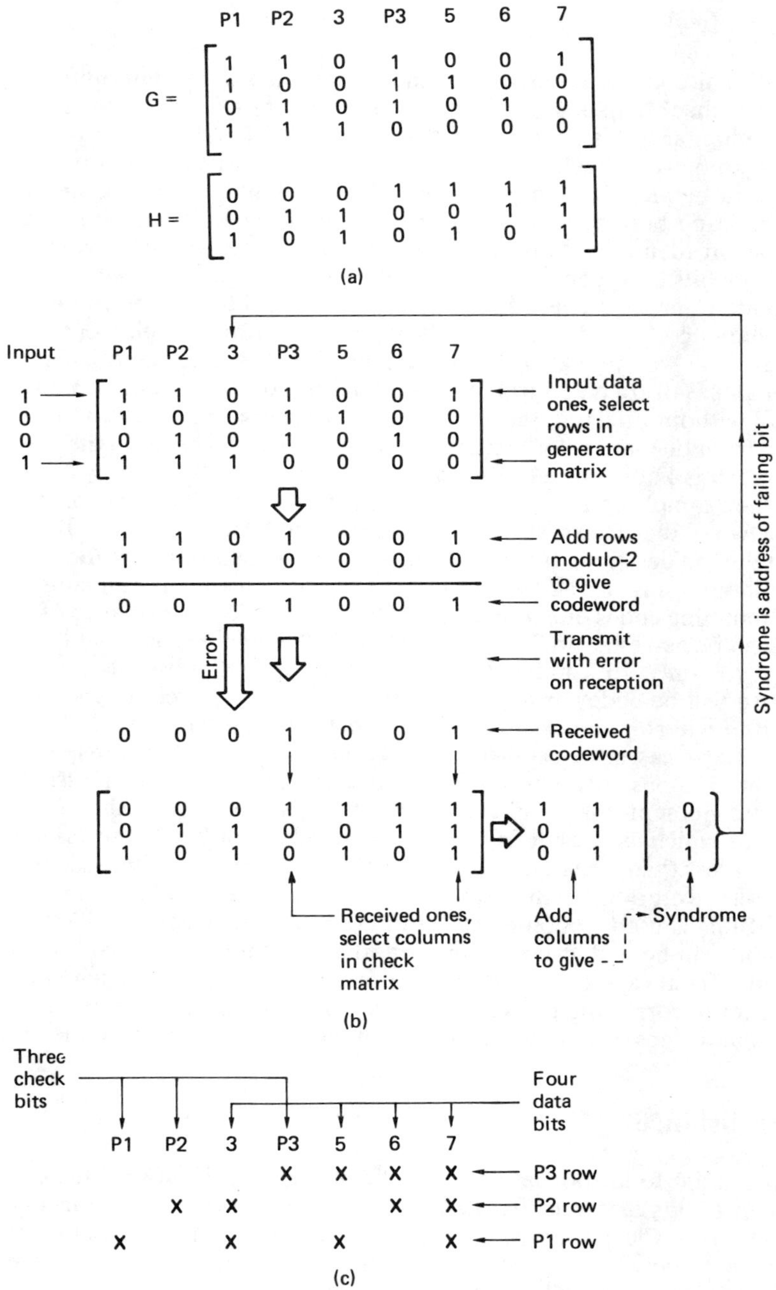

Figure 5.8 (a) The generator and check matrices of a Hamming code. The data and check bits are arranged as shown because it causes the syndrome to be the binary address of the failing bit. (b) An example of Hamming-code generation and error correction. (c) Another way of looking at Hamming code is to say that the rows of crosses in this chart are calculated to have even parity. If bit 3 fails, parity check P3 is not affected, but parity checks P1 and P2 both include bit 3 and will fail

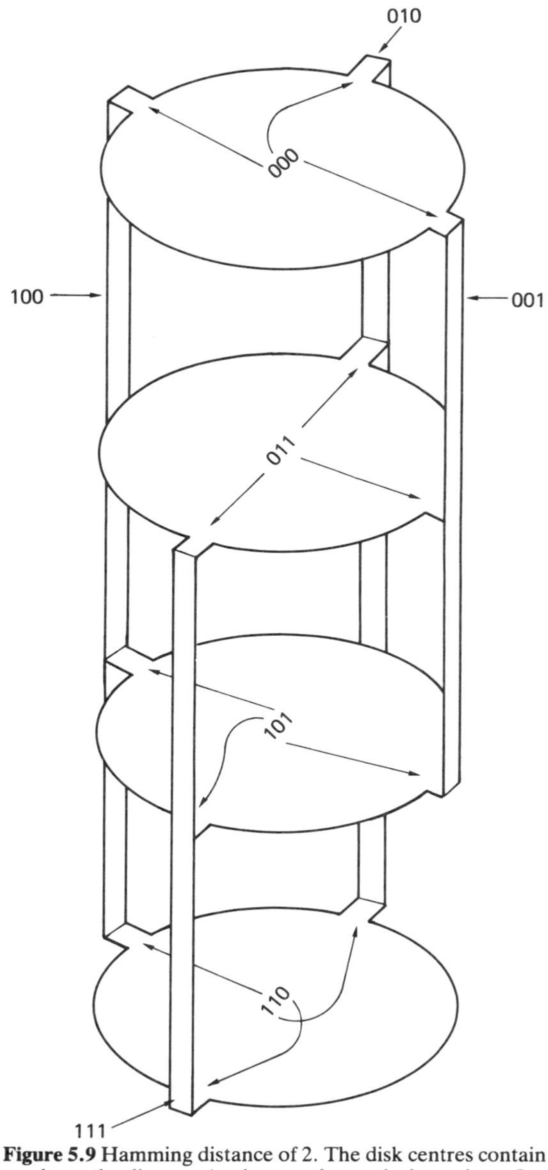

Figure 5.9 Hamming distance of 2. The disk centres contain codewords. Corrupting each bit in turn produces the distance 1 values on the vertical members. In order to change one codeword to another, 2 bits must be changed, so the code has a Hamming distance of 2

rows of the truth table in Figure 5.4. At the perimeter of the disks are shown the received words which would result from a single-bit error, i.e. they have a Hamming distance of 1. It will be seen that the same received word (on the vertical bars) can be obtained from a different single-bit corruption of any three codewords. It is thus not possible to tell which codeword was corrupted and so, although all single-bit errors can be detected, correction is not possible. This diagram should be compared with that of Figure 5.10, which is a Venn diagram where there is a set in which the MSB is 1 (upper circle), a set in which the middle bit is 1 (lower left circle) and a set in which the LSB is 1 (lower right circle). Note that in crossing any boundary only one bit changes, and so each boundary represents a Hamming distance change of one. The four codewords of Figure 5.9 are repeated here, and it will be seen that single-bit errors in any codeword produce a non-codeword, and so single-bit errors are always detectable.

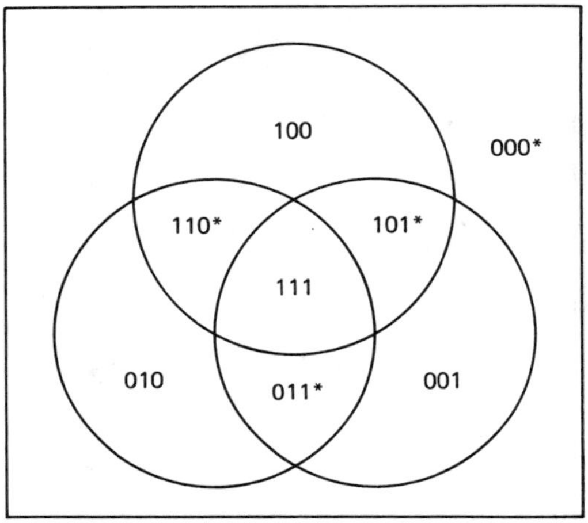

Figure 5.10 Venn diagram shows a 1 bit change crossing any boundary which is a Hamming distance of 1. Compare with Figure 5.9. Codewords marked*

Correction is possible if the number of non-codewords is increased by increasing the number of redundant bits. This means that it is possible to spread out the actual codewords in Hamming distance terms.

Figure 5.11(a) shows a distance 2 code, where there is only one redundancy bit, and so half of the possible words will be codewords. There will be non-codewords at distance 1 which can be produced by altering a single bit in either of two codewords. In this case it is not possible to tell what the original codeword was in the case of a single-bit error.

Figure 5.11(b) shows a distance 3 code, where there will now be at least two non-codewords between codewords. If a single-bit error occurs in a code word, the resulting non-codeword will be at distance 1 from the original codeword. This same non-codeword could also have been produced by changing 2 bits in a different codeword. If it is known that the failure mechanism is a single bit, it can be *assumed* that the original codeword was the one which is closest in Hamming distance to the received bit pattern, and so correction is possible. If, however, our assumption about the error mechanism proved to be wrong, and in fact a 2 bit error had occurred, this assumption would take us to the wrong codeword, turning the event into a 3 bit error. This is an illustration of the knee in the graph of Figure 5.6, where if the power of the code is exceeded it makes things worse.

Figure 5.11(c) shows a distance 4 code. There are now three non-codewords between codewords, and clearly single-bit errors can still be corrected by choosing

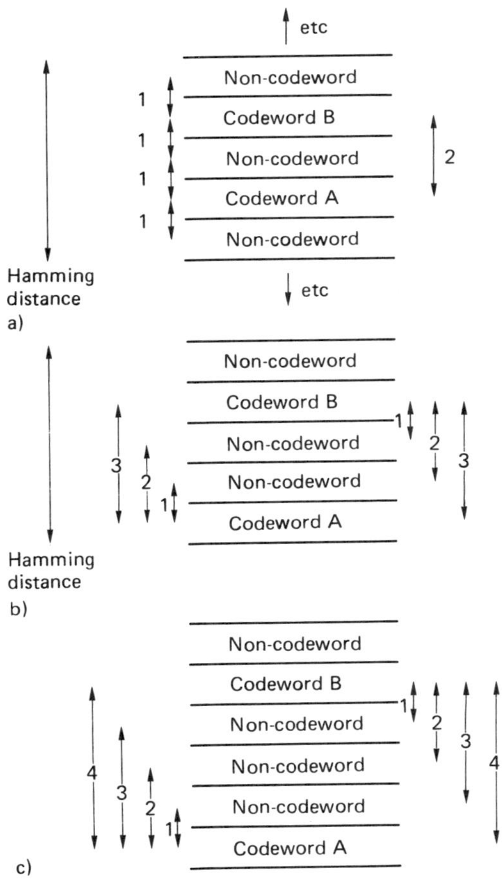

Figure 5.11 (a) Distance 2 code, non-codewords are at distance 1 from *two* possible codewords so it cannot be deduced what the correct one is. (b) Distance 3 code, non-codewords which have *single-bit errors* can be attributed to the nearest codeword. Breaks down in presence of double-bit errors. (c) Distance 4 code, non-codewords which have single-bit errors can be attributed to the nearest codeword, AND double-bit errors form *different* non-codewords, and can thus be detected but not corrected

the nearest codeword. Double-bit errors will be detected, because they result in non-codewords equidistant in Hamming terms from codewords, but it is not possible to say what the original codeword was.

Figure 5.12 shows a more complex arrangement of the 16 possible combinations of 4 bits. Travelling along any of the ties between discs is the equivalent of a Hamming distance of 1.

The correction of one bit is of little use in the presence of burst errors, but a Hamming code can be made to correct burst errors by using interleaving. Figure 5.13 shows that if several codewords are calculated beforehand and woven together as shown before they are sent down the channel, then a burst of errors which corrupts several bits will become a number of single-bit errors in separate codewords upon de-interleaving. Interleaving is used extensively in digital recording and will be discussed in greater detail later in this chapter.

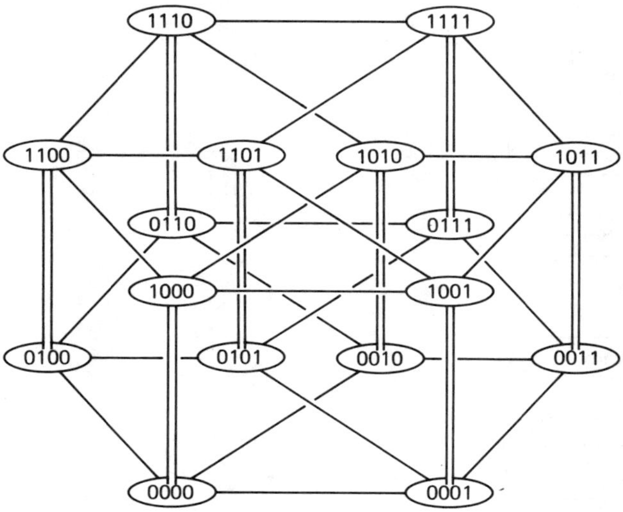

Figure 5.12 Hamming distance diagram for 4 bits. Each bond between the disks is unit Hamming distance

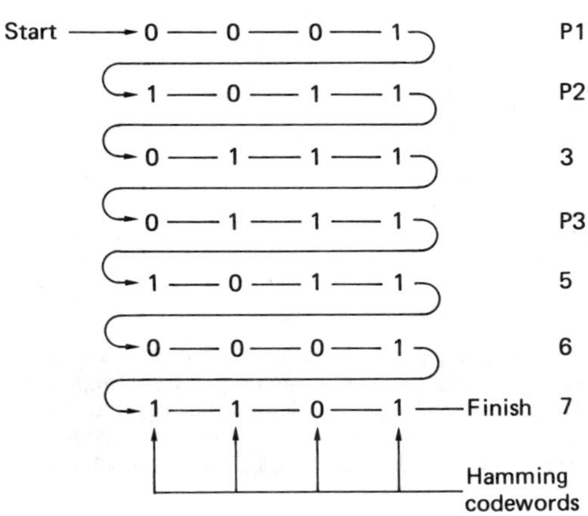

Figure 5.13 The vertical columns of this diagram are all codewords generated by the matrix of Figure 5.8 which can correct a single-bit error. If these words are recorded in the order shown, a burst error of up to 4 bits will result in one single-bit error in each codeword, which is correctable. Interleave requires memory and causes delay. De-interleave requires the same

5.9 Cyclic codes

The implementation of a Hamming code can be made very fast using parity trees, which is ideal for memory applications where access time is increased by the correction process. However, in most video applications, the data are stored serially on a magnetic track, and it is desirable to use relatively large data blocks to reduce the amount of the medium devoted to preambles, addressing and synchronizing. Where large data blocks are to be handled, the use of a lookup table or tree has to be abandoned because it would become impossibly large. The principle of generator and check matrices will still be employed, but they will be matrices which can be generated algorithmically by an equation. The syndrome will then be converted to the bit(s) in error not by looking them up, but by solving an equation.

Where data can be accessed serially, simpler circuitry can be used because the same gate will be used for many XOR operations. Unfortunately the reduction in component count is only paralleled by an increase in the difficulty of explaining what takes place.

The circuit of Figure 5.14 is a kind of shift register, but with a peculiar feedback arrangement which leads it to be known as a twisted-ring counter. If seven message bits A–G are applied serially to this circuit, and each one of them is clocked, the outcome can be followed in the diagram. As bit A is presented and the system is clocked, bit A will enter the left-hand latch. When bits B and C are presented, A moves across to the right. Both XOR gates will have A on the upper input from the right-hand latch, the left one has D on the lower input and the right one has B on the lower input. When clocked, the left latch will thus be loaded with the XOR of A and D, and the right one with the XOR of A and B. The remainder of the sequence can be followed, bearing in mind that when the same term appears on both inputs of an XOR gate, it goes out, as the XOR of something with itself is nothing. At the end of the process, the latches contain three different expressions. Essentially, the circuit makes three parity checks through the message, leaving the result of each in the three stages of the register. In the figure, these expressions have been used to

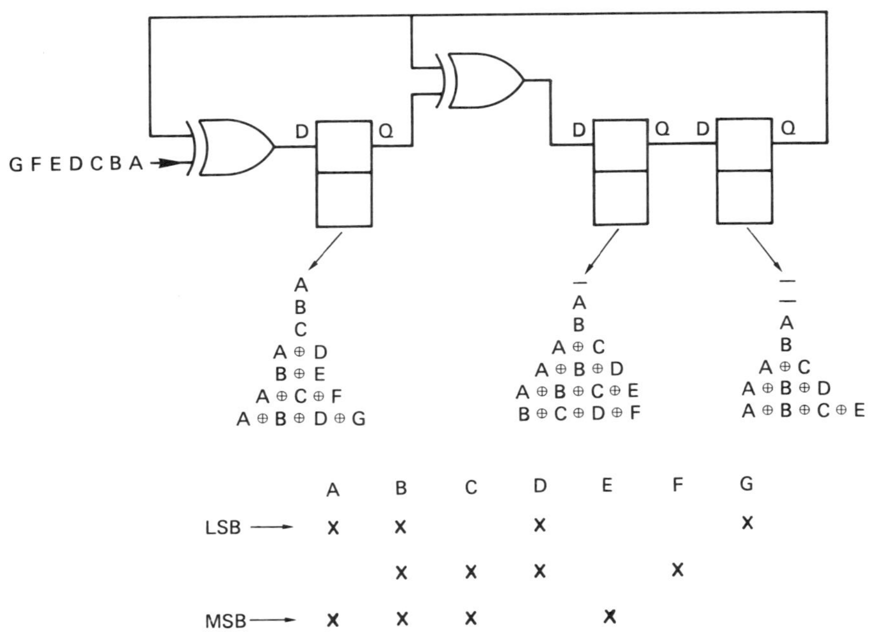

Figure 5.14 When seven successive bits A–G are clocked into this circuit, the contents of the three latches are shown for each clock. The final result is a parity-check matrix

draw up a check matrix. The significance of these steps can now be explained. The bits A B C and D are four data bits, and the bits E F and G are redundancy. When the redundancy is calculated, bit E is chosen so that there are an even number of ones in bits A B C and E; bit F is chosen such that the same applies to bits B C D and F, and similarly for bit G. Thus the four data bits and the three check bits form a 7 bit codeword. If there is no error in the codeword, when it is fed into the circuit shown, the result of each of the three parity checks will be zero and every stage of the shift register will be cleared. If a bit in the codeword is corrupted, there will be a non-zero result. For example, if bit D fails, the check on bits A B D and G will fail, and a one will appear in the left-hand latch. The check on bits B C D F will also fail, and the centre latch will set. The check on bits A B C E will not fail, because D is not involved in it, making the right-hand bit zero. There will be a syndrome of 110 in the register, and this will be seen from the check matrix to correspond to an error in bit D. Whichever bit fails, there will be a different 3 bit syndrome which uniquely identifies the failed bit. As there are only three latches, there can be eight different syndromes. One of these is zero, which is the error-free condition, and so there are seven remaining error syndromes. The length of the codeword cannot exceed 7 bits, or there would not be enough syndromes to correct all of the bits. This can also be made to tie in with the generation of the check matrix. If 14 bits, A–N, were fed into the circuit shown, the result would be that the check matrix repeated twice, and if a syndrome of 101 were to result, it could not be determined whether bit D or bit K failed. Because the check repeats every 7 bits, the code is said to be a cyclic redundancy check (CRC) code.

In Figure 5.8 an example of a Hamming code was given. Comparison of the check matrix of Figure 5.14 with that of Figure 5.8 will show that the only difference is the order of the matrix columns. The two different processes have thus achieved exactly the same results, and the performance of both must be identical. This is not true in general, but these examples have been selected to allow parallels to be seen. In practice Hamming code blocks will generally be much smaller than the blocks used in CRC codes.

It has been seen that the circuit shown makes a matrix check on a received word to determine if there has been an error, but the same circuit can also be used to generate the check bits. To visualize how this is done, examine what happens if only the data bits A B C and D are known, and the check bits E F and G are set to zero. If this message, ABCD000, is fed into the circuit, the left-hand latch will afterwards contain the XOR of A B C and zero, which is of course what E should be. The centre latch will contain the XOR of B C D and zero, which is what F should be and so on. This process is not quite ideal, however, because it is necessary to wait for three clock periods after entering the data before the check bits are available. Where the data are simultaneously being recorded and fed into the encoder, the delay would prevent the check bits being easily added to the end of the data stream. This problem can be overcome by slightly modifying the encoder circuit as shown in Figure 5.15. By moving the position of the input to the right, the operation of the circuit is advanced so that the check bits are ready after only four clocks. The process can be followed in the diagram for the four data bits A B C and D. On the first clock, bit A enters the left two latches, whereas on the second clock, bit B will appear on the upper input of the left XOR gate, with bit A on the lower input, causing the centre latch to load the XOR of A and B and so on.

The way in which the correction system works has been described in engineering terms, but it can be described mathematically if analysis is contemplated.

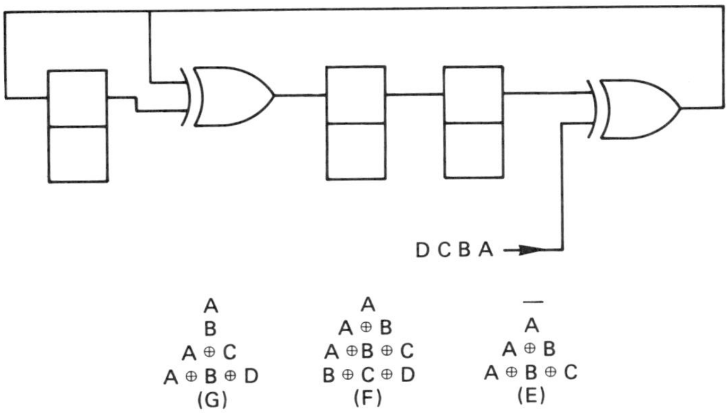

Figure 5.15 By moving the insertion point three places to the right, the calculation of the check bits is completed in only four clock periods and they can follow the data immediately. This is equivalent to premultiplying the data by x^3

Just as the position of a decimal digit in a number determines the power of ten (whether that digit means one, ten or a hundred), the position of a binary digit determines the power of two (whether it means one, two or four). It is possible to rewrite a binary number so that it is expressed as a list of powers of two. For example, the binary number 1101 means 8 + 4 + 1, and can be written:

$2^3 + 2^2 + 2^0$

In fact, much of the theory of error correction applies to symbols in number bases other than two, so that the number can also be written more generally as:

$x^3 + x^2 + 1 \; (x^0 = 1)$

which also looks much more impressive. This expression, containing as it does various powers, is of course a polynomial, and the circuit of Figure 5.14 which has been seen to construct a parity-check matrix on a codeword can also be described as calculating the remainder due to dividing the input by a polynomial using modulo-2 arithmetic. In modulo-2 there are no borrows or carries, and addition and subtraction are replaced by the XOR function, which makes hardware implementation very easy. In Figure 5.16 it will be seen that the circuit of Figure 5.14 actually divides the codeword by a polynomial which is:

$x^3 + x + 1$ or 1011

This can be deduced from the fact that the right-hand bit is fed into several lower-order stages of the register at once. Once all the bits of the message have been clocked in, the circuit contains the remainder. In mathematical terms, the special property of a codeword is that it is a polynomial which yields a remainder of zero when divided by the generating polynomial. If an error has occurred it is considered that this is due to an error polynomial which has been added to the codeword polynomial. If a codeword divided by the check polynomial is zero, a non-zero syndrome must represent the error polynomial divided by the check polynomial. Some examples of modulo-2 division are given in Figure 5.16 which can be compared with the parallel computation of parity checks according to the matrix of Figure 5.14.

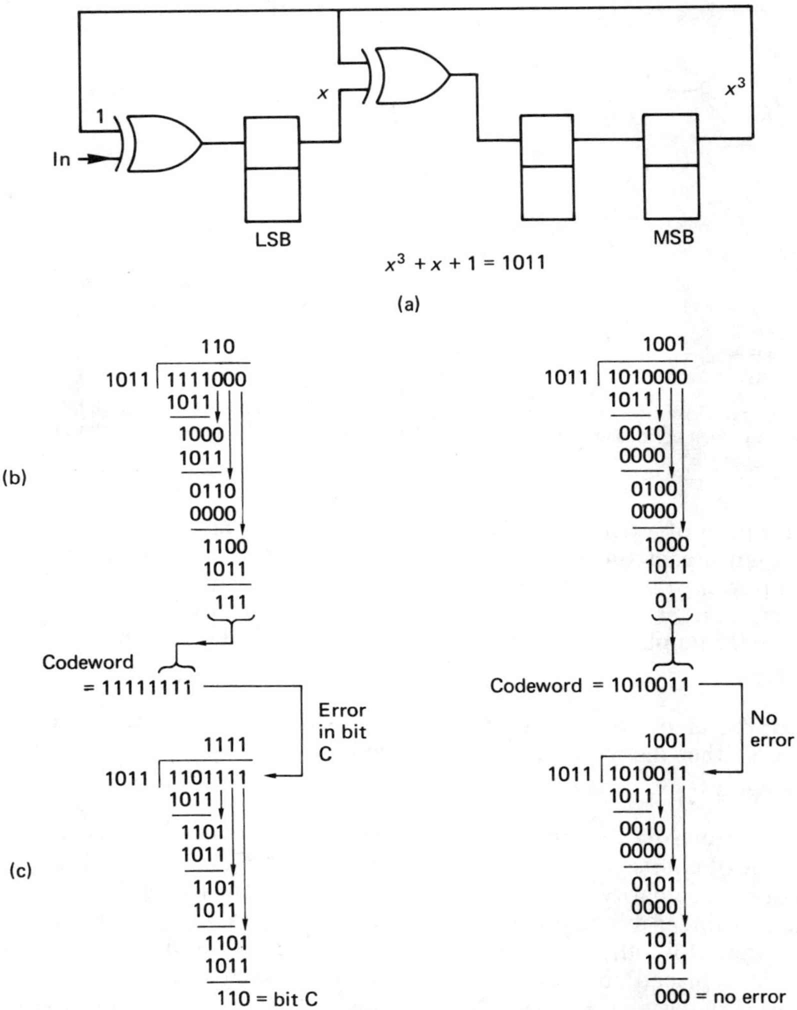

Figure 5.16 Circuit of Figure 5.14 divides by $x^3 + x + 1$ to find remainder. At (b) this is used to calculate check bits. At (c) left, there is an error, non-zero syndrome 110 points to bit C. At (c) right, zero syndrome, no error

The process of generating the codeword from the original data can also be described mathematically. If a codeword has to give zero remainder when divided, it follows that the data can be converted to a codeword by adding the remainder when the data are divided. Generally speaking the remainder would have to be subtracted, but in modulo-2 there is no distinction. This process is also illustrated in Figure 5.16. The four data bits have three zeros placed on the right-hand end to make the wordlength equal to that of a codeword, and this word is then divided by the polynomial to calculate the remainder. The remainder is added to the zero-extended data to form a codeword. The modified circuit of Figure 5.15 can be described as premultiplying the data by x^3 before dividing.

It is also interesting to study the operation of the circuit of Figure 5.14 with no input. Whatever the starting condition of the 3 bits in the latches, the same state will always be reached again after seven clocks, except if zero is used. The states of the latches form an endless ring of non-sequential numbers called a Galois field after the French mathematical prodigy Evariste Galois who discovered them. The states of the circuit form a maximum length sequence because there are as many states as are permitted by the wordlength. As the all-zero case is disallowed, the length of a maximum length sequence generated by a register of m bits cannot exceed $(2^m - 1)$ states. This figure will also be the length of the codeword in simple cyclic codes. The Galois field, however, includes the zero term. It is interesting to explore the bizarre mathematics of Galois fields which use modulo-2 arithmetic (exclusively!). Familiarity with such manipulations is useful when studying the Reed–Solomon codes used in D-2.

As the circuit of Figure 5.14 divides the input by the function $F(x) = x^3 + x + 1$, and there is no input in this case, the operation of the circuit has to be described by:

$$x^3 + x + 1 = 0$$

Each 3 bit state of the circuit can be described by combinations of powers of x, such as:

$$x^2 = 100$$

$$x = 010$$

$$x^2 + x = 110, \text{ etc.}$$

To avoid confusion, the 3 bit state of the field will be called a, which is a primitive element.

Now,

$$a^3 + a + 1 = 0$$

In modulo-2,

$$a + a = a^2 + a^2 = 0$$

$$a = x = 010$$

$$a^2 = x^2 = 100$$

$$a^3 = a + 1 = 011$$

$$a^4 = a \times a^3 = a\,(a + 1) = a^2 + a = 110$$

$$a^5 = a^2 + a + 1 = 111$$

$$\begin{aligned} a^6 &= a \times a^5 = a(a^2 + a + 1) \\ &= a^3 + a^2 + a = a + 1 + a^2 + a \\ &= a^2 + 1 = 101 \end{aligned}$$

$$\begin{aligned} a^7 &= a(a^2 + 1) = a^3 + a \\ &= a + 1 + a = 1 = 001 \end{aligned}$$

In this way it can be seen that the complete set of elements of the Galois field can be expressed by successive powers of the primitive element. Note that the twisted-ring circuit of Figure 5.14 simply raises a to higher and higher powers as it is clocked; thus the seemingly complex multibit changes caused by a single clock of the register become simple to calculate using the correct primitive and the appropriate power.

CRC codes are of primary importance for detecting errors, and several have been standardized for use in digital communications. The most common of these are:

$x^{16} + x^{15} + x^2 + 1$ (CRC-16)

$x^{16} + x^{12} + x^5 + 1$ (CRC-CCITT)

5.10 Burst correction

The concept of Hamming distance has been introduced in the context of single-bit correction. However, it can be extended to explain how more than 1 bit can be corrected. In Figure 5.17 the example of 2 bits in error is given. If a codeword 4 bits

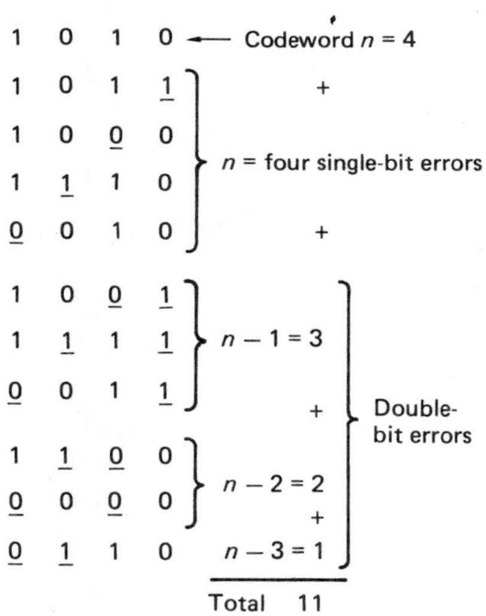

Figure 5.17 Where double-bit errors occur, the number of patterns necessary is $(n - 1) + (n - 2) + (n - 3) + \ldots$ etc. Total necessary is $1 + n + (n - 1) + (n - 2) + (n - 3) + \ldots$ etc. Example here is of 4 bits, and all possible patterns up to Hamming distance of 2 are shown (errors underlined)

long suffers a single-bit error, it could produce one of four different words. If it suffers a 2 bit error, it could produce one of 3 + 2 + 1 different words as shown in the figure (the error bits are underlined). The total number of possible words of Hamming distance 1 or 2 from a 4 bit codeword is thus:

$$4 + 3 + 2 + 1 = 10$$

If the 2 bit error is to be correctable, no other codeword can be allowed to become one of this number of error patterns because of a 2 bit error of its own. Thus every codeword requires space for itself plus all possible error patterns of Hamming distance 2 or 1, which is 11 patterns in this example. Clearly there are only 16 patterns available in a 4 bit code, and thus no data can be conveyed if 2 bit protection is necessary.

The number of different patterns possible in a word of n bits is:

$$1 + n + (n-1) + (n-2) + (n-3) + \ldots$$

and this pattern range has to be shared between the ranges of each codeword without overlap. For example, an 8 bit codeword could result in $1 + 8 + 7 + 6 + 5 + 4 + 3 + 2 + 1 = 37$ patterns. As there are only 256 patterns in 8 bits, it follows that only 256/37 pieces of information can be conveyed. The nearest integer below is six, and the nearest power of two below is four, which corresponds to two data bits and six check bits in the 8 bit word. The amount of redundancy necessary to correct *any* 2 bits in error is large, and as the number of bits to be corrected grows, the redundancy necessary becomes enormous and impractical. A further problem is that the more redundancy is added, the greater the probability of an error in a codeword. Fortunately, in practice errors occur in bursts, as has already been described, and it is a happy consequence that the number of patterns that result from the corruption of a codeword by *adjacent* 2 bit errors is much smaller.

It can be deduced that the number of redundant bits necessary to correct a burst error is twice the number of bits in the burst for a perfect code. This is done by working out the number of received messages which could result from corruption of the codeword by bursts of from 1 bit up to the largest burst size allowed, and then making sure that there are enough redundant bits to allow that number of combinations in the received message.

5.11 Reed–Solomon code

The Reed–Solomon codes (Irving Reed and Gustave Solomon)[3] are inherently burst correcting and have the extremely useful feature that the number of bursts which are correctable can be chosen at the design stage by the amount of redundancy.

The concept of the Galois field has been introduced earlier in this chapter. In the Reed–Solomon codes, data bits are assembled into words, or symbols, which become elements of the Galois field upon which the code is based. The number of bits in the symbol determines the size of the Galois field and hence the number of symbols in the codeword. A symbol size of eight is commonly used because it fits in conveniently with both 16 bit audio samples, 8 bit video samples and byte-oriented computers. A Galois field with 8 bit symbols has a maximum sequence length of $2^8-1 = 255$. As each symbol contains 8 bits, the codeword will be: $255 \times 8 = 2040$ bits long.

As further examples, 5 bit symbols could be used to form a codeword 31 symbols long, and 3 bit symbols would form a codeword seven symbols long. This latter size is small enough to permit some worked examples and will be used further here.

In Section 5.11 it was shown that the circuit of Figure 5.14 generated a Galois field when clocked with no input. The primitive element a will do this when raised to sequential powers. In Reed–Solomon coding, each symbol will be multiplied by some power of such a primitive element. It is necessary to construct hardware which will perform this multiplication. Figure 5.18 shows some examples, and Table 5.1 shows that a truth table can be drawn up for a Galois field multiplier by simply adding the powers of the inputs.

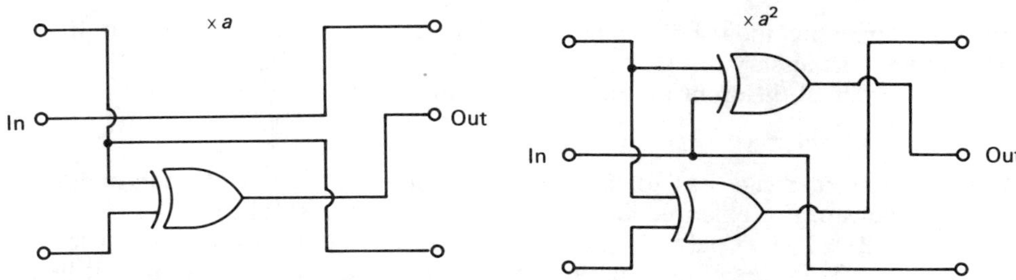

Figure 5.18 Some examples of GF multiplier circuits

Table 5.1 The truth table for Galois field multiplication of GF (2^3). $F(x) = x^3 + x + 1$. Primitive element $a = 010$.

Element: Bits:	0 000	a 010	a^2 100	a^3 011	a^4 110	a^5 111	a^6 101	$a^7 = 1$ 001
0 000	0 000	0 000	0 000	0 000	0 000	0 000	0 000	0 000
a 010	0 000	a^2 100	a^3 011	a^4 110	a^5 111	a^6 101	1 001	a 010
a^2 100	0 000	a^3 011	a^4 110	a^5 111	a^6 101	1 001	a 010	a^2 100
a^3 011	0 000	a^4 110	a^5 111	a^6 101	1 001	a 010	a^2 100	a^3 011
a^4 110	0 000	a^5 111	a^6 101	1 001	a 010	a^2 100	a^3 011	a^4 110
a^5 111	0 000	a^6 101	1 001	a 010	a^2 100	a^3 011	a^4 110	a^5 111
a^6 101	0 000	1 001	a 010	a^2 100	a^3 011	a^4 110	a^5 111	a^6 101
$a^7 = 1$ 001	0 000	a 010	a^2 100	a^3 011	a^4 110	a^5 111	a^6 101	1 001

For example:

$a^2 = 100$, $a^3 = 011$, so $100 \times 011 = a^5 = 111$

Note that the results of a Galois multiplication are quite different from binary multiplication. Because all products must be elements of the field, sums of powers which exceed seven wrap around by having seven subtracted:

$a^5 \times a^6 = a^{11} = a^4 = 110$

It has been stated that the effect of an error is to add an error polynomial to the message polynomial. The number of terms in the error polynomial is the same as the number of errors in the codeword. In a simple CRC system, the effect of the error is detected by ensuring that the codeword can be divided by a polynomial. In the Reed–Solomon codes, several errors can be isolated by ensuring that the codeword will divide by a number of first-order polynomials. If it is proposed to correct for a number of symbols in error given by t, the codeword must be divisible by $2t$ different polynomials of the form $(x + a^n)$ where n takes all values up to $2t$. a is the primitive element discussed earlier.

Figure 5.19 and Figure 5.20 show some examples of Reed–Solomon coding processes. The Galois field of Figure 5.14 has been used, and the primitive element

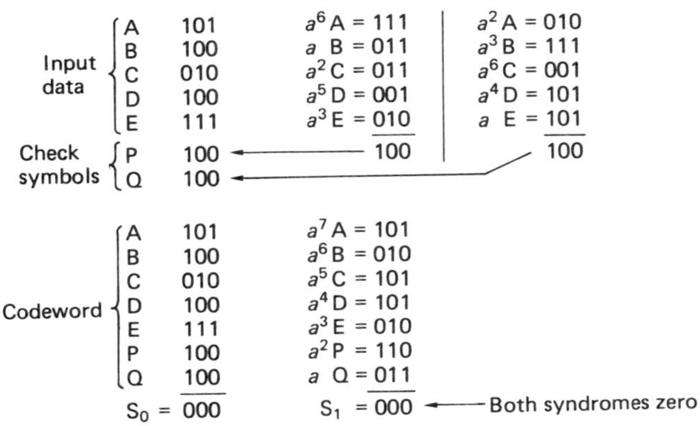

Figure 5.19 Five data symbols A–E are used as terms in the generator polynomials derived in Appendix 5.1 to calculate two redundant symbols P and Q. An example is shown at the top. Below is the result of using the codeword symbols A–Q as terms in the checking polynomials. As there is no error, both syndromes are zero

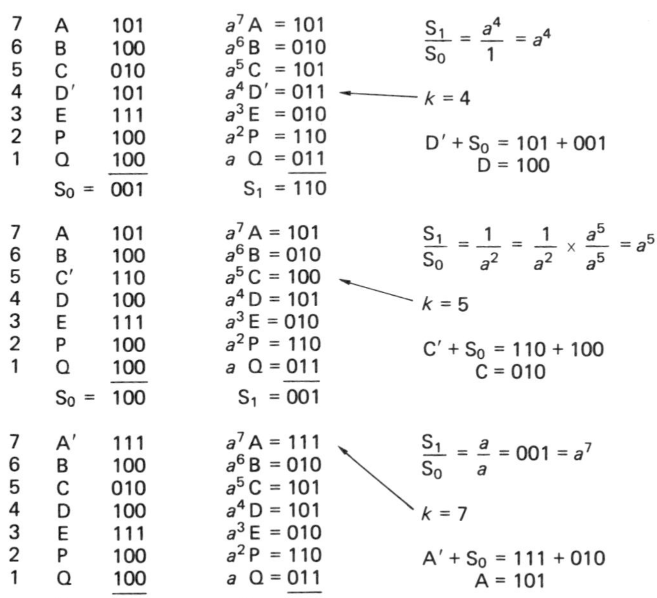

Figure 5.20(a) Three examples of error location and correction. The number of bits in error in a symbol is irrelevant; if all three were wrong, S_0 would be 111, but correction is still possible

A	101	a^7A =	101	
B	100	a^6B =	010	$S_0 = C \oplus D$
C	000	a^5C =	000	$S_1 = a^5 C \oplus a^4 D$
D	000	a^4D =	000	
E	111	a^3E =	010	
P	100	a^2P =	110	
Q	100	a Q =	011	
S_0	= 110	S_1 =	000	

$$S_1 = a^5S_0 \oplus a^5D \oplus a^4D = a^5S_0 \oplus D$$

$$\therefore D = S_1 \oplus a^5S_0 = 000 \oplus 100 = \underline{100}$$

$$S_1 = a^5C \oplus a^4C \oplus a^4S_0 = C \oplus a^4S_0$$

$$\therefore C = S_1 \oplus a^4S_0 = 000 \oplus 010 = \underline{010}$$

Figure 5.20(b) When the position of errors is known by some other means, two errors can be corrected for the same amount of redundancy. The technique is to set the symbols in error to zero, as shown, and calculate what they should have been. This gives the name of erasure correction to the system

a = 010 from Table 5.1 will be used. Five symbols of 3 bits each, A–E, are the data, and two redundant symbols, P and Q, will be used because this simple example will locate and correct only a single symbol in error. It does not matter, however, how many bits in the symbol are in error.

The two check symbols are solutions to the following equations:

$$A \oplus B \oplus C \oplus D \oplus E \oplus P \oplus Q = 0$$

$$a^7A \oplus a^6B \oplus a^5C \oplus a^4D \oplus a^3E \oplus a^2P \oplus aQ = 0$$

By some tedious mathematics, which is shown for reference in Appendix 5.1, it is possible to derive the following expressions which must be used to calculate P and Q from the data in order to satisfy the above equations. These are:

$$P = a^6A \oplus aB \oplus a^2C \oplus a^5D \oplus a^3E$$

$$Q = a^2A \oplus a^3B \oplus a^6C \oplus a^4D \oplus aE$$

In Figure 5.19 the redundant symbols have been calculated. In order to calculate P, the symbol A is multiplied by a^6 according to Table 5.1, B is multiplied by a, and so on, and the products are added modulo-2. A similar process is used to calculate Q. The entire codeword now exists and can be recorded. In Figure 5.19 it is also demonstrated that the codeword satisfies the checking equations.

Upon replaying the information, two checks must be made on the received message to see if it is a codeword. This is done by calculating syndromes using the following expressions, where the prime implies the received symbol which is not necessarily correct:

$$S_0 = A' \oplus B' \oplus C' \oplus D' \oplus E' \oplus P' \oplus Q'$$

(This is in fact a simple parity check.)

$$S_1 = a^7A' \oplus a^6B' \oplus a^5C' \oplus a^4D' \oplus a^3E' \oplus a^2P' \oplus aQ'$$

In Figure 5.20(a) three examples of errors are given, where the erroneous symbol is marked with a dash. As there has been an error, the syndromes S_0 and S_1 will not be zero. The syndrome calculation is performed with Table 5.1 as before.

The parity syndrome S_0 determines the error bit pattern, and the syndrome S_1 is the same error bit pattern, but it has been raised to a different power of a dependent on the position of the error symbol in the block. If the position of the error is in symbol k, then:

$$S_0 \times a^k = S_1$$

Hence:

$$a^k = \frac{S_1}{S_0}$$

The error symbol can be located by multiplying S_0 by various powers of a (which is the same as multiplying by successive elements of the Galois field) until the product is the same as S_1. The power of a necessary is known as the locator, because it gives the position of the error. The process of finding the error position by experiment is known as a Chien search.[4] Once the locator has identified the erroneous symbol, the correct value is obtained by adding S_0 (the corrector) to it.

In the examples of Figure 5.20(a), two redundant symbols have been used to locate and correct one error symbol. If the positions of errors are known by some separate mechanism (see product codes, Section 5.12), the number of symbols which can be corrected is equal to the number of redundant symbols. In Figure 5.20(b) two errors have taken place, and it is known that they are in symbols C and D. Since S_0 is a simple parity check, it will reflect the modulo-2 sum of the two errors. Hence:

$$S_0 = EC \oplus ED$$

The two errors will have been multiplied by different powers in S_1, such that:

$$S_1 = a^5EC \oplus a^4ED$$

It is possible to solve these two equations, as shown in the figure, to find EA and EB, and the correct value of the symbol will be obtained by adding these correctors to the erroneous values. It is, however, easier to set the values of the symbols in error to zero, and the correct values are then found more simply as shown in Figure 5.20(b). This setting of symbols to zero gives rise to the term erasure.

The necessary circuitry for encoding the examples given is shown in Figure 5.21. The P and Q redundancy is computed using suitable Galois field multipliers to obtain the necessary powers of the primitive element according to Table 5.1. In Figure 5.22 the circuitry for calculating the syndromes is shown. The S_0 circuit is a simple parity checker which produces the modulo-2 sum of all symbols fed to it. The S_1 circuit is more subtle, because it contains a Galois field multiplier in a feedback loop, such that early symbols fed in are raised to higher powers than later symbols because they have been recirculated through the GF multiplier more often. It is possible to compare the operation of these circuits with the examples of Figure 5.19 and Figure 5.20 to confirm that the same results are obtained.

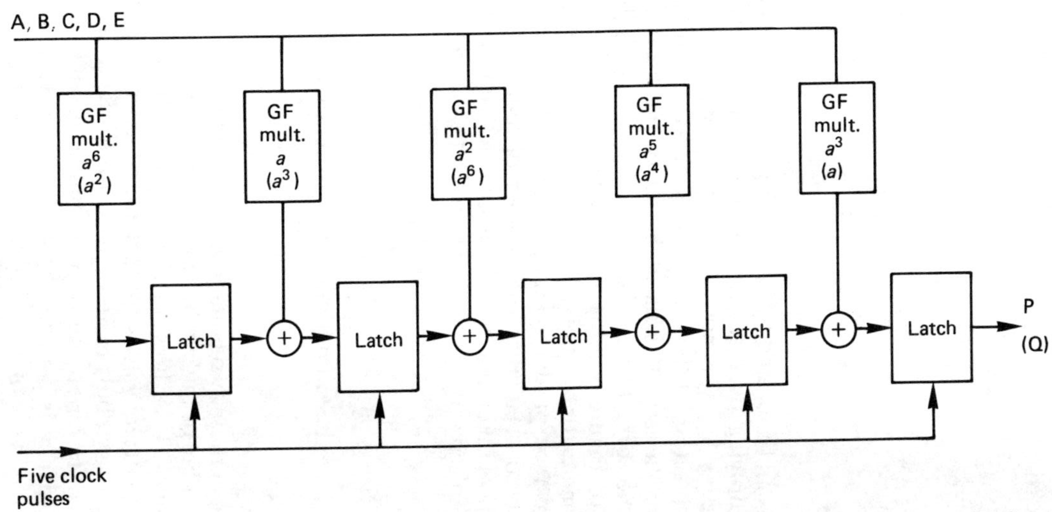

Figure 5.21 If the five data symbols of Figure 5.19 or Figure 5.20(b) are supplied to this circuit in sequence, after five clocks, one of the check symbols will appear at the output. Terms without brackets will calculate P, bracketed terms calculate Q

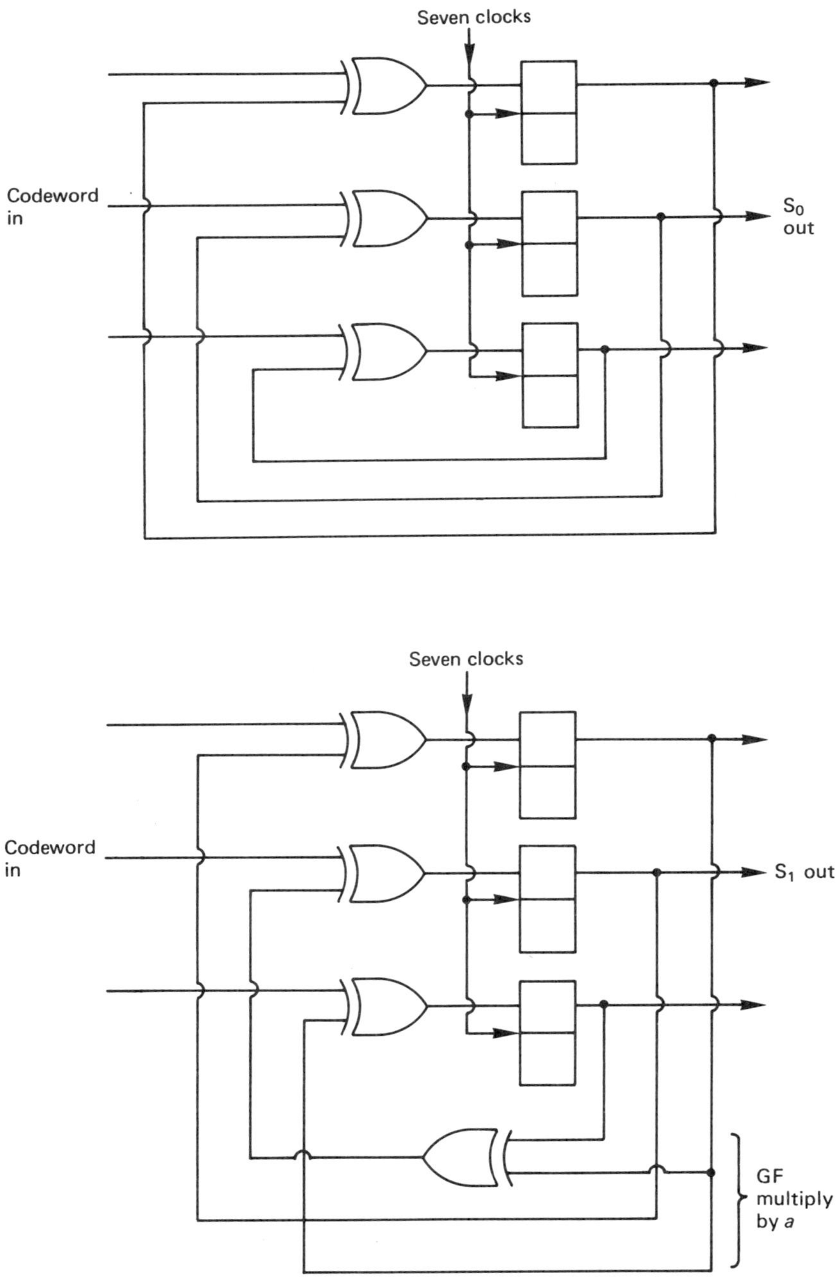

Figure 5.22 Circuits for parallel calculation of syndromes S_0, S_1. S_0 is a simple parity check. S_1 has a GF multiplication by *a* in the feedback, so that A is multiplied by a^7, B is multiplied by a^6, etc., and all are summed to give S_1

Where two symbols are to be corrected without the help of erasure pointers, four redundant symbols are necessary, and the codeword polynomial must then be divisible by:

$$(x + a^0)(x + a^1)(x + a^2)(x + a^3)$$

Upon receipt of the message, four syndromes must be calculated, and the two error patterns and their positions are determined by solving four simultaneous equations. This generally requires an iterative procedure, and a number of algorithms have been developed for the purpose.[5–7] Clearly a double-error-correcting Reed–Solomon (R–S) code will be capable of four-symbol correction if erasure pointers are available. Such techniques are very powerful, because the amount of overhead necessary can be made quite small without sacrificing output error rate. The D-2 format uses erasure techniques with R–S coding extensively. The primitive polynomial commonly used with GF (256) is:

$$x^8 + x^4 + x^3 + x^2 + 1$$

The larger Galois fields require less redundancy, but the computational problem increases. LSI chips have been developed specifically for R–S decoding in D-2 and in other formats.[8]

5.12 Interleaving

The concept of bit interleaving was introduced in connection with a single-bit correcting code to allow it to correct small bursts. With burst-correcting codes such as Reed–Solomon, bit interleave is unnecessary. In most channels, particularly high-density recording channels used for digital audio, the burst size may be many bytes rather than bits, and to rely on a cyclic code alone to correct such errors would require a lot of redundancy. The solution in this case is to employ word interleaving, as shown in Figure 5.23(a). Several codewords are encoded from input data, yet these are not recorded in the order they were input, but are physically reordered in the channel, so that a real burst error is split into smaller bursts in several codewords. The size of the burst seen by each codeword is now determined primarily by the parameters of the interleave, and Figure 5.24 shows that the probability of occurrence of bursts with respect to the burst length in a given codeword is modified. The number of bits in the interleave word can be made equal to the burst-correcting ability of the code in the knowledge that it will be exceeded only very infrequently.

There are a number of different ways in which interleaving can be performed. Figure 5.25, shows that, in block interleaving, words are reordered within blocks which are themselves in the correct order. This approach is attractive for digital video cassette recorders, such as D-2, because the blocks fit into the segmented frame structure of the television waveform, and editing is easy. The block interleave is achieved by writing samples into a memory in sequential address locations from a counter, and reading the memory with non-sequential addresses from a sequencer. The effect is to convert a one-dimensional sequence of samples into a two-dimensional structure having rows and columns.

Rotary-head recorders naturally interleave spatially on the tape. Figure 5.26 shows that a single, large tape defect becomes a series of small defects owing to the geometry of helical scanning.

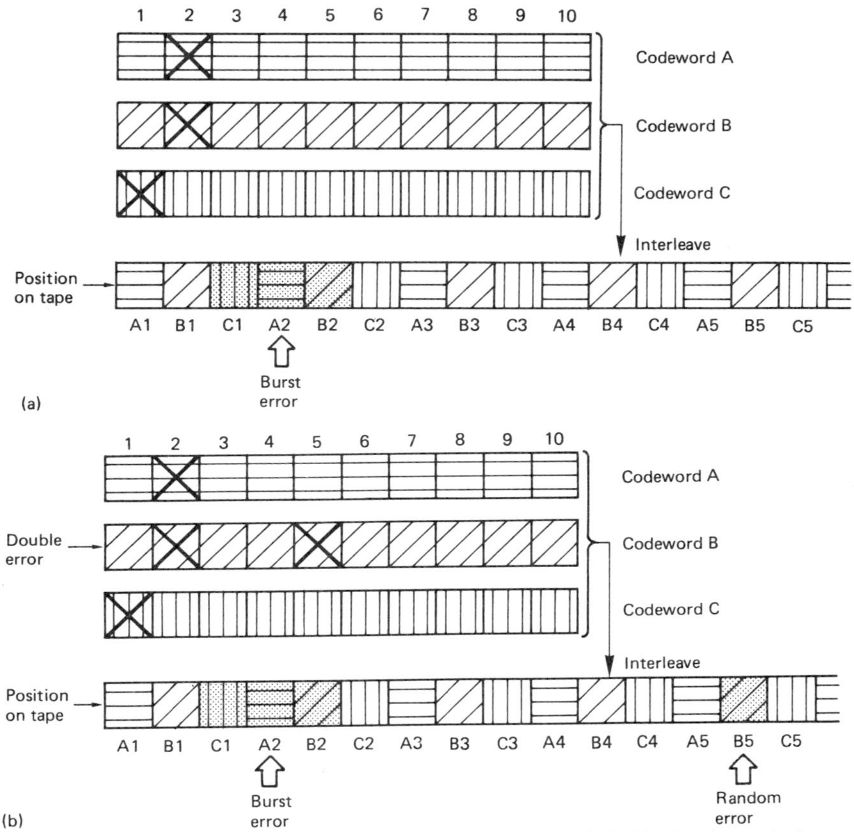

Figure 5.23 At (a), interleave controls the size of burst errors in individual code words, but at (b) the system falls down when a random error occurs adjacent to a burst

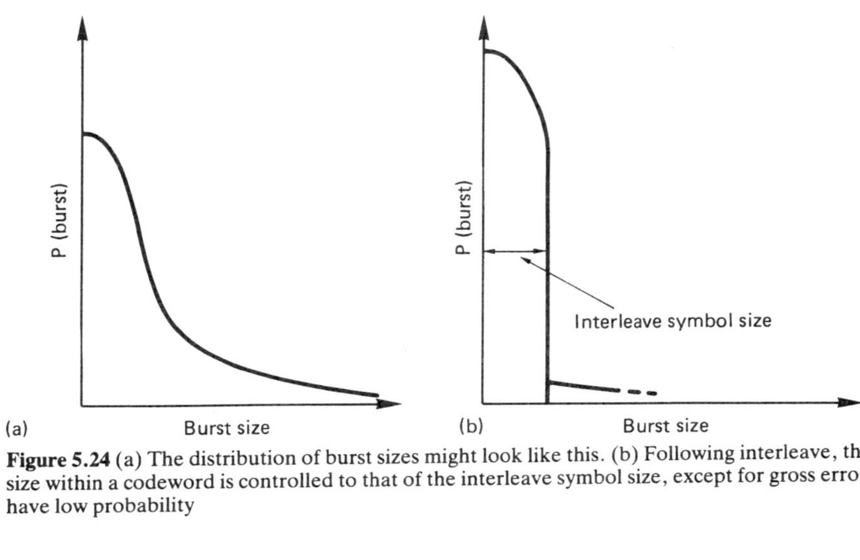

Figure 5.24 (a) The distribution of burst sizes might look like this. (b) Following interleave, the burst size within a codeword is controlled to that of the interleave symbol size, except for gross errors which have low probability

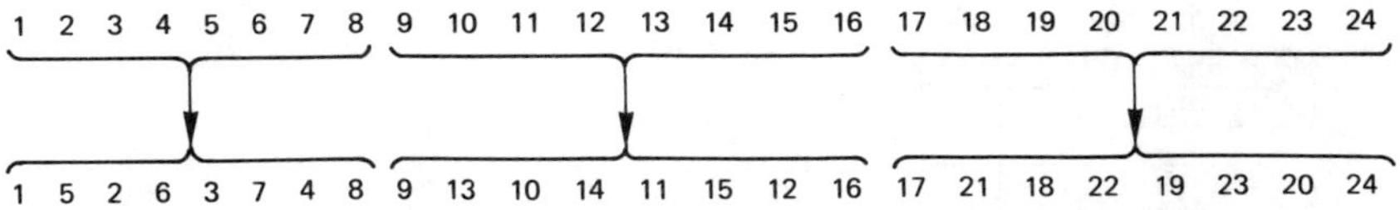

Figure 5.25 In block interleaving, data are scrambled within blocks which are themselves in the correct order

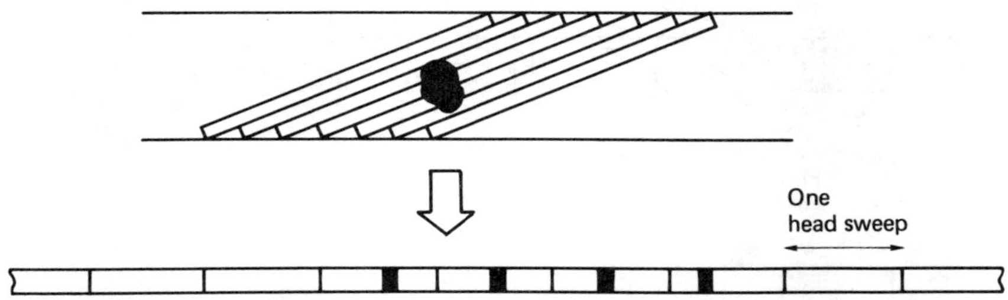

Figure 5.26 Helical scan recorders produce a form of mechanical interleaving, because one large defect on the medium becomes distributed over several head sweeps

5.13 Erasure and product codes

In the presence of burst errors alone, the system of interleaving works very well, but it is known that in many channels there are also uncorrelated errors of a few bits due to noise. Figure 5.23(b) shows that a noise error in the vicinity of a burst error will cause two errors in one codeword, which may not be correctable. The solution to this problem is to use a system where codewords are formed both before and after the interleave process. In block interleaving, this results in a product code.

In fact, the power of product codes goes beyond the ability to deal with real-life errors. The fact that there is a two-dimensional structure allows the position of an error to be discovered because it will be at the intersection of two codewords. If the position of the error can be established geometrically, then it is not necessary to find it by using a code which needs redundancy. The overhead required for a product code is actually less than any other system for a given performance, because the detection of an error in one codeword before de-interleave can be used to generate erasure pointers which help a further codeword after de-interleave. The combination of codewords with interleaving in several dimensions yields an error-protection strategy which is truly synergistic, in that the end result is more powerful than the sum of the parts. Needless to say, the technique is used extensively in D-2.

5.14 Introduction to error correction in D-2

The interleave and error-correction systems of D-2 will now be discussed. Figure 5.27 shows a conceptual block diagram of the system. D-2 uses a product code

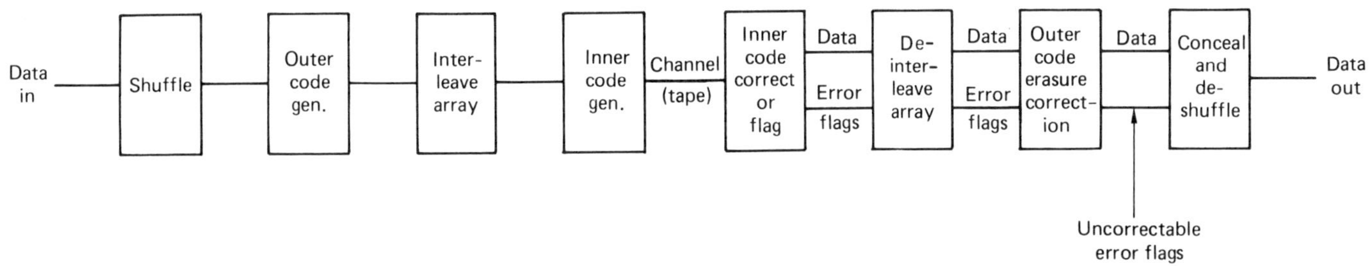

Figure 5.27 Block diagram of error-correction strategy of D-2 DVTR. See text for details

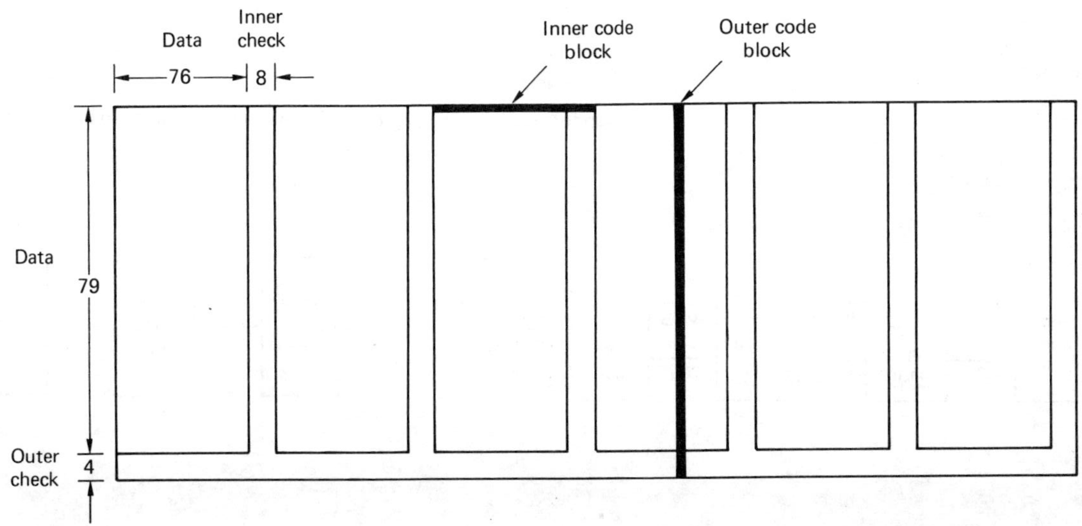

Figure 5.28 Video product-block Reed–Solomon code structure for D-2 DVTR (PAL parameters shown)

formed by producing Reed–Solomon codewords at right angles across an array. The array is formed in a memory, and the layout used in PAL can be seen in Figure 5.28. Incoming samples (bytes) are written into the array in columns. Each column is then made into a codeword by the addition of 4 bytes of redundancy. These are the outer codewords. Each row of the array is then formed into six inner codewords by the addition of 8 bytes of redundancy to each. In order to make a recording, the memory is read in rows, and two inner codewords fit into one block along the track. Every block in an eight-field sequence is given a unique header address. This process continues until the entire array has been laid on tape to form a segment. The process then repeats for the other record-head pair. Writing the memory in columns and reading in rows achieves the necessary interleave.

On replay, a combination of random errors and burst errors will occur.

Data will come from tape in the sequence of the inner codewords. In D-2 each of these has 8 bytes of redundancy, so it would be possible to locate and correct up to 4 bytes in error in the codeword. This, however, is not done, because it is better to limit the correction power to 3 bytes and use the remaining redundancy to decrease the probability of miscorrection. In this way, random errors are corrected with a high degree of confidence. Errors exceeding 3 bytes cause the entire inner codeword to be declared corrupt, and it is written into the appropriate row of the de-interleave array as all zeros, with error flags set. When all of the rows of the array are completed, it is possible to begin processing the columns. A burst error on tape which destroys one or more inner codewords will result in single-byte errors with flags in many different outer codewords. The presence of the error flags means that the outer code does not need to compute the position of the errors, so the full power of the outer code redundancy is available for correction. The four redundancy bytes of D-2 can correct four error bytes. The process of using error flags to assist a code is called erasure. It will be evident that enormous damage must be done to the tape track before 4 bytes are corrupted in a single outer codeword, which gives high resistance to dropouts and tape scratches.

5.15 Concealment and shuffle

In the event of a gross error, correction may not be possible, and some bytes may be declared uncorrectable. In this case the errors must be concealed by interpolation from adjacent samples. Owing to the regular structure of the product code block, uncorrectable burst errors would show up as a regular pattern of interpolations on the screen. The eye is extremely good at finding such patterns, so the visibility is higher than might be expected. The solution to this problem is to perform a two-dimensional pseudo-random shuffle on the video samples before generating the product code. The reverse shuffle is performed on replay. The positions of interpolated samples arising from an uncorrected error now form a pseudo-random pattern with much decreased visibility.

The details of the D-2 format and the shuffle mechanism will be discussed in more detail in Chapter 7.

Appendix 5.1 Calculation of Reed–Solomon generator polynomials

For a Reed–Solomon codeword over $GF(2^3)$, there will be seven 3 bit symbols. For location and correction of one symbol, there must be two redundant symbols P and Q, leaving A–E for data.

The following expressions must be true, where a is the primitive element of $x^3 + x + 1$ and + is XOR throughout.

$$A + B + C + D + E + P + Q = 0 \qquad (1)$$
$$a^7A + a^6B + a^5C + a^4D + a^3E + a^2P + aQ = 0 \qquad (2)$$

Dividing eq. (2) by a:

$$a^6A + a^5B + a^4C + a^3D + a^2E + aP + Q = 0 = A + B + C + D + E + P + Q$$

Cancelling Q, and collecting terms:

$$(a^6 + 1)A + (a^5 + 1)B + (a^4 + 1)C + (a^3 + 1)D + (a^2 + 1)E = (a + 1)P$$

Using Table 5.1 to calculate $(a^n + 1)$, e.g. $a^6 + 1 = 101 + 001 = 100 = a^2$

$$a^2A + a^4B + a^5C + aD + a^6E = a^3P$$
$$a^6A + aB + a^2C + a^5D + a^3E = P \qquad (3)$$

Multiplying eq. (1) by a^2 and equating to eq. (2):

$$a^2A + a^2B + a^2C + a^2D + a^2E + a^2P + a^2Q = 0$$
$$= a^7A + a^6B + a^5C + a^4D + a^3E + a^2P + aQ$$

Cancelling a^2P and collecting terms (remember $a^2 + a^2 = 0$):

$$(a^7 + a^2)A + (a^6 + a^2)B + (a^5 + a^2)C + (a^4 + a^2)D + (a^3 + a^2)E = (a^2 + a)Q$$

Adding powers according to Table 5.1, e.g. $a^7 + a^2 = 001 = 101 = a^6$:

$$a^6A + B + a^3C + aD + a^5E = a^4Q$$
$$a^2A + a^3B + a^6C + a^4D + aE = Q$$

References

1. SHANNON, C. E. A mathematical theory of communication. *Bell Syst. Tech. J.*, **27**, 379 (1948)
2. HAMMING, R. W. Error-detecting and error-correcting codes. *Bell Syst. Tech. J.*, **29**, 147–160 (1950)
3. REED, I. S. and SOLOMON, G. Polynomial codes over certain finite fields. *J. Soc. Ind. Appl. Math.*, **8**, 300–304 (1960)
4. CHIEN, R. T., CUNNINGHAM, B. D. and OLDHAM, I. B. Hybrid methods for finding roots of a polynomial – with application to BCH decoding. *IEEE Trans. Inf. Theory*, **IT-15**, 329–334 (1969)
5. BERLEKAMP, E. R. *Algebraic Coding Theory*, New York: McGraw-Hill (1967). Reprint edn: Laguna Hills, CA: Aegean Park Press (1983)
6. SUGIYAMA, Y. *et al.* An erasures and errors decoding algorithm for Goppa codes. *IEEE Trans. Inf. Theory*, **IT-22** (1976)
7. PETERSON, W. W. and WELDON, E. J. *Error Correcting Codes*, 2nd edn. Cambridge, MA: MIT Press (1972)
8. *Ampex Data Sheets* V5125 and V5126, Ampex Data Systems Division, Redwood City, CA

Chapter 6

D-2 tape transports and control systems

There is more to digital video recording than electronics. It depends heavily on mechanical engineering of the highest quality and several extraordinarily accurate servo systems. This chapter looks at the techniques necessary to build and control a rotary-head D-2 transport.

6.1 Why rotary heads?

The helical scan tape deck has more in common with the helicopter than just the geometrical terminology. The rotary-head recorder is to the fixed-head recorder what the rotary-winged flying machine is to the fixed-wing machine.

In both cases the rotary machine is more complex, more expensive, consumes more power, needs more maintenance and adjustment, and makes more noise than its fixed counterpart. If these were our only criteria, this comparison would amount to a condemnation. The truth of the matter is that all of these characteristics have to be accepted in order to obtain one particular attribute which cannot be obtained in any other way. In the case of the helicopter, the ability to stop dead in mid-air is vital for many applications. In the case of the helical recorder, the attribute obtained is the ability to record exceptional bandwidths with high density and to be able to replay at varying speed. If these are the requirements, there is no competition from simpler hardware. The problems have to be overcome by matching each one with the appropriate mechanical design, materials and control systems.

The attractions of the helical recorder are that the head-to-tape speed and hence bandwidth are high, whereas the linear tape speed is not. The space between tracks is determined by the linear tape speed, not by multitrack head technology. Chapter 4 showed how rotary-head machines using azimuth recording make better use of the tape area. The high head speed raises the frequency of offtape signals, and since output is proportional to frequency, playback signals are raised above head noise even with very narrow tracks.

6.2 Helical geometry

Figure 6.1 shows the two fundamental approaches to helical scan. In the rotating upper drum system, one or more heads are mounted on the periphery of a

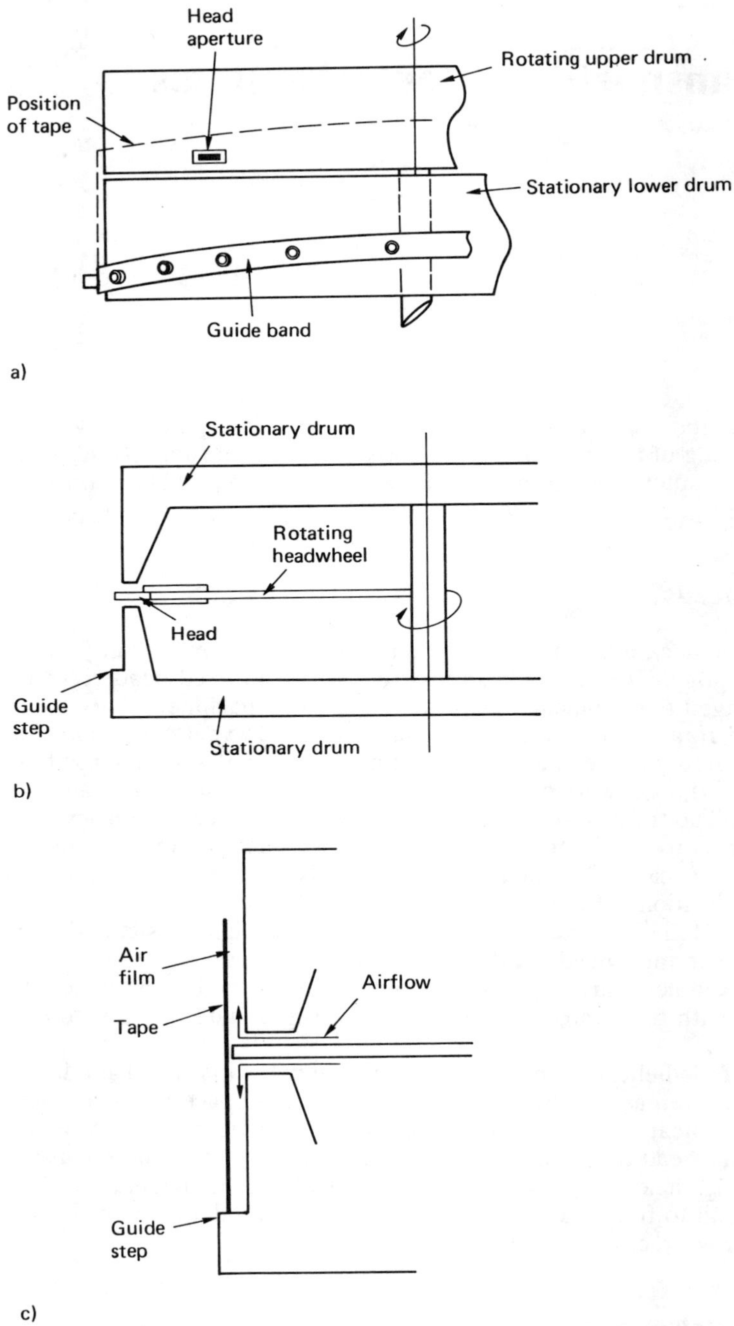

Figure 6.1 (a) Rotating-top-type scanner. The upper drum is slightly smaller than the lower drum due to the air film it develops. Helical band guides lower edge of tape. (b) Stationary-top scanner, where headwheel rotates in a slot between upper and lower drums. (c) Airflow in stationary-top scanner. Headwheel acts as a centrifugal pump, producing air film between tape and drums

revolving drum or scanner. The fixed base of the scanner carries a helical ramp in the form of a step or, for greater wear resistance, a hardened band which is suitably attached. Alternatively, both top and bottom of the scanner are fixed, and the headwheel turns in a slot between them. Both approaches have advantages and disadvantages. The rotating upper drum approach is simpler to manufacture than the fixed upper drum, because the latter requires to be rigidly and accurately cantilevered out over the headwheel. The rotating part of the scanner will produce an air film which raises its effective diameter. The rotor will thus need to be made slightly smaller than the lower drum, so that the tape sees a constant diameter as it rises up the scanner. There will be plenty of space inside the rotor to install individually replaceable heads.

The fixed upper drum approach requires less power, since there is less air resistance. The headwheel acts as a centrifugal pump and supplies air to the periphery of the slot where it lubricates both upper and lower drums, which are of the same diameter. This approach gives head contact which is more consistent over the length of the track, but it makes the provision of replaceable heads more difficult, and it is generally necessary to replace the entire headwheel as an assembly.

In both cases, a pair of fixed guides, known as the entry and exit guides, lead the tape to and from the ramp. The tape may wrap around the scanner by various amounts. Some early machines used a complete circuit around the scanner which led the tape to cross itself in the so-called Alpha wrap. The C-format uses almost a complete circuit of the scanner where the tape turns sharply around the entrance and exit guides in the shape of an Omega. The tape may only pass half-way around the scanner in a U-shape, hence the name U-matic. The total angle of scanner rotation for which the heads touch the tape is called the *wrap angle*.

The scanner can rotate either with or against the direction of linear tape motion; in D-2 it turns against the tape motion. It will be evident that as the tape is caused to climb up the scanner axis by the ramp, it actually takes on the path of a helix. The head rotates in a circular path, and so will record diagonal tracks across the width of the tape. So it is not really the scanning which is helical, it is the tape path. If the tape is stationary, the head will constantly retrace the same track, and the angle of the track can be calculated by measuring the rise of the tape along the scanner axis and the circumferential distance over which this rise takes place. The latter can be obtained from the diameter of the scanner and the wrap angle. The tangent of the *helix angle* is given by the rise over the distance. It can also be obtained by measuring the distance along the ramp corresponding to the wrap angle; this dimension will give the sine of the helix angle.

When the tape moves, the angle between the tracks and the edge of the tape will not be the helix angle. Figure 6.2 shows how this occurs. When the head contact commences, the tape will be at a given location, but when the head contact ceases, the tape will have moved a certain linear distance. To obtain the *track angle* it is necessary to take into account the tape motion.

The length of the track resulting from scanning a stationary tape, which will be at the helix angle, can be resolved into two distances at right angles as shown in Figure 6.2. One of these is across the tape width, the other is along the length of the tape. The tangent of the helix angle will be the ratio of these lengths. If the length along the tape is corrected by adding or subtracting the distance the tape moves during one scan, depending upon whether tape motion aids or opposes scanner rotation, the new ratio of the lengths will be the tangent of the track angle. In practice this

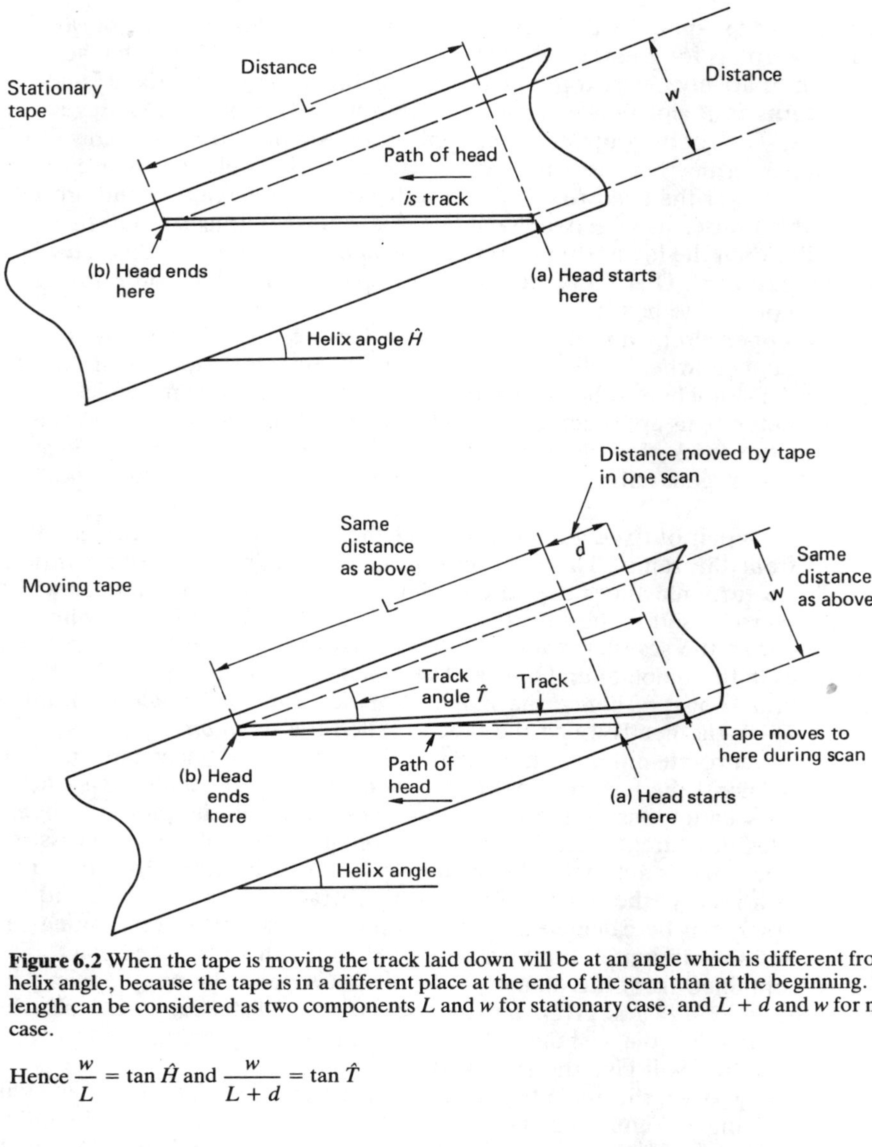

Figure 6.2 When the tape is moving the track laid down will be at an angle which is different from the helix angle, because the tape is in a different place at the end of the scan than at the beginning. Track length can be considered as two components L and w for stationary case, and $L + d$ and w for moving case.

Hence $\dfrac{w}{L} = \tan \hat{H}$ and $\dfrac{w}{L + d} = \tan \hat{T}$

process will be reversed, because the track angle of D-2 is standardized, and the transport designer has to find a helix angle which will produce it.

Provided that the tape linear speed and scanner speed remain the same, the theoretical track angle will remain the same. In practice this will only be the case if the tape tension is constant. Figure 6.3 shows that if the back tension changes, the effective length of the tape will also change, and with it the track angle, a phenomenon known as *skew*. All digital rotary-head recorders need some form of tape tension servo to maintain the tape tension around the scanner constant irrespective of the size of the pack on the supply hub. In practice tension servos can

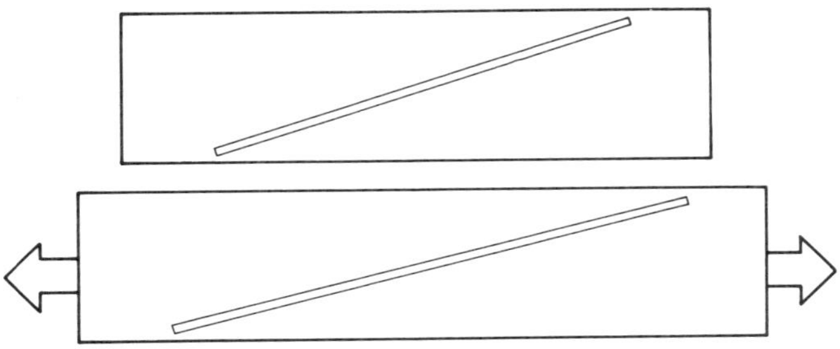

Figure 6.3 Tape is flexible, so change of tension will result in change of length. This causes a change in track angle known as skew shown exaggerated here

only control the tension of the tape entering the scanner. Since there will be friction between the tape and the scanner, the tape tension will gradually increase between the entrance and exit guides, and so the tape will be extended more towards the exit guide. When the tape subsequently relaxes, it will be found that the track is actually curved. The ramp can be made to deviate from a theoretical helix to counteract this effect, or all scanners can be built to the same design, so that effectively the track curvature becomes part of the format. In any case azimuth recording tolerates residual tracking errors.

When the tape direction reverses, the sense of the friction in the scanner will also reverse, so the back tension has to increase to keep the average tension the same as when going forward.

When verifying the track angle produced by a new design, it is usual to develop a tape which it has recorded with magnetic fluid and take measurements under a travelling microscope. With the very small track widths of digital recorders, it is usually necessary to compensate for skew by applying standard tension to the tape when it is being measured, or by computing a correction factor for the track angle from the modulus of elasticity of the tape.

The *track pitch* is the distance, measured at right angles to the tracks, from a given place on one track to the same place on the next. The track pitch is a function of the linear tape speed, the head passing frequency and the helix angle. It can be seen in Figure 6.4 that the linear speed and the helix angle determine the rise rate

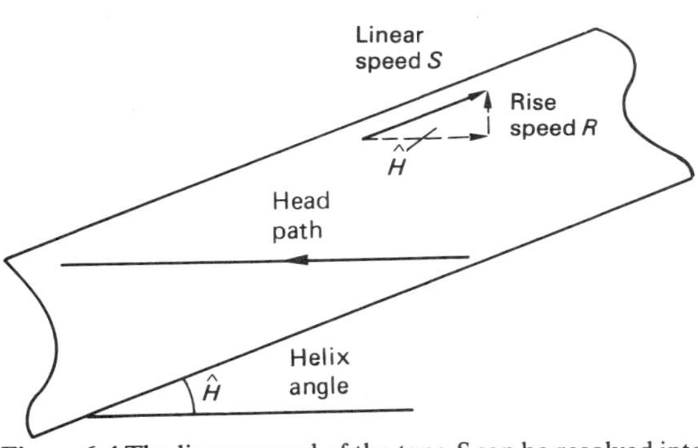

Figure 6.4 The linear speed of the tape S can be resolved into the speed of the tape along the scanner axis which is the rise speed R. $R = S \sin \hat{H}$. Dividing the rise speed by the head passing frequency gives the track pitch. Head passing frequency is simply the number of heads multiplied by the scanner rotational frequency. Thus track pitch is proportional to tape speed

up the scanner axis, and knowledge of the rotational period of the scanner will allow the rise in one revolution to be calculated. If the rise is divided by the number of active heads on the scanner, the result will be the track pitch. If everything else remains equal, the track pitch is proportional to the linear tape speed. Note that the track angle will also change with tape speed.

Figure 6.5 shows that, in guard band recording, the track pitch is equal to the width of the track plus the width of the guard band. The track width is determined by the width of the head poles, and the linear tape speed will be high enough so that the desired guard band is obtained.

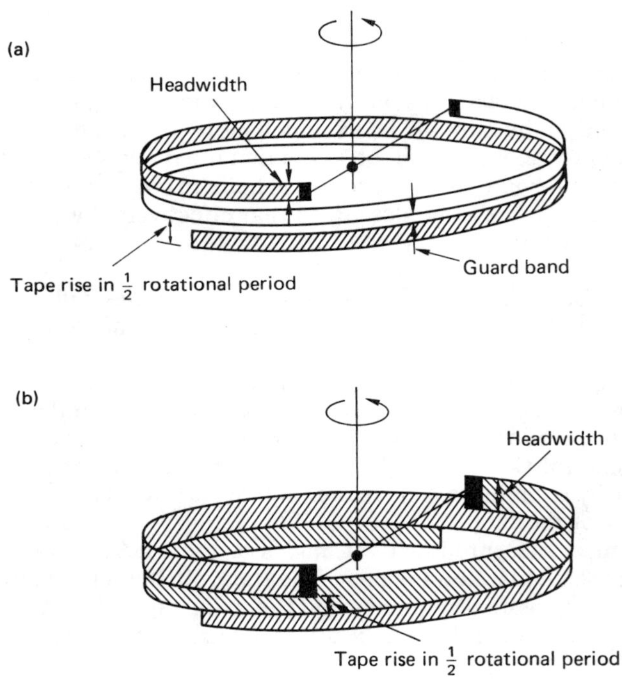

Figure 6.5 (a) Where tape rise is more than head width, a guard band between the tracks results. (b) Where tape rise is less than head width, overwrite takes palce and the tracks are narrower than the heads. Shown here for two-headed scanner, the minimum configuration for azimuth recording

In azimuth recording the reverse situation is used. The head width is greater than the track pitch, so that the tape does not rise far enough for one track to clear the previous one. Part of the previous track will be overwritten, so that the track width and the track pitch become identical.

The length of the track laid on the tape is a function of the scanner diameter and the wrap angle. Figure 6.6 shows a number of ways in which the same length of track can be put on the tape.

A small scanner with a large wrap angle can lay down the same track pattern as a large scanner with a small wrap angle. If the tape speed, and hence the track rate is constant, and for the purposes of this comparison the number of heads on the

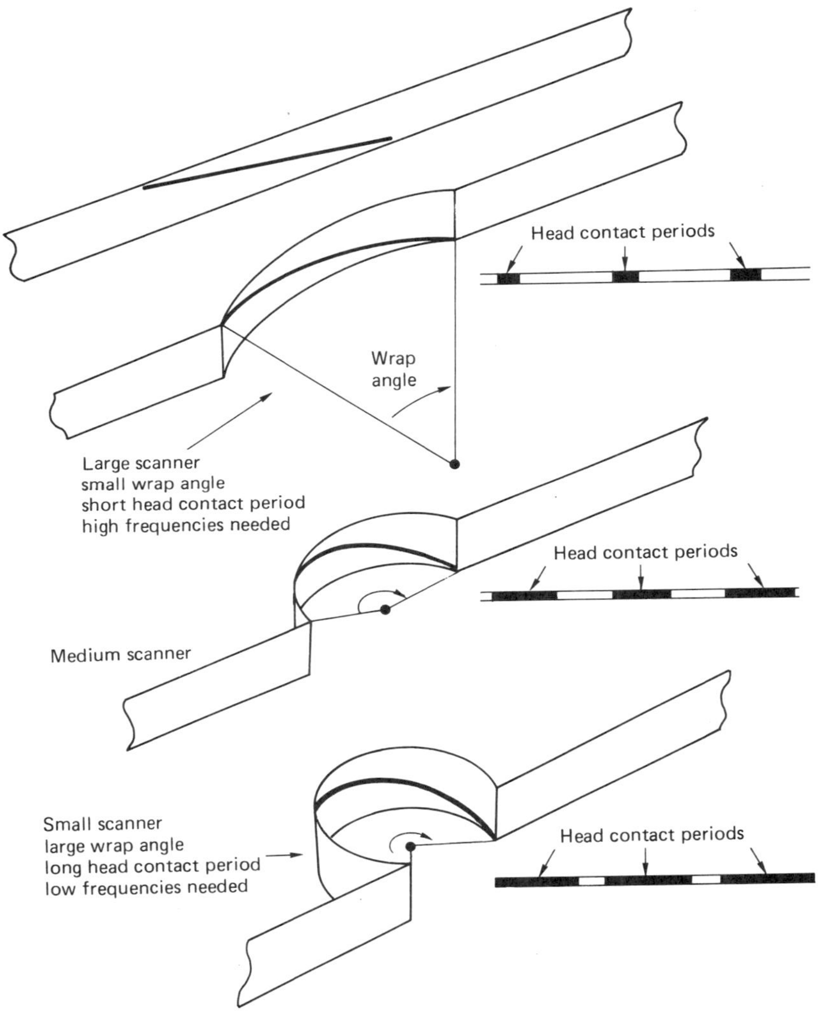

Figure 6.6 A standard track can be put on tape with a variety of scanners. A large wrap allows the lowest frequencies at the head but increases complexity of threading mechanism and increases tape path friction. The helix angle will be different in each case in order to match the tape rise speed to the head peripheral speed – giving the same track angle

scanners remains constant, the heads on the smaller scanner will take a longer time to traverse the track than the heads on the larger scanner. This means that the helix angle will need to be different so that the same track angle is achieved in all cases. In practice, a larger scanner can be used to mount more heads which work in turn, so that the scanner speed will then be less than before. Alternatively the space inside the larger scanner may be used to incorporate track-following heads. Another advantage of the large scanner is that the reduced wrap angle is easier to thread up. A small scanner is attractive for a portable machine since it is compact and requires less power. The greater wrap angle then has to be accepted.

The different scanner sizes in the example of Figure 6.6 traced standard tracks in different periods of time, as shown in the diagram. In analog recorders such as the C-format and U-matic, the time period had to be one television field so that the television waveform could be directly fed to the heads, and design of the scanner was extremely inflexible.

Using time compression, the actual duration of the head sweep and the real time in the recording which it represents are independent. The principle can be used in both analog and digital recorders. Component camcorders time-compress the colour difference signals to record both sequentially on one track. Consumer machines such as Betamovie and Compact VHS use time compression so that real-time video can be recorded with a smaller than usual scanner for portable applications.

In the digital domain there is even more freedom. The video frame is expressed by a given number of samples, and as long as these appear on replay in the right sequence, their actual position in the recording or the exact time at which they were recorded is of no consequence. Clearly the record and replay processes must complement one another.

6.3 Track layout and segmentation

The number of bits to be recorded for one field will be a direct result of the sampling scheme chosen. The minimum wavelength which can be reliably recorded will then determine the length of track that is necessary.

With $4 \times F_{sc}$ composite sampling, the figures will be about 3 Mbits and 2 Mbits per field for PAL and NTSC respectively.

Since current wavelengths are restricted to about 0.9 μm (which records 2 bits) then the track length needed will be about 1.7 metres and 1.1 metres respectively. Clearly no one in their right mind is going to design a scanner large enough to put the whole of one field on one track. A further consideration is that current head technology does not permit operation at a high enough frequency to pass all the data through one head. Eddy current losses in the head get worse at high frequencies.

The solution to both problems is *segmentation* where the data for one field are recorded in a number of head sweeps. With segmentation it is easy to share the data between more than one head. In the D-2 format, two heads are necessary to support the data rate.

With segmentation, the contents of a field can be divided into convenient-sized pieces for recording. This allows the width of the tape to be chosen. If the tape is wide, then tracks of a given length can be placed at a greater angle to the edge of the tape, and so the tape will move more slowly for a given track rate. This means that the effective rewinding time will be reduced because a given recording will occupy a shorter length of tape. Taken to extremes, this would result in very wide tape on small reels.

This is not the only criterion, however. A wide tape needs a physically larger scanner assembly and requires greater precision in guidance. It is seldom possible to engineer a tape transport with total precision, so the tape will always be expected to accommodate small inaccuracies by flexing. The stiffness of the tape against bending in its own plane is proportional to the moment of inertia of its cross-section. This rises disproportionally with tape width. The selection of a narrower tape will ease transport accuracy requirements slightly. Another factor to be considered is the recording density. A high-density recording requires less tape area per unit time, and so the effective rewinding speed will be proportional to the recording density. Consideration of these factors led to the choice of ¾ inch wide tape for the digital video cassette. The choice was basically between that and 1 inch tape, since choice of a non-standard width would have unnecessarily raised the cost of the medium as existing slitters could not have been used.

Having chosen the tape width, the number of segments per field can be established. A large number of short, steep tracks will give the same overall track length as a smaller number of long, shallow tracks. For normal speed operation, there is little difference between the two approaches, but for variable speed, the smaller the number of segments the better, since the magnitude of head jumps will be reduced. Where two TV standards have to be recorded, as in D-2, the number of segments will be chosen so that a whole number results in both standards.

The wide disparity between the subcarrier frequencies of NTSC and PAL leads to different segmentation being used in each case. The fundamental unit in the D-2 format is the segment, which consists of a pair of tracks, 0 and 1, one of each azimuth type. These can be seen in Figure 6.7. The first recorded track in a field is always a type 0 track, in which the azimuth is rotated clockwise by 14.97° from the track edge when looking at the magnetic coating side of the tape. In a type 1 track, the azimuth is rotated 15.03° anticlockwise.

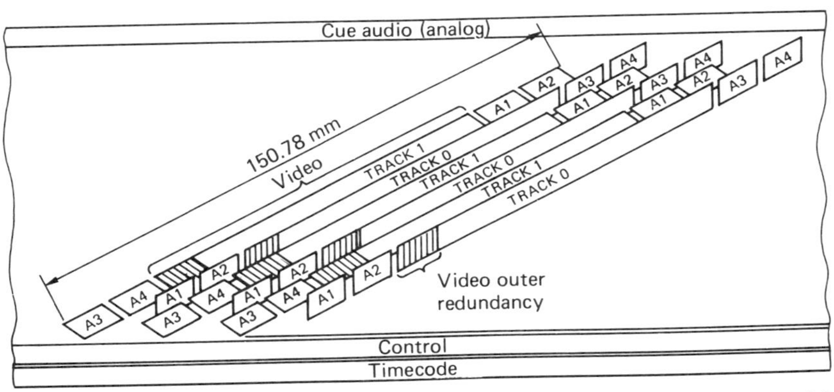

Figure 6.7 Track layout of D-2. Type 0 track has clockwise azimuth rotation, type 1 track has anti-clockwise azimuth rotation. Note video outer redundancy at beginning of video sector. This can be omitted in some modes to give the heads more time to settle on track. Audio channels are in separate sectors with gaps so they can be independently edited. Note second copy of audio on opposite side of tape and on different head

Although the heads are physically mounted at ±15° to the scanner axis, the tape moves as the heads turn, and, as Figure 6.2 shows, the track angle and the helix angle are then different. This accounts for the discrepancy of 0.03° in the azimuth angle with respect to the track.

Figure 6.8(a) shows that in PAL, four segments of 76 lines each record a total of 304 lines per field,[1] whereas in NTSC, shown in Figure 6.8(b), three segments of 85 lines each record a total of 255 lines per field.[2] The segment and track numbering is also shown. The control track necessary for segmented operation must convey

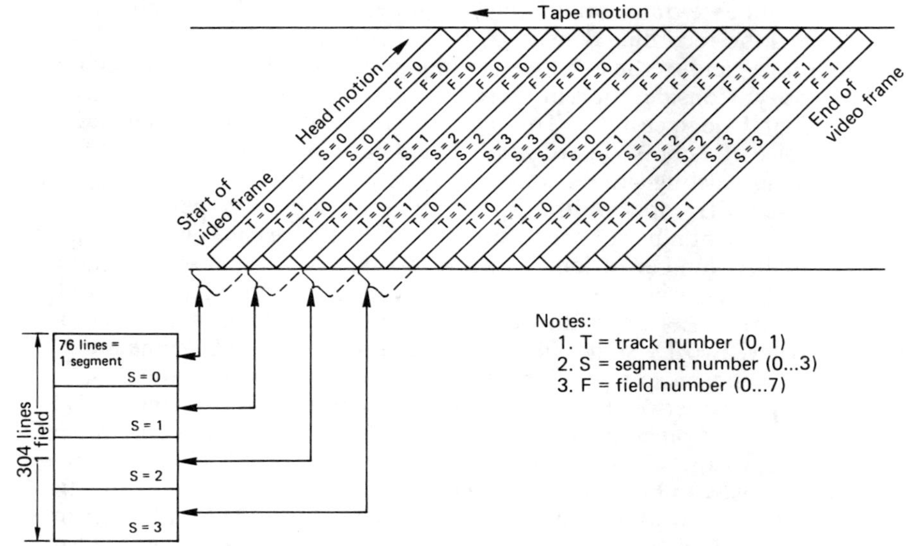

Figure 6.8(a) PAL segmentation in D-2. Four segments of 76 lines are necessary to record one field. The numbering scheme needed to recreate correctly framed pictures is shown here. The field number reflects the eight-field sequence of PAL. Audio blocks are omitted for clarity

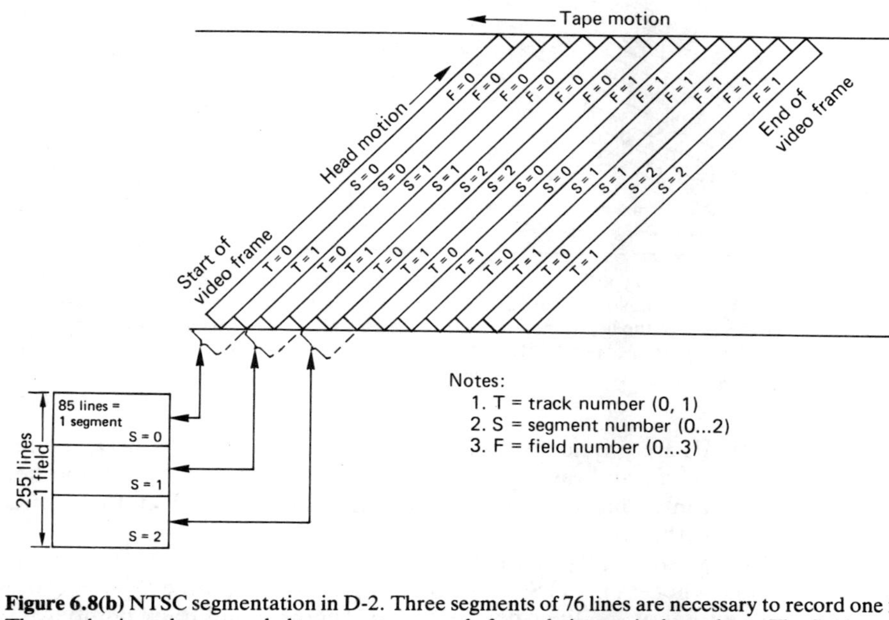

Figure 6.8(b) NTSC segmentation in D-2. Three segments of 76 lines are necessary to record one field. The numbering scheme needed to recreate correctly framed pictures is shown here. The field number reflects the four-field sequence of NTSC. Audio blocks are omitted for clarity

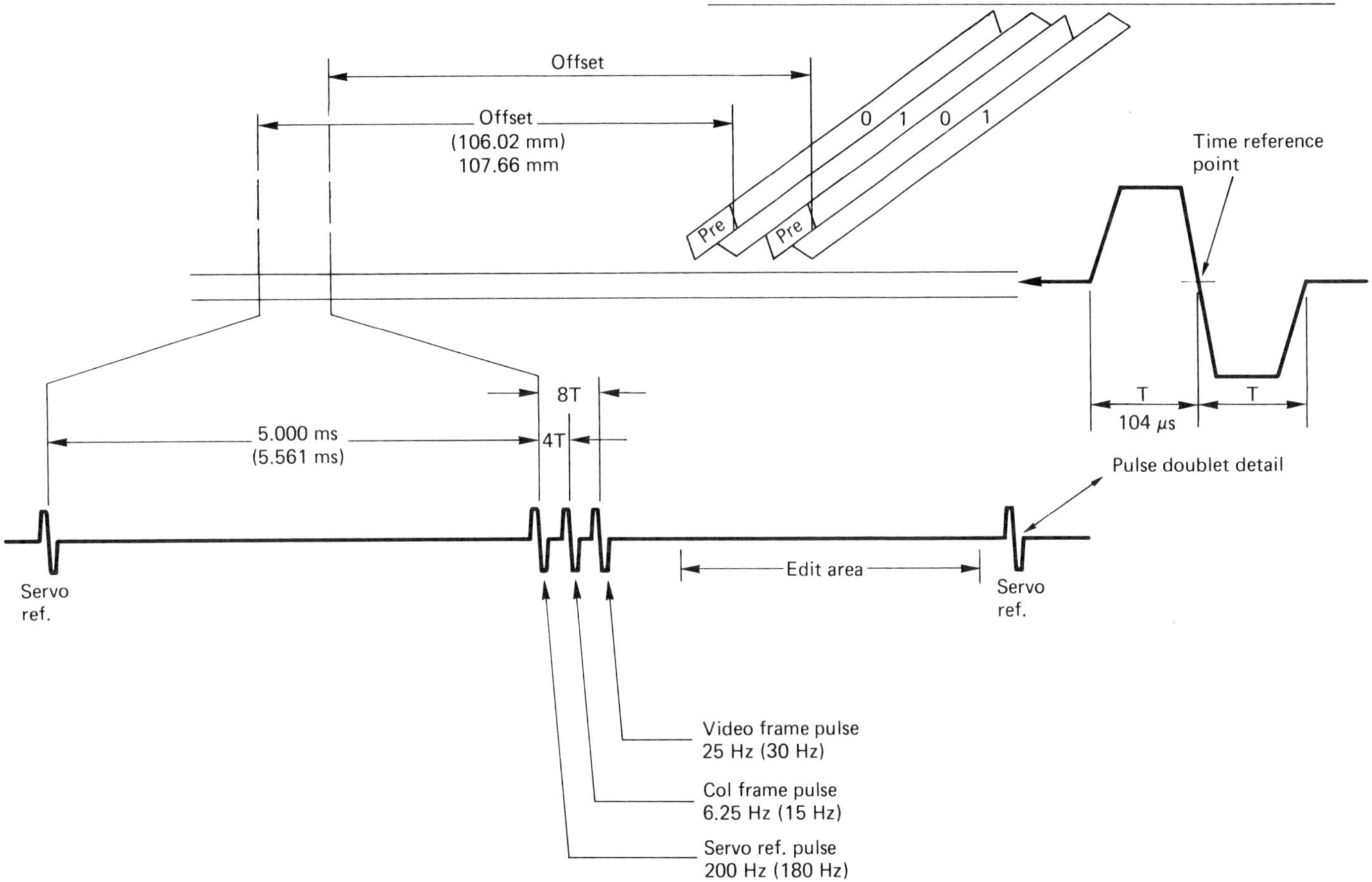

Figure 6.8(c) The control track of D-2 contains a servo reference pulse at segment rate, but the pulses are offset from the beginning of the segment, which is at the end of the preamble in the video sector. Servo pulse may be followed by colour frame pulse or video frame pulse as appropriate. Figures are for PAL, NTSC in brackets

segment, colour frame and frame information. Figure 6.8(c) shows that there is an event in the control track once per segment, and there are three positions in which a control track pulse can occur. A pulse is always present in the first position and is used to control the capstan servo so that the segments will be properly traced on replay. The second pulse will occur once per colour frame, and the third pulse will occur once per frame. When tape travel initially starts, the capstan speed will slip until a control track colour frame pulse coincides with a pulse from the reference colour framer. At this point the capstan phase-locked loop will be closed, and the chroma phase of the replay signal will match that of the reference. The frame pulses allow the picture to be reassembled from segments correctly when playback is not colour framed, for example in TSO (Tape Speed Override) mode.

6.4 Head configurations

There are some practical matters to consider when choosing the positions of heads in the scanner. Figure 6.9 shows a scanner which carries four heads spaced evenly around the periphery. All of the heads are at the same height on the scanner axis, which means that they would all pass the same point in space as the scanner rotates.

One revolution of such a scanner would lay down four evenly spaced tracks on the tape and would play them back equally well. With a wrap angle of 180°, two heads would be in contact with the tape at any one time, and so the data rate of an individual head will be half the total.

If it is necessary to play back at variable speed, each of the heads will need to be movable and will need to be individually controlled, so the cost of implementation will be rather high. The approach used in actual machines is shown in Figure 6.9(b) where the heads are mounted in pairs at 180°. The same tape format can be recorded, provided that the heads in each pair are mounted in different planes. Two rotations of such a scanner will be needed for one PAL field, and one and a half rotations for NTSC, giving scanner speeds of 100 Hz and 90/1.001 Hz respectively. The tape speed does not differ between the two TV standards, so that playing time remains the same for both. The different scanner speeds result in the two formats having different track angles. The track width is in inverse proportion to the scanner speed, so the PAL track will be 90% of the width of an NTSC track. The separation needed between the planes of the two heads in each pair depends on the track pitch and the angular separation betwen the heads. If this geometry is correct, the same format will result, but the time at which the various tracks are laid down will be different, as the figure shows. No problem will arise provided that the record and reproduce electronics are designed to expect to transfer data at the appropriate time. The main advantage of this approach is that there is now only a requirement for two head positioners, which is a great improvement over four.

The stagger between the heads is necessary whether the heads are intended for azimuth recording or guard band recording. In both cases the head is physically larger than the track it writes, owing to the poles being milled away in the area of the gap. It is not possible to put the heads side by side without a large guard band resulting. Figure 6.10 shows that staggering the heads allows the guard band to be any size, even negative in azimuth applications.

The bit error rate from the trailing head of a staggered head pair is generally slightly worse than that of the leading head because it is working in the shock wave pattern set up by the first head and flies over any particles of debris which the first head loosens.

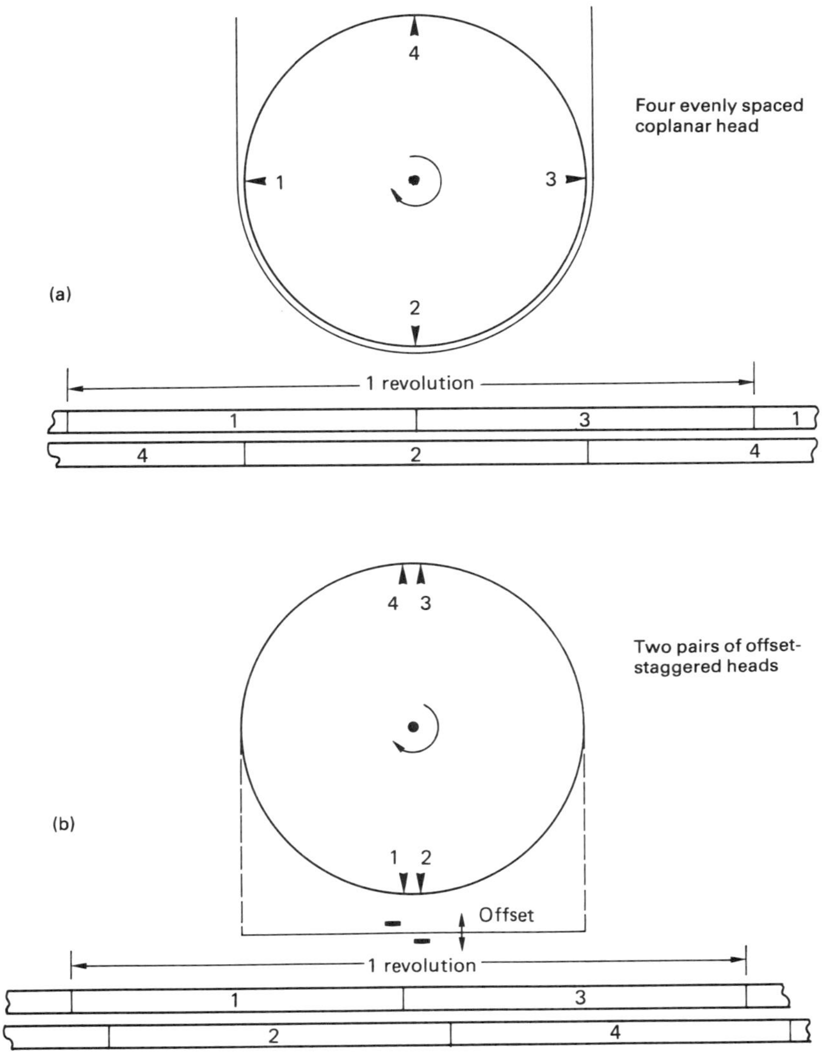

Figure 6.9 (a) Evenly spaced heads in a single plane require individual actuators for variable speed. (b) Two of the heads are slid along the tape track to give pairs of offset-staggered heads. Same tape format results if signal timing to heads is suitably modified

The heads which record the track pairs are staggered 4.22° around the scanner, so that the trailing type 1 head partially overwrites the edge of the track written by the leading type 0 head. This is shown in Figure 6.11(a). The leading head of the next head pair then trims the edge of the track written by the trailing head. In this way there can be no gaps between the tracks, and so it is not possible to replay unwanted signals from previous recordings while mistracking, a condition necessary because there is no separate erase process. Additionally, the tracks are narrower than the heads which wrote them, since the track width is controlled by

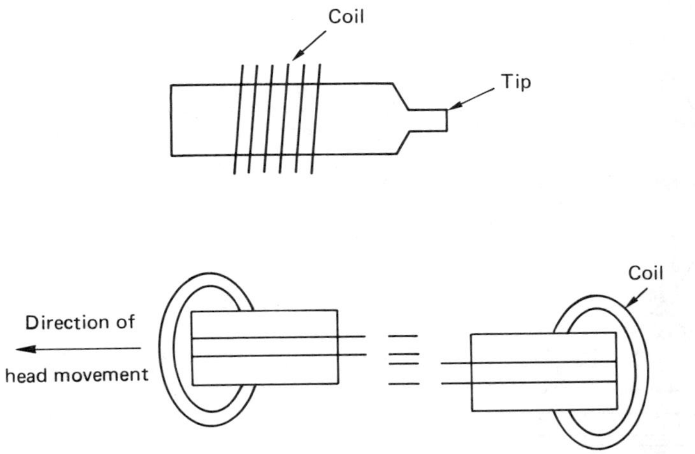

Figure 6.10 Staggering the heads allows any relative track spacing to be used

the linear tape speed rather than the pole width of the head. The amount of overwrite is kept small so that tracks inserted in the centre of an existing recording in certain edit modes do not unnecessarily overwrite adjacent existing tracks. The longitudinal offset of 4.22° is necessary to allow the two heads to overlap physically, but it also means that the transverse displacement between the two heads is not equal to the track pitch. Figure 6.11(b) shows that the longitudinal offset between the heads must take into account the fact that the tape is moving up the scanner as the heads pass. By the time the trailing head has reached a given point on a track made by the leading head, that track will be higher on the scanner than it was when it was written, and so the trailing head must be positioned higher on the scanner to put the next track in the correct place.

It is easy to calculate the adjustment necessary. In both formats, the tape rises by two track widths in half a rotation of the scanner. The correction is obtained by calculating the rise which takes place in 4.22° of rotation, which is as follows:

(a) for PAL

$$\text{tape rise} = \frac{35.2 \times 2 \times 4.22}{180}\ \mu\text{m} = 1.7\ \mu\text{m}$$

(b) for NTSC

$$\text{tape rise} = \frac{39.1 \times 2 \times 4.22}{180}\ \mu\text{m} = 1.8\ \mu\text{m}$$

Subtracting these correction factors from the track pitch results in the head displacement along the scanner axis which is necessary to produce the standard format. The displacement is 33.5 μm for PAL and 37.3 μm for NTSC.

The angular separation between the heads in a pair means that compensation is also necessary in the timing with which one head begins to write with respect to the other. In helical recording the start points of successive tracks are staggered, and the timing difference takes into account the track stagger and the head stagger. The

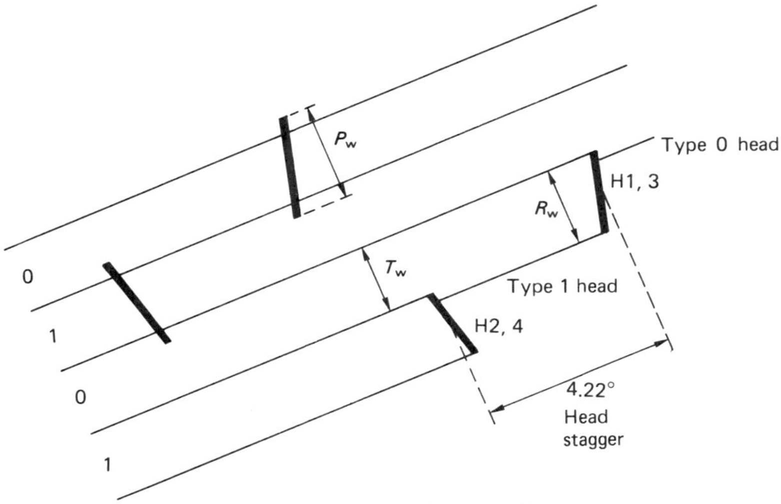

	R_w	T_w	P_w μm
PAL	37-41	35.2	52
NTSC	41-45	39.1	58

Figure 6.11(a) A pair of heads, one of each azimuth type, follow one another across the tape. The resulting track width, T_w, is less than the width of the record heads, R_w, because of partial overwrite. The replay head width, P_w, can be usefully made 50% greater than the track width

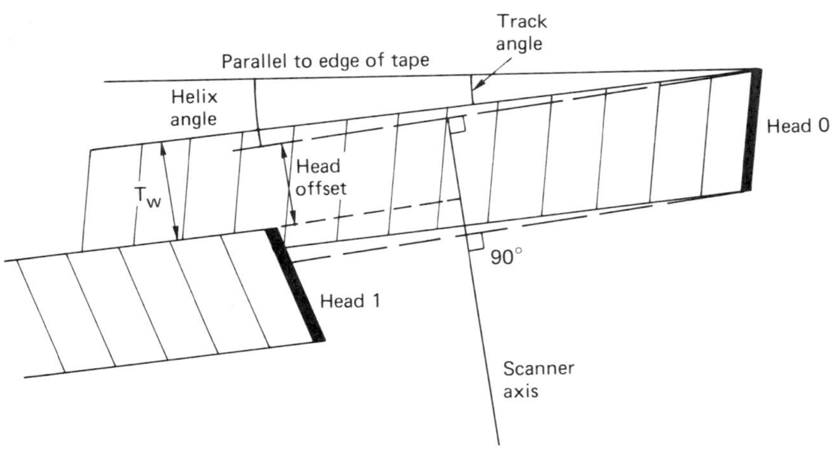

Figure 6.11(b) The movement of the tape causes the track angle to be different to the helix angle of the scanner. Thus the track width T_w, measured on the track angle, is not the same as the displacement between the heads which is measured at right angles to the helix angle – which is along the scanner axis. The compensation value is derived in the text

data to be recorded by each head will be available in an interleave memory, and the timing difference is easily accommodated by reading the memory at different times for each head.

In most machines there will be a separate pair of replay heads 90° of scanner rotation away from the record heads. The use of azimuth recording allows the replay-head width to be somewhat larger than both the record head and the track, so there is considerable immunity to tracking error on replay and a raised probability of recovering data when tracks are crossed in shuttle.

Prior to an edit, the machine will also read tape tracks with the record heads in order to arrive at the correct mechanical relationship with the existing recording when the new recording commences.

Pole widths for record heads are shown in Figure 6.11 for the two formats, and considerable tolerance exists in this dimension owing to the nature of azimuth recording. The replay pole widths are suggestions, since the format only specifies the track width, and a manufacturer can choose any suitable replay-head width which meets his requirements.

6.5 The basic rotary-head transport

Figure 6.12 shows the important components of a rotary-head helical scan tape transport. There are four servo systems which must correctly interact to obtain all modes of operation: two reel servos, the scanner servo and the capstan servo. The capstan and reel servos together move the tape, and the scanner servo moves the heads. For variable-speed operation a further servo system will be necessary to deflect the heads.

There are two approaches to capstan drive: those which use a pinch roller and those which do not. In a pinch roller drive, the tape is held against the capstan by pressure from a resilient roller which is normally pulled towards the capstan by a solenoid. The capstan only drives the tape over a narrow speed range, generally the range in which broadcastable pictures are required. Outside this range, the pinch

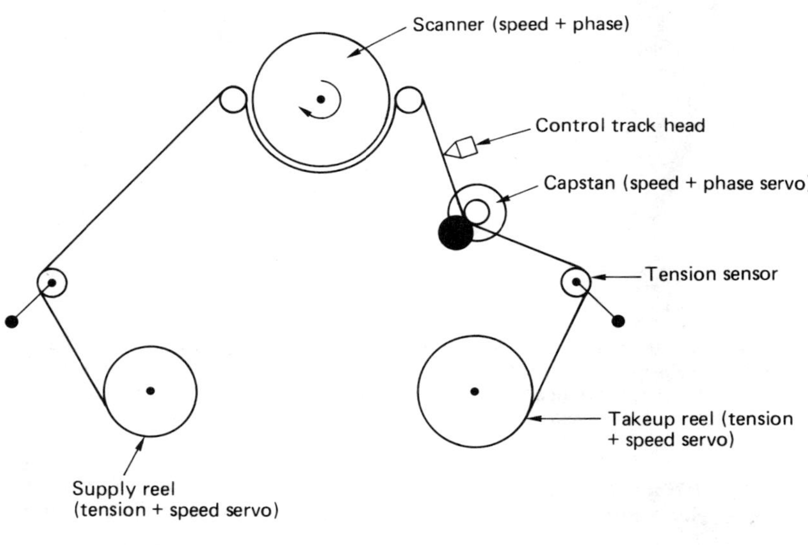

Figure 6.12 The four servos essential for proper operation of a helical scan DVTR. Cassette-based units will also require loading and threading servos, and for variable speed a track-following servo will be necessary

roller retracts, the tape will be driven by reel motors alone, and the reel motors will need to change their operating mode.

In a pinch-roller-less transport, the tape is wrapped some way around a relatively large capstan to give a good area of contact. A drilled capstan connected to a suction pump can also be used to obtain extra adhesion. The tape is always in contact with the capstan, irrespective of operating mode, and so the reel servos never need to change mode. A large capstan has to be used to give sufficient contact area and to permit high shuttle speed without excessive motor rotational speed. This means that at play speed it will be turning slowly and must be accurately controlled and free from cogging. A multipole ironless-rotor pancake-type brush motor is often used, or a sinusoidal-drive brushless motor.

The simplest operating mode to consider is the first recording on a blank tape. In this mode, the capstan will rotate at constant speed and drive the tape at the linear speed specified for the format. The scanner must rotate at a precisely determined speed, so that the correct number of tracks per unit distance will be laid down on the tape. Since in a segmented recording each track will be a constant fraction of a television field, the scanner speed must ultimately be determined by the incoming video signal to be recorded. To take the example of a PAL D-2 recorder having two record head pairs, eight tracks or four segments will be necessary to record one field, and so the scanner must make exactly two complete revolutions in one field period, requiring it to run at 100 Hz. In the case of NTSC D-2, there are six tracks or three segments per field, and so the scanner must turn at one and a half times field rate, or a little under 90 Hz. The phase of the scanner depends upon the time delay necessary to shuffle and interleave the video samples, which was seen to be necessary in Chapter 5. This will be at least one segment.

6.6 Controlling motor speed

In various modes of operation, the capstan and/or the scanner will need to have accurate control of their rotational speed. During crash record where the tape is deemed to be blank, the capstan must run at an exact and constant speed. When the scanner is first started, it must be brought to the correct speed before phase lock can be attempted. The principle of speed control commonly used will be examined here.

Figure 6.13(a) shows that the motor whose speed is to be controlled is fitted with a toothed wheel or slotted disk. For convenience, the number of slots will usually be some power of two. A sensor, magnetic or optical, will produce one pulse per slot, and these will be counted by a binary divider. A similar counter is driven by a reference frequency. This may often be derived by multiplying the input video field rate in a phase-locked loop. The operation of a phase-locked loop was described in Chapter 4.

The outputs of the two counters are taken to a full adder, whose output drives a DAC which in turn drives the motor. The bias of the motor amplifier is arranged so that a DAC code of one-half of the quantizing range results in zero drive to the motor, and smaller or larger codes will result in forward or reverse drive.

If the count in the tacho divider lags behind the count in the reference divider, the motor will receive increased power, whereas if the count in the tacho divider leads the count in the reference divider, the motor will experience reverse drive,

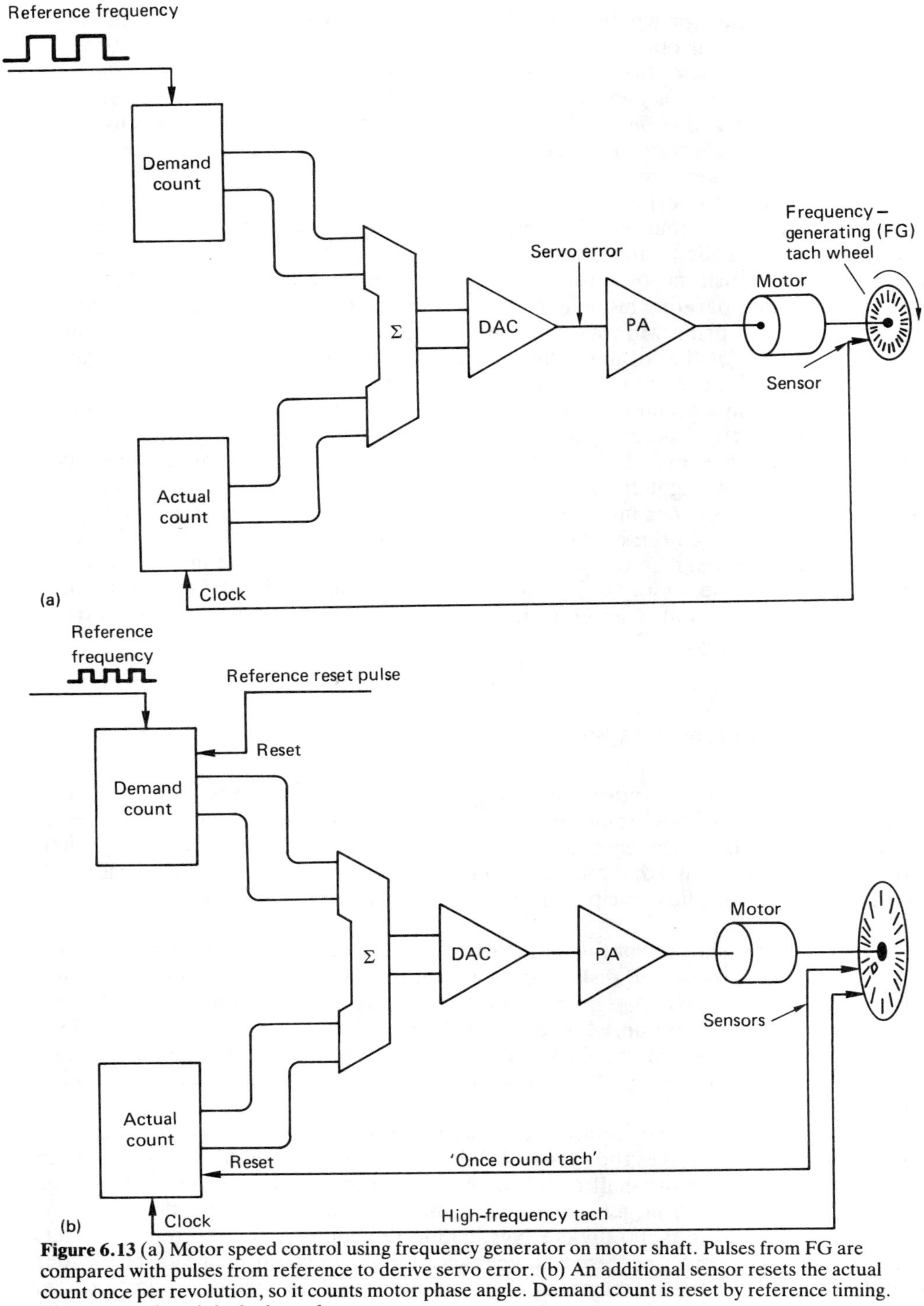

Figure 6.13 (a) Motor speed control using frequency generator on motor shaft. Pulses from FG are compared with pulses from reference to derive servo error. (b) An additional sensor resets the actual count once per revolution, so it counts motor phase angle. Demand count is reset by reference timing. Thus motor phase is locked to reference

which slows it down. The result is that the speed of the motor is exactly proportional to the reference frequency. In principle the system is a phase-locked loop, where the voltage-controlled oscillator has been replaced by a motor and a frequency-generating (FG) wheel.

6.7 Phase-locked servos

In a DVTR the rotational phase of the scanner and capstan must be accurately controlled. In the case of the scanner, the phase must be controlled so that the heads reach the beginning of a track at a time which is appropriate to the station reference video fed to the machine.

A slightly more complex version of the speed control system of Figure 6.13(a) is required, as will be seen in Figure 6.13(b). In addition to a toothed wheel or slotted disk, the motor carries a reference slot which produces a rotational phase reference commonly called *once-round tach*. This reference presets the tach divider, so that the tach count becomes an accurate binary representation of the actual angle of rotation.

A similar divider is fed by a reference clock as before and preset at appropriate intervals. The reference clock has the same frequency as the tooth-passing frequency of the tacho at normal speed. In a PAL D-2 machine which needs two scanner rotations per field, the counter will need to be preset twice per field, or, more elegantly, the counter will have an additional high-order bit which does not go to the adder, and then it can be preset once per field, since disconnection of the upper bit will cause two repeated counts of half a field duration each, with an overflow between.

The adder output will increase or decrease the motor drive until the once-round tach occurs exactly opposite the reference preset pulse, because in this condition the sum of the two inputs to the adder is always zero.

The binary count of the tach counter can be used to address a rotation phase PROM. This will be programmed to generate signals which enable the different sectors of the recorded format to be put in the correct place on the track. For example, if it is desired to edit one audio channel without changing any other part of a recording, the record head must be enabled for a short period at precisely the correct scanner angle. The scanner phase PROM will provide the timing information needed.

When the tape is playing, the phase of the control track must be locked to reference segment phase in order to achieve accurate tracking.

Figure 6.14 shows that a similar configuration to the scanner servo is used, but there is no once-round tach on the capstan wheel. Instead, the tach counter is reset by the segment pulses obtained by replaying the control track. Since the speed of the control track is proportional to the capstan speed, resetting the tach count in this way results in a count of control track phase. The reference counter is reset by segment rate pulses, which can be obtained from the scanner, and so the capstan motor will be driven in such a way that the phase error between control track pulses and reference pulses is minimized. In this way the rotary heads will accurately track the diagonal tracks.

During an assemble edit, the capstan will phase-lock to control track during the preroll, but must revert to constant speed mode at the in-point, since the control

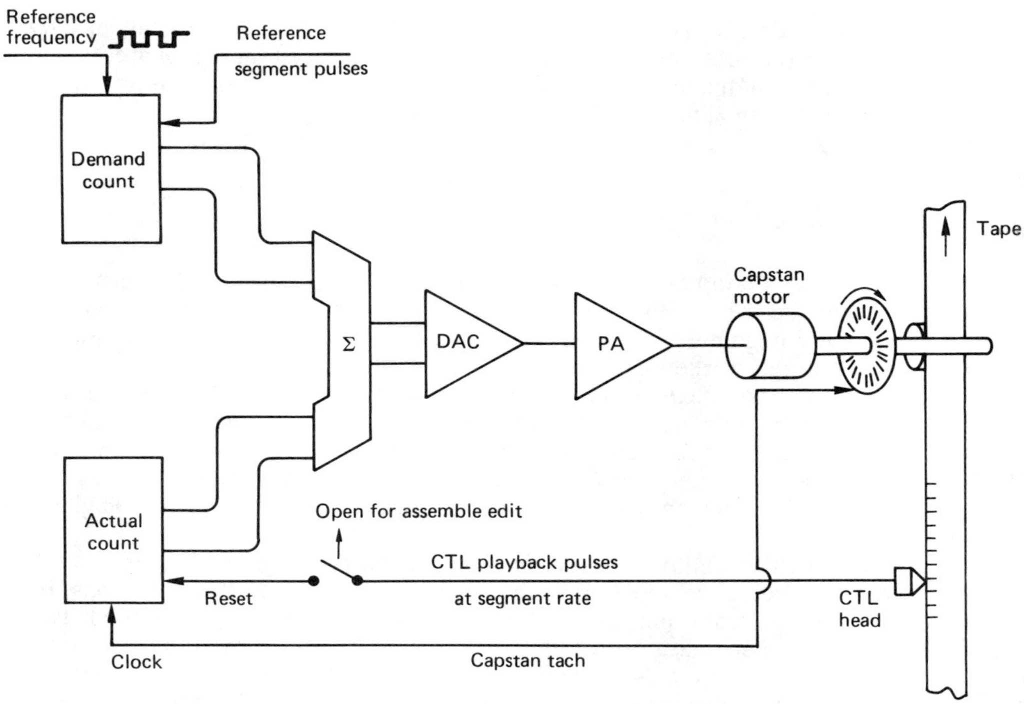

Figure 6.14 If the encoder counter is reset by CTL pulses, it will count linear tape phase, i.e. distance between CTL pulses. Controlling demand counter from reference segment pulses phase-locks tape CTL track to segment rate resulting in correct tracking. At an assemble edit the reset is disabled and the capstan servo makes a smooth transition to the velocity control mode of Figure 6.13(a)

track will be recorded from that point. This transition can be obtained by simply disabling the capstan tach counter reset, which causes the system to revert to the speed control servo of Figure 6.13.

6.8 Tension servos

It has been shown that tape tension control is critical in helical scan machines, primarily to ensure interchange and the correct head contact. Tension control will also be necessary for shuttle.

Figure 6.15 shows a typical back tension control system. A swinging guide called a tension arm is mounted on a position sensor which can be optical or magnetic. A spring applies a force to the tension arm in a direction which would make the tape path longer. The position sensor output is connected to the supply reel motor amplifier which is biased in such a way that when the sensor is in the centre of its travel the motor will receive no drive. The polarity of the system is arranged so that the reel motor will be driven in reverse when the spring contracts. The result is that the reel motor will apply a reverse torque which will extend the spring as it attempts to return the tension arm to the neutral position. If the system is given adequate

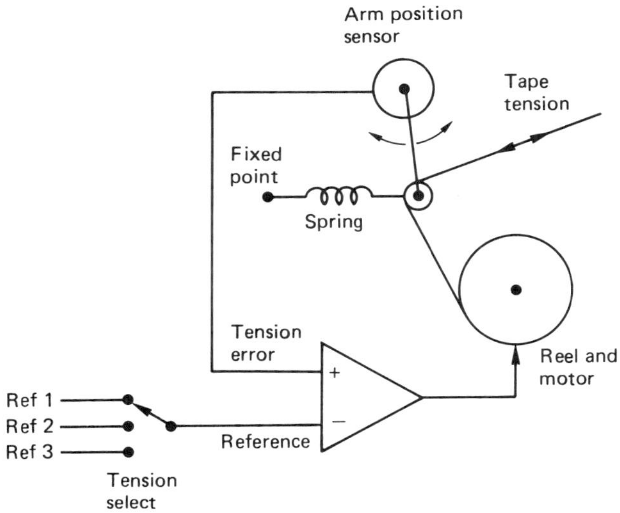

Figure 6.15 Tension servo controls reel motor current so that motor extends sprung arm. Sensor determines extension of spring and thus tape tension

loop gain, a minute deflection of the arm will result in full motor torque, so the arm will be kept in an essentially constant position. This means that the spring will have a constant length, and so the tape tension will be constant. Different tensions can be obtained by switching offsets into the feedback loop, which result in the system null being at a different extension of the spring.

The error in the servo will only be small if the reel can accelerate fast enough to follow tape motion. In a machine with permanent capstan engagement, the capstan can accelerate much faster than the reels can. If not controlled, this would lead to the tension arm moving to one or other end of its travel, resulting in tape stretch, or a slack loop which will subsequently snatch tight with the same results. A fixed acceleration limit on the capstan servo would prevent the problem, but it would have to be set to allow a full reel on a large-size cassette to keep up. The motion of a small cassette with vastly reduced inertia would be made artificially ponderous. The solution is to feed back the displacement of the tension sensors to the capstan servo. If the magnitude of the tension error from either tension arm is excessive, the capstan acceleration is cut back. In this way the reels are accelerated as fast as possible without the tension becoming incorrect.

In a cassette-based recorder, the tension arms usually need to be motorized so that they can fit into the mouth of the cassette for loading and pull tape out into the threaded position. It is possible to replace the spring which provides tape tension with a steady current through the arm motor. By controlling this current, the tape tension can be programmed.

6.9 Tape-remaining sensing

It is most important in a cassette to prevent the tape running out at speed. The tape is spliced to a heavy leader at each end, and this leader is firmly attached to the reel hub. In an audio cassette, there is sufficient length of this leader to reach to the other reel, and so the impact of a run-off can be withstood. This approach cannot be used with a video cassette, because the heavy leader cannot be allowed to enter the scanner, as it would damage the rotating heads.

The transport must compute the tape remaining in order to prevent running off at speed. This will be done by measuring the linear speed of the tape and comparing it with the rotational speed of the reel. The linear speed will be the capstan speed in a permanent capstan drive transport, while a timer roller will be necessary in a pinch-roller-type transport. The rotational speed of the reels will be obtained from a frequency generator on the reel motors. Figure 6.16 shows a simple method of computing the tape remaining. Pulses from the capstan FG cause a counter to

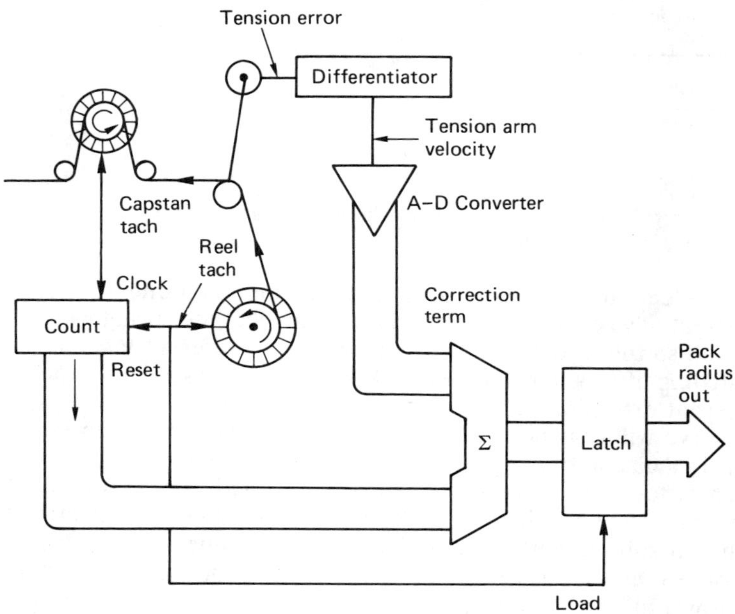

Figure 6.16 Tape remaining is calculated by comparing capstan tach with reel tach. If tension arm is moving, tape speed at reel and capstan will be different, so correction term from tension arm sensor is added.

Pinch-roller-type transports disengage the capstan in shuttle, and so the timer roller tacho would be used instead

increment. When a reel FG pulse occurs, the count will be transferred to a latch, and the counter will be reset. When the reel is full, there will be many capstan FG pulses between reel FG pulses, but as the radius of the tape pack falls, the count transferred to the latch will also fall. It will be possible to determine the limit count from knowledge of the capstan and reel hub diameters and the number of teeth on their generators. The latch count is not quite the pack radius, because if the tension arm is moving as a result of an acceleration, the linear speed of the tape at the capstan will not be the same as the linear speed at the reel. To prevent false shutdowns when accelerating near the end of the tape, the tension arm signal is differentiated to give an arm velocity signal, and this can be digitized and used to produce a correction factor for the radius parameter.

To prevent run-off, the reel pack radius of the reel which is unwinding profiles the allowable shuttle speed, so that as the pack radius falls, the tape speed falls with it. Thus when the tape finally runs out, it will be travelling at very low speed, and the photoelectric sensor which detects the leader will be able to halt the transport without damage.

6.10 Brushless DC motors

The reliability of digital circuitry is of little consequence if a breakdown is caused by the failure of an associated mechanical component. Video recorders, including digital video recorders, are a complex alliance of mechanical, magnetic and electronic technology, and the electronic wizardry has to be matched by some good mechanical engineering. As the cost of electronics falls, it becomes possible to replace certain mechanical devices cost-effectively, and there is an impetus to do this if a wear mechanism can be eliminated. The brushless motor is one such device. The conventional DC brush motor is ubiquitous, and life would be infinitely less convenient without it, for it allows reasonable efficiency, a wide speed range, reversing and relatively compact size. The weak point of all motors of this kind is the brush/commutator system, which serves to distribute current to the windings which are in the best position relative to the field to generate torque. There is a physical wear mechanism, and the best brush material developed will still produce a conductive powder which will be distributed far and wide by the need to have airflow through the motor for cooling. The interruption of current in an inductive circuit results in sparking, which can cause electrical interference and erosion of the brushes and commutator. The brushless motor allows all of the benefits of the DC motor, without the drawbacks of commutation.

Figure 6.17 shows that a brushless motor is not unlike a normal motor turned inside out. In a normal motor, the field magnet is stationary and the windings

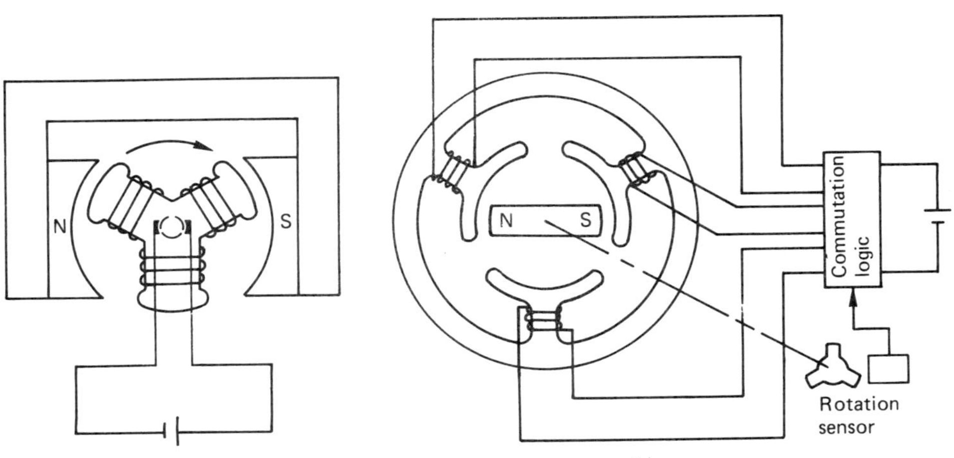

Figure 6.17 At (a) a conventional brush motor has rotating windings. Current is fed through stationary brushes and commutation is automatic. At (b) the motor has been turned inside-out, the magnet revolves and the windings are now stationary so they can be directly connected to an electronic commutating circuit. The rotation sensor on the magnet shaft tells the commutator when to switch

rotate. In a brushless motor, the windings are stationary and the magnet rotates. The stationary windings eliminate the need to supply current to the rotor, and with it the need for brushes. The commutating action is replaced by electronic switches. These need to be told the rotational angle of the shaft, a function which is intrinsic in the conventional commutator. A rotation sensor performs this function.

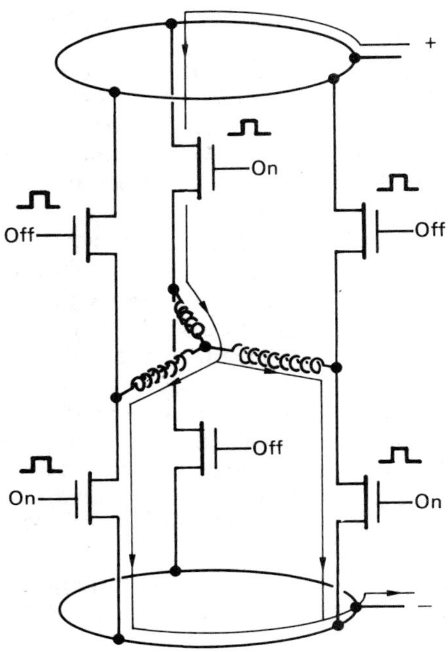

Figure 6.18 Circuit of a brushless DC motor is basically a three-phase bridge. Six different flux directions can be obtained by switching combinations of FETs. One example is shown

Figure 6.18 shows the circuit of a typical brushless motor. The three-phase winding is driven by six switching devices, usually power FETs. By switching these on in various combinations, it is possible to produce six resultant field directions in the windings. The switching is synchronized to the rotation of the magnet, such that the magnetic field of the rotor always finds itself at right angles to the field from the windings, and so produces the most torque. In this condition the motor will produce the most back EMF, and the normal way of adjusting the rotation sensor is to allow the motor to run on no load and to adjust the angular sensor position mechanically until the motor consumes minimum current.

The rotating magnet has to be sufficiently powerful to provide the necessary field, and in early brushless motors it was heavy, which gave the motor a high inertial time constant. Whilst this was of no consequence in steady speed applications, it caused problems in servos. The brushless servo motor was basically waiting for high-energy permanent magnets, which could produce the necessary field without excessive mass. These rare earth magnets are currently more expensive than the common ferrite magnet, and so the first applications of brushless motors have been in machines where the motor cost is a relatively small proportion of the high overall cost and where reliability is paramount. Professional recorders come into this category. Down time for brush replacement and labour costs eclipse the cost of the brushes. The scanner motor of a video recorder in a modern post-production suite has to run at speed hour after hour, day after day, and is a natural application of a brushless motor.

The development of the power FET also helped brushless motors, since all DC motors regenerate, which is to say that when the torque opposes the direction of rotation, the current reverses. FETs are not perturbed by current direction, whereas bipolar devices were at best inefficient or would at worst fail in the presence of reverse current. Power FETs can handle currents which would destroy brush gear, and so large transient accelerations are easily obtained, overcoming the inertia of the rotating magnet.

A typical brushless motor appears to the drive amplifier just like a normal DC motor. It has only two power terminals, has back EMF, and will regenerate. The only major difference is the way that reversing is achieved. Brush motors reverse by reversing drive polarity, whereas brushless motors reverse by changing the phase of the commutation.

The number of motor poles directly influences the complexity of the switching circuitry, and brushless motors tend to have relatively few poles. Conventional motors suffer from cogging or a cyclic variation in torque, and those with few poles suffer most. Whilst it is of little consequence in a scanner motor, which runs at high speed and has the inertia of the scanner to smooth out rotation, it can be important in direct-drive capstan motors, which turn at low speed and have low inertia.

Many transports retain multipole brush motors for the capstan, but it is possible to construct a cogging-free motor. The commutation circuit is no longer switching, but consists of a number of analog amplifiers which will feed the motor with sinusoidal waveforms in fixed phase relationships. In the case of a two-phase motor, this will be two waveforms at 90°; for a three-phase motor, it will be three waveforms at 120°.

The drive signals for the amplifiers are produced by a number of rotation sensors, which directly produce analog sinusoids. These can conveniently be Hall effect sensors which detect the fields from rotating magnets.

6.11 Switched-mode motor amplifiers

The switched-mode motor amplifier has much in common with the switched-mode power supply, since both have the same objectives, namely an increase in electrical efficiency through reduced dissipation, and reduction in size and weight. In studio-based machines reduced dissipation will help to increase reliability, whereas in portable machines the efficiency is paramount since existing battery technology is still a handicap.

The linear amplifier must act as a variable resistance, and so current flowing through it will cause heat to be developed. Linear amplifiers are particularly inefficient for driving tape reel motors, since to obtain fast shuttle speed the supply voltage must be high, and this means that when the machine plays, the reels are turning slowly and the motor voltage will be very small, so that most of the power supply voltage has to be dropped by the amplifier, which is extremely inefficient. Some transports using linear motor amplifiers will program a switched-mode power supply to provide just enough voltage for the instantaneous requirements, helping to raise efficiency.

The principle of a switched-mode amplifier is quite simple. The current is controlled by an electronic switch which is either fully on or fully off. In both cases dissipation in the switch is minimal. The only difficulty consists of providing a variable output despite the binary switching. This will be done by a combination of duty cycle control and filtering.

Figure 6.19 shows a typical switched-mode motor amplifier. It consists of a normal bridge configuration with a number of additional components, a pair of inductors and a current sense resistor in series with the motor, and a number of reverse-connected flywheel diodes.

If it is desired to make current flow through the motor from left to right, then field-effect transistors T_1 and T_4 will be turned fully on. Owing to the presence of the inductors, the current through the motor increases gradually, and when the desired current is reached, transistor T_1 is switched off. The current flowing in the inductive circuit must continue flowing, and it will do so through D_3, as shown in (b). Greater efficiency will be obtained if T_3 is then switched on, as the forward drop of the diode is then bypassed. Transistor T_4 is then switched off, and the current loop will be blocked, causing the flux in the inductors to decay and raising their terminal voltage. This will result in current flowing through D_2 back to the power supply. Greater efficiency will be obtained if T_2 is turned on. The current is now being opposed by the power supply and will decay, and after a short time transistor T_2 must be switched off and transistor T_4 will be switched on again. In this way, the current rises and falls about the desired value. If the switching rate is made high, the variations in current will be small with respect to the average current. The inertia of the motor will in any case prevent any mechanical response at the switching frequency. It will be noted that during this process current flowed in reverse through two of the transistors, requiring the use of field-effect devices.

Clearly if current flowing from right to left is required, the bridge switching will change over, but the principle remains the same.

The torque generated by the motor is proportional to the current, and torque can be applied with or against rotation. For example, when the transport is in play mode, the take-up motor is turning in the same direction as its torque, whereas the supply motor is providing back tension and is turning against its torque. In this case, the motor EMF will cause current to flow in the flywheel circuit, and the switching will take energy out of the current flywheel back to the power supply. This is the principle of regenerative braking. If the motor is being turned by the tape, it will generate, and a current will flow as shown in (c). The current will rise gradually owing to the presence of the inductors. When the current reaches the value needed to give the correct back tension, T_3 will be switched off, and the flywheel current will find itself blocked. The result will be a flux collapse causing the inductors to raise their terminal voltage until a current path is found. When the voltage exceeds the power supply voltage, current will flow in D_1, back to the supply. Again greater efficiency will be obtained if T_1 is switched on, since the diode drop is bypassed. Since the power supply voltage is opposing the motor EMF, the current will fall, and when it has fallen sufficiently, T_1 will be switched off and T_3 will be switched on again.

The switching sequence of the transistors is important, since turning on the wrong pair will short out the power supply! One method of controlling the switching is to produce a current error by comparing the current sensed with the desired current. The current error then changes the DC level of a triangle wave, which is compared with four thresholds, each of which controls one of the

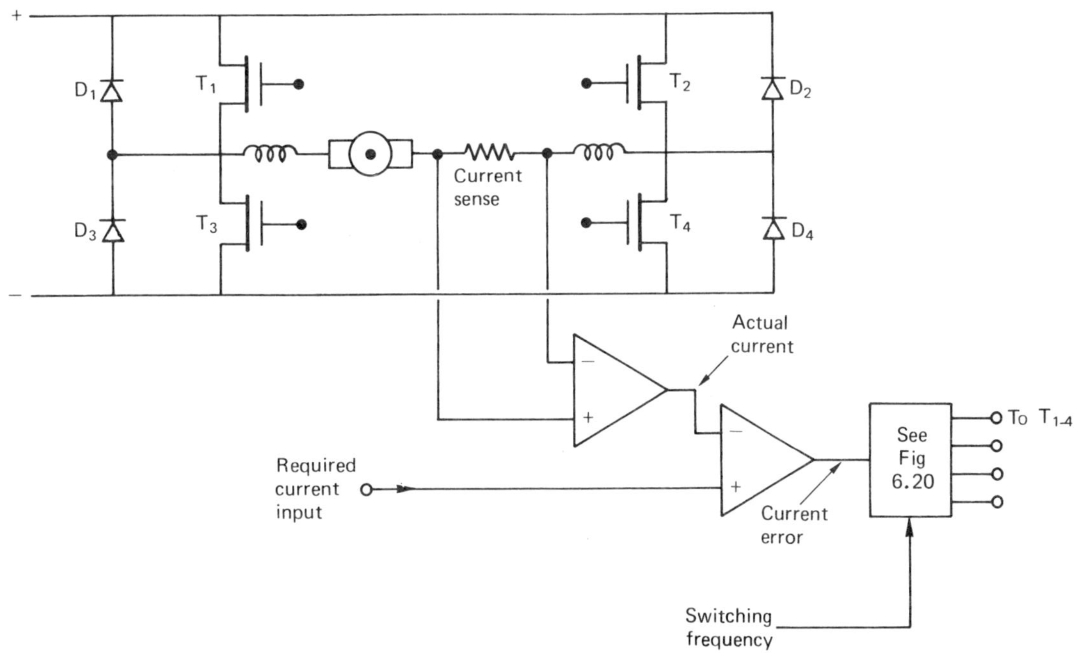

Figure 6.19 General arrangement of switched-mode power amplifier. Devices T_1–T_4 are power FETs which are on or off. Inductors in series with motor inductance limit rate of change of current

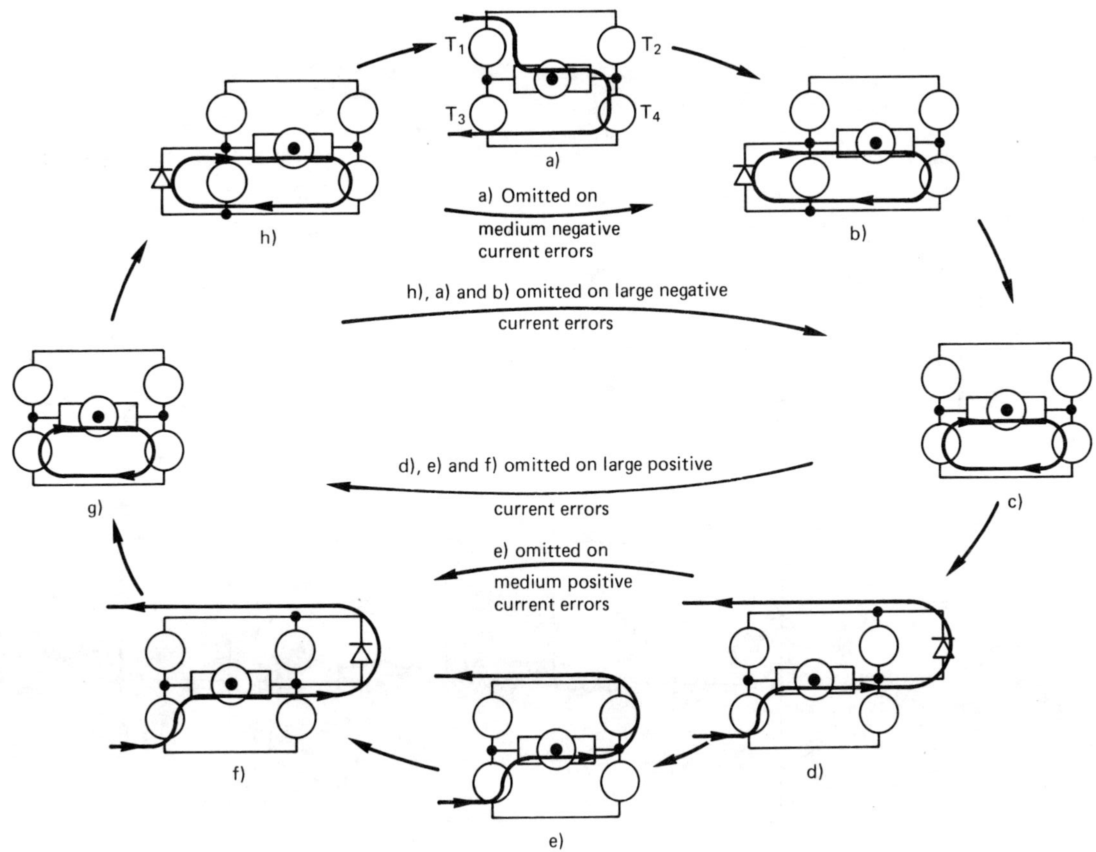

Figure 6.19 (*Continued*). Switching sequence begins at (a) with current building up, a transitional phase (b), a flywheel phase (c), transitional phase (d), and at (e) the system regenerates or opposes the power supply. This causes current to fall. System then passes through remaining phases back to (a). To increase current, system spends more time in phase (a), whereas to reduce current, system spends more time in phase (e). Figure 6.20 shows how this is done

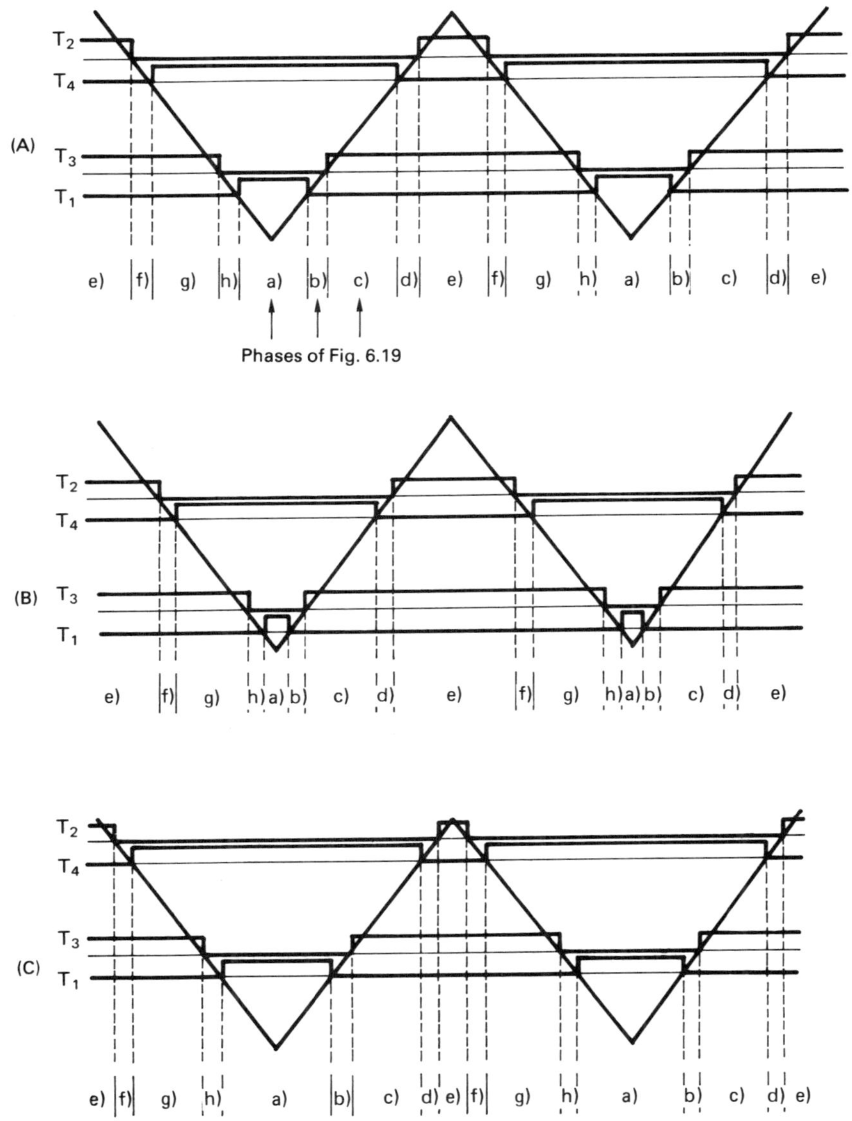

Figure 6.20 The power FETs of Figure 6.19 are switched by comparing the current error with a sawtooth signal produced from the switching frequency. Four offset voltages are added to the current error, and four comparators are used, one for each power device. The offsets are constant, so all four current errors move together.

At (A) above, the current error is zero, so the switching is symmetrical because time period (a) has the same length as time period (e) and the current is neither increased nor reduced. At (B) the current error is negative, and phase (a) is very short whereas phase (e) is much longer, thus current reduces because the motor is regenerating. At (C) the current error is positive and phase (a) is now longer than phase (e) so current increases. If the current error increases further, some phases will no longer take place. An inverse of the error in (C) will result in phase (e) disappearing since (d) and (f) are identical. Further increase will cause (d), (e) and (f) to disappear giving the sequence a, b, c, g, h, …

transistors. Figure 6.20 shows the switching sequence which results. Note that when the current error is zero, the system alternates equally between opposing and assisting the current, and so there is no change.

A motor which is producing back tension will return power to the supply, which will be used by the motor which is taking up, and by the scanner motor. If the machine is braking from a high shuttle speed, both reel motors may together produce more power than the rest of the machine can use, and the power supply voltage will rise. In AC-powered machines, it is necessary to shunt the supply with a resistor in this condition, whereas in battery-powered machines, the regenerative current simply recharges the battery. Clearly a power failure in shuttle is no problem, for as long as the reels are being decelerated, power is generated. This can be augmented if necessary by decelerating the scanner.

6.12 The digital video cassette

The D-2 format cassette uses the same mechanical parts and dimensions as the D-1 cassette, even though the kind of tape and the track pattern are completely different.

The main advantages of a cassette are that the medium is better protected from contamination whilst out of the transport, and that an unskilled operator or a mechanical elevator can load the tape.

It is not true that cassettes offer any advantage for storage except for the ease of handling. In fact a cassette takes up more space than a tape reel, because it must contain two such reels, only one of which will be full at any one time. In some cases it is possible to reduce the space needed by using flangeless hubs and guiding liner sheets, as is done in the Compact Cassette and RDAT, or pairs of hubs with flanges on opposite sides, as in U-matic. Whilst such approaches are acceptable for consumer and industrial products, they are inappropriate for the digital video cassette, because as a professional unit it will be expected to wind at high speeds in automation and editing systems. Accordingly the digital cassette contains two fully flanged reels side by side. The use of a cassette causes a further drawback which is that it is not as easy to provide a range of sizes as it is with reels. Simply putting smaller reels in a cassette with the same hub spacing does not produce a significantly smaller cassette. The only solution is to specify different hub spacings for different sizes of cassette. This gives the best volumetric efficiency for storage, but it does mean that the transport must be able to reposition the reel drive motors if it is to play more than one size of cassette.

If the small, medium and large digital video cassettes are placed in a stack with their throats and tape guides in a vertical line, the centres of the hubs will be seen to fall on a pair of diagonal lines going outwards and backwards. This arrangement was chosen to allow the reel motors to travel along a linear track in a machine which accepts more than one size.

The large-size cassette was the largest which could be accommodated in a transport which would still fit in a 19 inch rack. This cassette will hold 1300 metres of 16 μm tape or 1650 metres of 13 μm tape. The medium-size cassette holds about half of these amounts, and the small cassette holds 190 metres or 250 metres.

All three sizes of cassette have basically the same structure, and detail differences will be noted. The cassette has a double-door arrangement shown in Figure 6.21. When the cassette is removed, both sides of the tape in the throat are

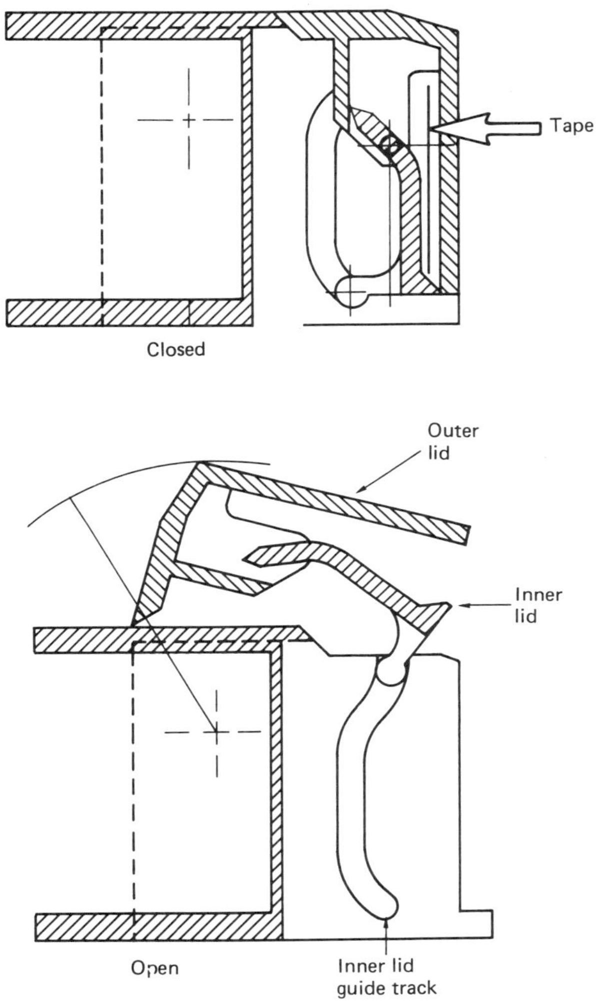

Figure 6.21 When the cassette lid is closed, both sides of the tape are protected by covers. As the outer lid swings up on a pivot, the inner lid is guided around the back of the tape by a curved track

covered. The inner door is guided by a curved track. The door extends to the edges of the small and medium cassettes, but this is unnecessary on the large cassette. The door has a lock which prevents accidental opening, and this is released by a pin when the cassette enters the transport. The lock release mechanism is as near the edge of the small cassette as possible, but it cannot be in the same place on the medium cassette, as it would foul the tape path to the larger reel. Three lock release pins are needed, one for each size, and the larger cassettes need dummy slots which clear the pins used by the smaller sizes.

The cassettes also have hub locks which prevent unwanted rotation in storage or transit. On the larger two sizes, the lock is released by a lever operated by the act of opening the door. There is insufficient room for this mechanism on the small cassette, and so the brake is released by a central post in the transport.

The cassettes are designed for front or side loading, and so have guiding slots running at right angles. Currently, all studio recorders are front loading, whereas some automated machines use side loading. The front-loading guide groove is centrally positioned and the threading throat forms a lead-in to it, helping to centralize the smaller cassettes in a loading slot which will accept a large cassette.

A number of identification holes are provided in the cassettes which are sensed by switches on the transport. Four of these are coding holes which are in the form of break-off tabs which will be set when the cassette is made, and four of them are resettable user holes, which can be controlled with a screwdriver. Table 6.1 shows the significance of the coding and user holes.

Table 6.1 (a) Manufacturers' coding holes (0 = tab removed). (b) User holes (0 = tab removed).

(a) Holes 1 and 2 shall be used in combination to indicate tape thickness according to the following logic table:

Hole number: 1 2	
0 0	= 16 μm tape
0 1	= 13 μm tape (D-2)
1 0	= Undefined/reserved
1 1	= Undefined/reserved

Holes 3 and 4 shall be used to indicate the coercivity of the magnetic recording tape:

Hole number: 3 4	
0 0	= 850 Oe (D-1)
0 1	= 1500 Oe (D-2)
1 0	= Undefined/reserved
1 1	= Undefined/reserved

(b) When a '0' state exists, the user holes shall identify the following conditions:

1. Total record lock out (audio/video/cue/timecode/control track)
2. Reserved and undefined
3. Reserved and undefined
4. Reserved and undefined

Areas are specified for the positioning of labels, and these are recessed to prevent fouling of elevators or guides. Automated machines will use the end location for a bar code which can be read by the elevator while the cassette is stacked.

6.13 Threading the tape

The sequence of operations when a cassette is inserted in the machine will be followed.

In many machines, all three cassette sizes can be used without any adjustment. The operator simply pushes the cassette into the aperture in the machine. The smaller sizes are guided into the centre of the aperture by a central slot with a tapered lead-in. The presence of the cassette is sensed optically, or by switches, and by the same means the machine can decide which size of cassette has been inserted. The cassette is drawn into the machine horizontally by motor-driven tractor belts which are positioned such that all sizes of cassette can be gripped. Alternatively the cassette fits into a cage which is driven by a toothed rack.

As the hub spacing differs between cassette sizes, the transport automatically moves the sliding reel motors to the correct position. In the case of a small cassette, the hub locks are released when the central guide reaches the end of the cassette slot and pushes the brake lever at the end of the horizontal travel into the machine. In all sizes, the final part of horizontal travel unlocks the cassette door. The cassette elevator is then driven downwards. The door is opened, and in the medium and large sizes this releases the hub brakes. The opened cassette is then lowered on to the reel motors, and locating dowels on the transport register the cassette body. The several guides and tension arms are positioned so that the throat of the cassette drops over them with the front run of tape between them and the head drum. In the Ampex transport the permanently engaged capstan also fits inside the cassette throat. A short movement of the tape guides allows the tape to wrap the capstan and the stationary timecode/erase head. In transports having pinch rollers, the latter will be positioned in the cassette throat during threading and will move a short distance to engage the fixed capstan. In both systems, this condition is known as coplanar mode, which means that the tape path is still entirely in the plane of the tape reels. In this mode tape can be shuttled at high speed, and it is possible to erase and timecode-stripe without causing wear to the rotary-head assembly. Timecode can also be read to position a tape without rotary-head wear.

For functions involving the rotary heads, the threading process proceeds further. It is inherent in helical scan recorders that the tape enters and leaves the head drum at different heights. In open reel recorders, the tape reels are simply mounted at different heights, but this is not possible in a cassette if it is to be of reasonable size.

In the Ampex VPR-300 the drum is positioned and angled so that tape leaving it is in the plane of the cassette, and all manipulation of the tape is done on the supply side. Tape entering the head drum has to be moved out of the plane of the cassette first. In the Sony transports, the scanner is tilted to put both the entrance and exit in the cassette plane, and the tape is manipulated on both sides of the drum to overcome the angular discrepancy. These two threading methods will be contrasted here.

Various methods exist to achieve the helical displacement of the tape. Some VTR transports have used conical posts, but these have the disadvantages that there is a considerable lateral force against the edge of the tape and that they cannot be allowed to revolve or the tape would climb off them. Angled pins do not suffer side force, but again cannot be allowed to revolve, although they can be air lubricated. The Sony transports use angled pins, but without lubrication. In the VPR-300, the tape in the plane of the cassette is slightly twisted on a long run and then passes around a conventional guide which causes it to leave the cassette plane at an angle. The twist is too small to approach the elastic limit of the tape, which is quite unharmed. A second guide set at the same angle as the drum axis passes the tape to the drum.[3] Figure 6.22 contrasts the tape twist and angled guide methods. As with any VCR, the guides start inside the cassette and move to various positions as threading proceeds. As the VPR-300 is designed for rapid threading and unthreading without tape damage, particular care was taken to ensure that the tape path geometry is perfect not just when threaded, but at all times during the threading process. There are two guides which primarily control threading the tape around the drum. The sequence can be followed in Figure 6.23. The drum guide moves in a circular path on a threading ring, which rotates concentric with the drum. This guide begins its motion on the wrong side of the tape, but as the threading ring turns, the guide pops up and engages the back of the tape. Then as

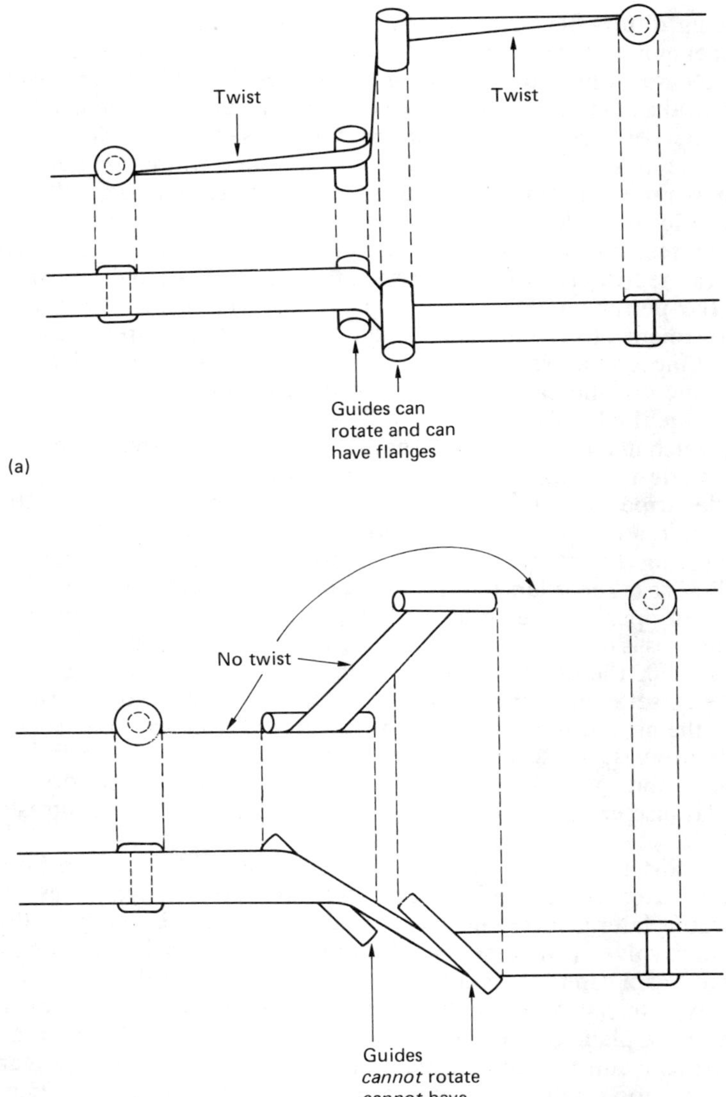

Figure 6.22 The contradictions of shifting the tape plane in helical scan. If the tape is twisted, the guides can rotate. If it is not, they cannot. Long gentle twist is good, but not compact enough for portable use

the threading ring rotates further, the guide is slowly lowered along the drum axis at the correct rate by a cam, such that the tape is laid on to the drum at the helix angle. When the drum guide reaches the end of its travel, its supporting shaft comes up against a fixed V-block which positions it repeatably. It is held in contact with the block by a solenoid.

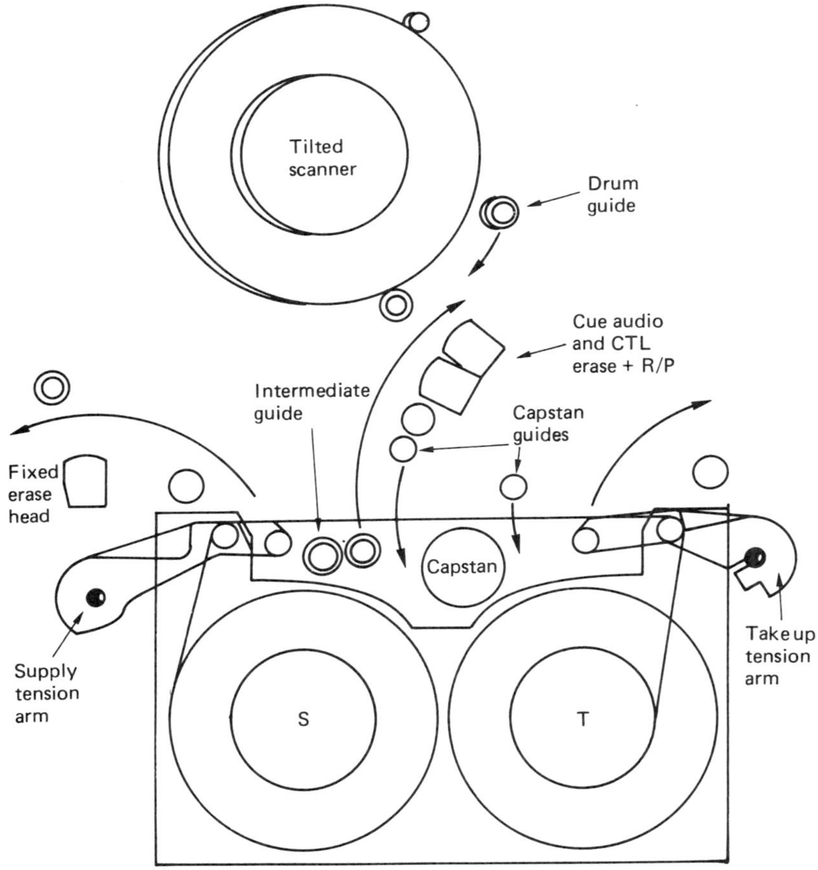

Figure 6.23(a) Ampex D-2 transport with cassette just lowered on to drive hubs. Note permanently engaged capstan in cassette throat

The intermediate guide is designed to move away from the cassette as the drum guide proceeds around the threading ring, and the tilt angle of this guide, which imparts the tape twist, is changed as it moves so that a cylindrical wrap is always maintained. This guide is mounted on three V-shaped rollers which are preloaded outwards and ride in a matching slot in the deck plate. The tilt of the guide is controlled by a spherical roller on an extension of the guide shaft, which bears against a cam profile below the guide track. The principle is shown in Figure 6.24. All of the cam profiles which achieve these operations were computer generated.

Since the total guide wrap angle of cassette transports inevitably amounts to several revolutions, tension buildup is often a problem. In this transport, all tape guides are lubricated by compressed air, except for the tilting threading guide, which is a roller. The drum guide has to travel a long distance, and it is difficult to provide a permanent air supply. The air feed is through the solenoid locking arm, so that air supply becomes available when the tape is fully threaded.

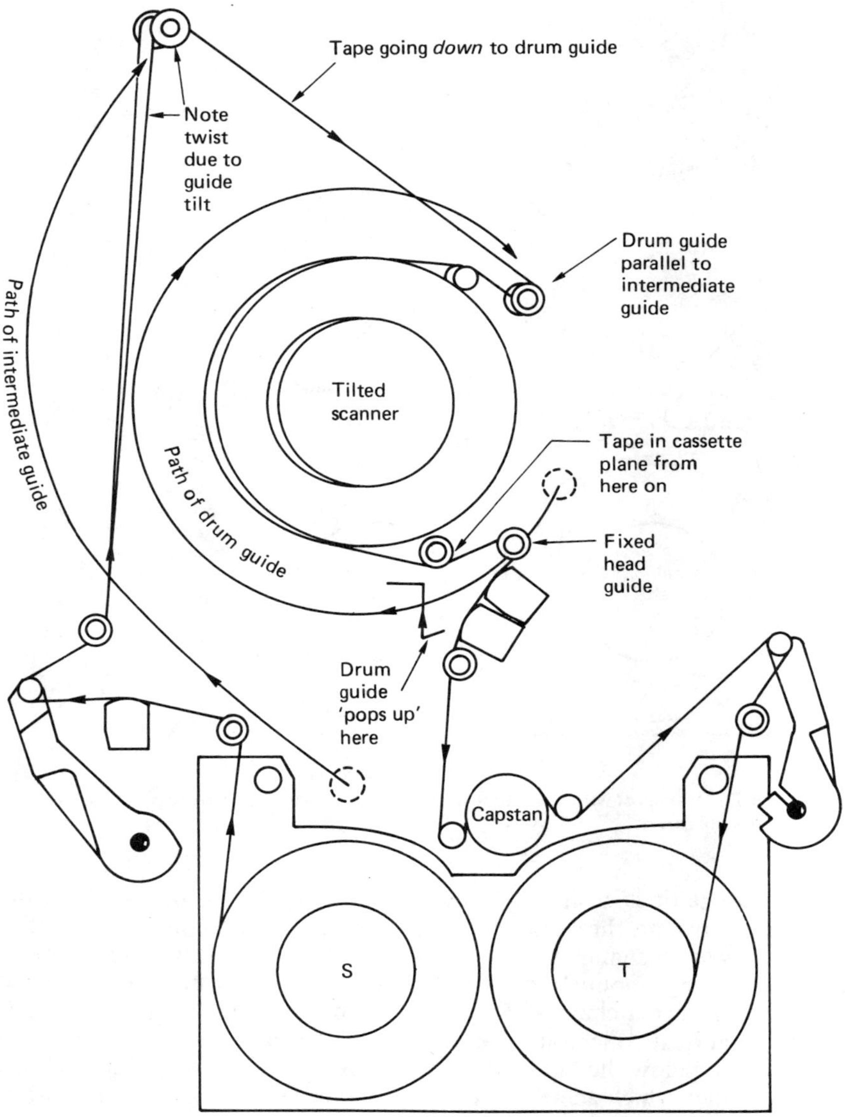

Figure 6.23(b) Fixed head guide swings out, and capstan guides swing in. Tension arms extend. Then drum and intermediate guides complete scanner wrap.

If the scanner guides are not operated, tape can be transported past stationary heads for timecode striping or pre-positioning tape without scanner wear

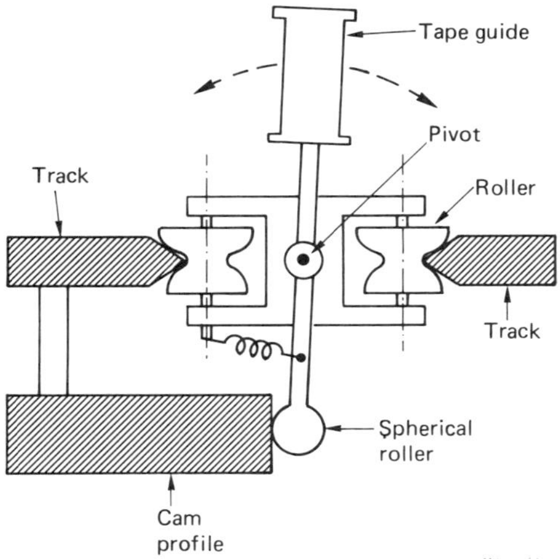

Figure 6.24 The spherical roller and cam profile tilt the tape guide axis to the correct angle as threading proceeds. Tape geometry can then be perfect at all points in the threading

The VPR-300 transport has much in common with the ACR-225 transport and is based on a massive casting which is CNC machined. The other major component is a substantial cast bridge holding fixed heads and guides.

In the Sony D-2 transports, a different approach is taken to the tape guidance. Figure 6.25(a) shows tape wrapping a scanner, with the usual elevation difference between entrance and exit. In (b) the scanner has been tilted so that entrance and exit guides are at the same height. The tape now approaches in the wrong plane. It will be found that if the wrap angle on the entry and exit guides is reduced, an angle can be found where the tape enters and leaves in the plane of the cassette, as shown in (c). At this stage the tape is in the cassette plane, but its surface is not at right angles to that plane – it is leaning over. The entry and exit guides are now tilted in the plane of the tape tangential to the scanner until the tape feeding them is at right angles to the cassette plane, as in (d). In this way, the tape is not subjected to any twisting as it always remains planar. The same approach is used in the RDAT transports built by Sony. The coplanar approach allows a very shallow transport construction which is highly suitable for portable operation.

6.14 Operating modes of a digital recorder

The digital video recorder is expected to do much more than record and replay in a modern installation. The main operating modes will be examined in the next sections, followed by the more obscure modes.

In *crash record* the tape is either blank or considered to be blank, and no reference will be made to any previous information on the tape. As with analog video recorders, if a crash recording is made in the middle of an existing recording, the playback will lose lock at the transition. In this mode the full width erase head will be active, and the entire tape format will be laid down, including audio blocks and control track.

There are a fixed number of segments in a field, and so the scanner speed will be locked to incoming video. The scanner phase will be determined by the delay

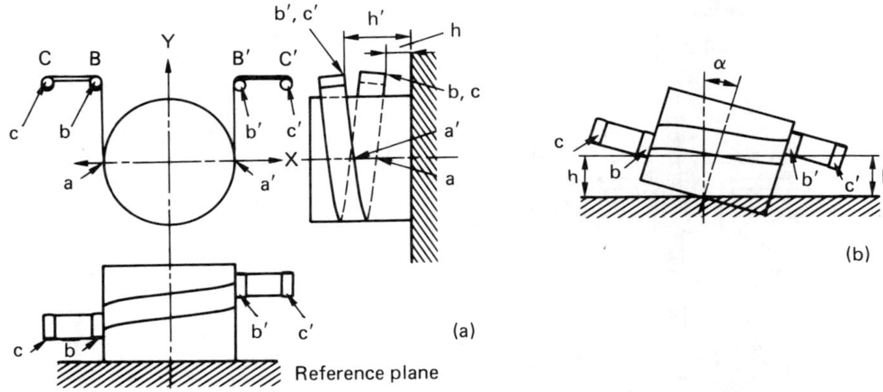

Figure 6.25 At (a) the scanner has a vertical axis which results in a height difference and a twist between the entrance plane BC and the exit plane B′ C′.

At (b) the scanner has been tilted through an angle α such that points b and b′ are both at height h

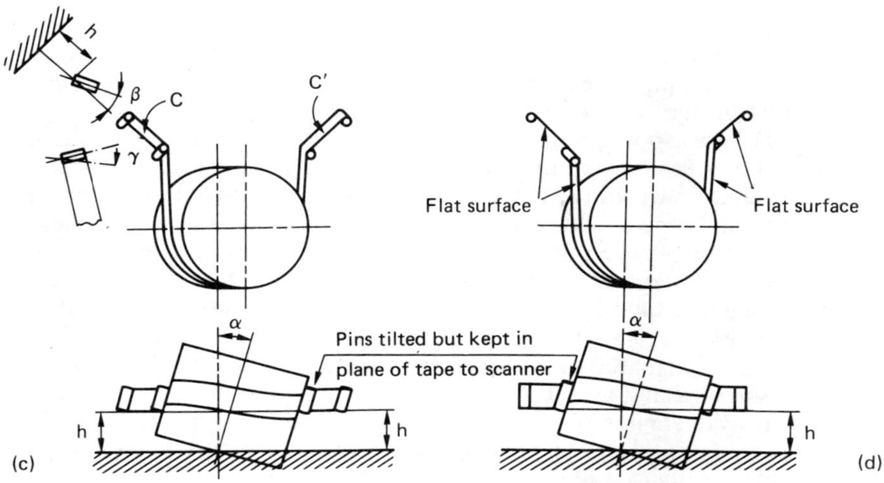

Figure 6.25 (c) The wrap angle on the guides nearest the scanner is reduced until the edge of the tape lies in the reference plane. The tape in planes C and C′ remains flat. The pins now incline at angle β to the reference surface.

At (d) the outer pins can be made vertical if the plane of C and C′ is brought perpendicular to the reference surface. This is done by tilting the inner pins but keeping them in the plane of the tape from inner pin to scanner. Tape entry and exit is now coplanar. (Based on drawings courtesy of Sony Broadcast)

needed to shuffle and encode the data in a segment. Owing to the encoding delay, the recording heads on the scanner will begin tracing a segment some time after the beginning of the input field. The capstan will simply rotate at constant speed, driving the tape at the speed specified in the format. As the rotating heads begin a diagonal track, a segment pulse will be recorded in the control track. Once per frame, the segment pulse will be accompanied by a frame pulse. There will also be a colour frame pulse recorded at appropriate intervals.

6.15 Colour framed playback

In colour framed playback, the number of frames played from the tape will be equal to the number of station reference frames. Ordinary framing requires that the segmentation on tape is properly undone, so that the first segment in a field arrives at the top of the picture monitor and that the interlace sequence is synchronous with reference, so that an odd field comes from tape when there is an odd field in the station reference. In colour framing, however, the offtape field will also be the same type of field in the four- or eight-field sequence as exists in the NTSC or PAL reference, respectively.

The error-correction and de-interleaving processes on replay require a finite time in which to operate, and so the transport must play back segments ahead of real time, in the same way that analog machines play back ahead of real time for timebase correction purposes. The scanner speed will be synchronous with reference, and the scanner phase will be determined by the decoding delay.

The capstan system will be responsible for tracking and framing and will need to vary the speed of the tape until framing is achieved. It will then vary the phase of the tape to optimize tracking.

The capstan and tape form part of a phase-locked loop where the phase of the control track pulses is compared with reference pulses. In order to achieve colour framing, the phase comparison must be between reference colour frame pulses and offtape colour frame pulses. In this way the linear motion of the tape is phased such that reference and offtape framing are the same. The capstan speed must vary during this part of the lock-up sequence, and the picture may be temporarily degraded unless the machine has provision for operating at non-normal speed.

Once framing is achieved, the capstan will switch to a different mode in order to obtain accurate tracking. The phase comparison will now be between offtape segment pulses and pulses generated by the rotation of the scanner. If the phase error between these is used to modify the capstan drive, the error can be eliminated, since the capstan drives the tape which produces the control track segment pulses. Eliminating this timing error results in the rotating heads following the tape tracks properly. Artificially delaying or advancing the reference pulses from the scanner will result in a tracking adjustment.

6.16 Assembly editing

In an assemble edit, new material is appended to an existing recording in such a way that the video timing structure continues unbroken over the edit point. Since this is a composite recording, the subcarrier phase will also need to be continuous.

The recording will begin at a vertical interval, and it will be necessary to have a preroll in order to synchronize tape and scanner motion to the signal to be recorded. The procedure during the preroll is similar to that of framed playback, except that the input video timing is used. The goal of the preroll process is to ensure that the record head is at the beginning of the first segment in a field exactly one encode delay after the beginning of the desired field in the input video signal. At this point the record head will be turned on. D-2 machines will erase by overwrite, so no flying erase is needed. After the assemble point, there will not necessarily be a control track, and the capstan must smoothly enter constant speed mode without a disturbance to tape motion. The control track will begin to record in a continuation of the phase of the existing control track. This process will then continue indefinitely until halted.

6.17 Insert editing

An insert differs from an assemble in that a short part of a recording is replaced by new material somewhere in the centre. An insert can only be made on a tape which has a continuous control track, because the in- and out-point edits must both be synchronous. The preroll will be exactly as for an assemble edit, and the video record process will be exactly the same, but the capstan control will be achieved by playing back the control track at all times, so that the new video tracks are laid down in exactly the same place as the previous recording. At the out-point, the record heads will be turned off at the end of the last segment in a field.

6.18 Tape speed override

One of the most useful features of a production recorder is the ability to operate at variable speed. The variable-speed modes variously available in D-2 are TSO (Tape Speed Override), slow motion/jog and shuttle.

In TSO, the speed is only changed by a maximum of 15% from normal in order to trim a programme to fit a broadcast time slot. Clearly the audio must be fully recovered, and so no replay-head jumping is permissible. The speed of the entire transport is changed by changing the transport timing reference with respect to station reference. The drum and the capstan change speed in proportion so that the heads still follow the helical tracks properly. All of the audio blocks are recovered, and full-quality digital audio is output. Owing to the speed change, the audio sampling rate will not be exactly 48 kHz, and a digital pitch changer or sampling-rate converter can be used if necessary. Clearly the offtape field rate will no longer match reference, and the difference is accommodated by a framestore/timebase corrector, which is written at offtape speed and read at reference speed. Fields will occasionally be skipped or repeated as TBC addresses lap one another.

In slow motion or jog, it is not common to recover the audio fully, but some machines have this feature which is detailed in Section 8.18.3.

6.19 Variable speed in D-2

The variable-speed range of C-format machines has essentially become the yardstick by which competing formats will be measured. It is important for the success of D-2 that at least an equal speed range is available, and the means to obtain this will be discussed here. Although the problem of variable speed is defined by the D-2 format, the solutions adopted by Ampex and Sony are quite different and will be contrasted.

It was seen in Section 6.2 that the movement of the tape results in the tracks having an angle different from the helix angle. The rotary head will only be able to follow the track properly if the tape travels at the correct speed. At all other speeds, the head will move at an angle to the tracks. If the head is able to move along the scanner axis as it turns, it will be possible to follow whole tracks at certain speeds by moving the head as a function of the rotational angle of the scanner. The necessary function can be appreciated by considering the situation when the tape is stopped. The tape track will, of course, be at the track angle, but the head will rotate at the helix angle. The head can be made to follow a stationary track by deflecting it at constant rate by one track pitch per sweep. In other words, a ramp or triangle waveform is necessary to deflect the head. The slope of the ramp is proportional to the speed *difference*, since at normal speed the difference is zero and no deflection is needed. Clearly the deflection cannot continue to grow for ever, because the head will run out of travel, and it must then jump to miss out some tracks and reduce the deflection.

There are a number of issues to be addressed in providing a track-following system in D-2. The use of a segmented format means that head jumps necessary to omit or repeat one or more fields must jump over several segments. This requires a mechanical head-positioning system which has the necessary travel and will work reliably despite the enormous acceleration experienced at the perimeter of the scanner. As there are two head pairs, two such systems are needed, and they must be independently controlled since they are mounted in opposition on the scanner and contact the tape alternately. Also required is a control system which will ensure that jumps only take place at the end of a field to prevent a picture from two different fields being displayed.

The degree of accuracy required is much higher than in C-format because the tracks are much narrower.

In PAL, the track pitch is 35 μm, and eight tracks or four segments are required in one field. The heads must jump a distance of 0.28 mm to omit or repeat a field. In NTSC the track pitch is 39 μm, and six tracks or three segments are required in one field. The corresponding jump distance is 0.23 mm.

When the tape speed is close to three times normal, most of the time a two-field jump will be necessary at the end of every field. Figure 6.26 shows the resulting ramp deflection waveform (for one head pair only). Since the control track phase is random, an offset of up to ±½ field will be superimposed on the deflection, so that if a two-field jump must always be possible, a total deflection of three fields must be available. This determines the mechanical travel of the heads, which will be 0.84 mm for PAL, which is the larger of the two.

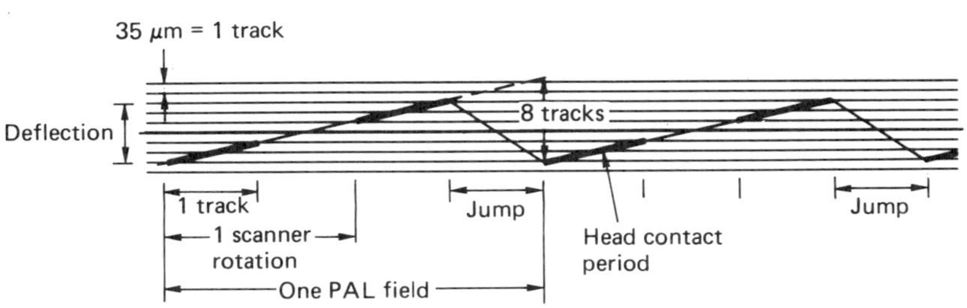

Figure 6.26 Head deflection waveform for PAL D-2 at +3× normal speed. This requires a jump of two fields between every field played, and this corresponds to 8 tracks. Since head pairs trace the tape alternately, the head has half a revolution in which to jump, so the actual jump is only 6 tracks. However, the vertical position of the above waveform is subject to an uncertainty of ±2 tracks because capstan phase is random during variable speed

6.20 The actuator

In Sony D-2 machines, the actuator used for head deflection is based on piezoelectric elements as used in numerous previous products. Figure 6.27 shows the construction of the Sony dynamic tracking head. A pair of parallel piezoelectric bimorphs is used, to ensure that head zenith is affected as little as possible by deflection. The basic principle of the actuator is that an applied voltage causes the barium titanate crystal to shrink. If two thin plates are bonded together to create a bimorph, the shrinkage of one of them will result in bending. A stale sandwich displays the same effect. Application of voltage to one or other of the elements allows deflection in either direction, although care must be taken to prevent reverse voltage being applied to an element since this will destroy the inherent electric field. Piezoelectric actuators display hysteresis, and some form of position feedback is necessary to allow a linear system. This is obtained by strain gauges which are attached to one of the bimorphs and which can be seen in the figure. When power is first applied to the transport, the actuator is supplied with a gradually decaying sinusoidal drive signal, which removes any unwanted set from the bimorph.

In Ampex D-2 machines, the piezoelectric actuators were not considered adequate for the larger travel demanded in a segmented format,[4] and a moving-coil actuator has been developed. This had been used experimentally in certain analog

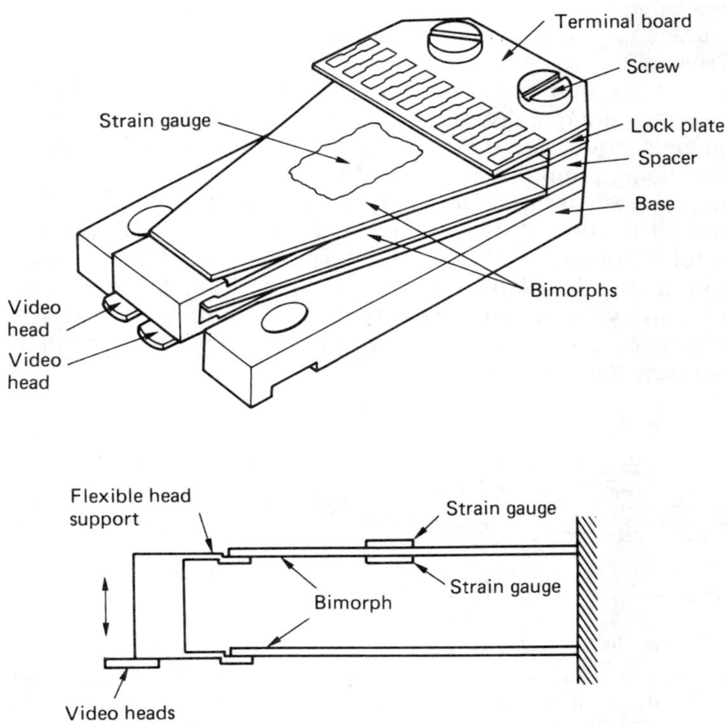

Figure 6.27 The dynamic tracking head of a Sony D-2 transport uses a pair of parallel bimorphs to maintain head zenith angle. Note use of strain gauges for feedback. (Courtesy Sony Broadcast)

recorders, but gave way to piezoelectric actuators in the C-format. Now that rare earth magnets are available, which offer high field strength with low mass, the moving-coil actuator becomes attractive again, because it allows a low-mass cantilever which has higher resonant frequencies and requires less force to deflect. The moving coil is inherently a low-impedance device requiring a current drive which is easier to provide than the high voltages needed by piezoelectric devices.

Figure 6.28(a) shows the concept of the moving-coil head and Figure 6.28(b) shows the appearance of the actual unit. The cantilever is folded from thin sheet metal which is perforated to assist the folding process. The resulting structure is basically a torsion box supported on a wide flexural pivot. This means that it can bend up and down, but it cannot twist, since twisting would introduce unwanted azimuth errors. The cantilever carries the actuator coil and is supported in a metal shoe which carries the magnet.

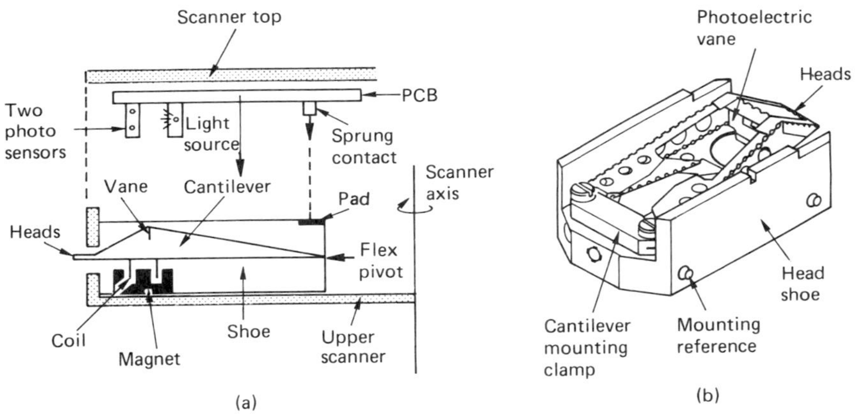

Figure 6.28 (a) Showing the concept of moving-coil head deflection. The cantilever is flexurally pivoted at the opposite end to the heads, and position feedback is obtained from a vane on the cantilever which differentially varies the light falling on two photosensors.
(b) Shows the appearance of an actual replaceable AST head assembly (courtesy AMPEX)

Positional feedback of the cantilever deflection is obtained by a photoelectric system. This has one light source and two sensors between which moves a blade which is integral with the cantilever. When the cantilever deflects, a differential signal results from the sensors. The photoelectric sensor is mounted inside the scanner top and automatically aligns with the moving blade when the top is fitted.

Whichever of these actuators is used, the combination of actuator, drive amplifier and position feedback gives a subassembly which will deflect the head in proportion to an applied voltage. It is then necessary to provide suitable drive signals to ensure that the deflection makes the head follow the track.

6.21 Detecting the tracking error

Track following is a means of actively controlling the relationship between the replay head and the track so that the track is traced more accurately than it would be by purely mechanical means.

This can be applied to systems operating at normal speed, in order to allow interchange in adverse conditions, but in D-2 track following is an option, and satisfactory interchange can be achieved without it.

The tracking error will be used on two different levels. First, the DC component of the tracking error will be used to set the average elevation of the head about which the ramp deflection will take place. Secondly, variations of tracking error during the track can be used to compensate for tracks which are not straight, owing to some relative misalignment between the machine which recorded the tape and the player. The tracking error detection systems differ between manufacturers and these differences will be noted.

Figure 6.29 shows three relationships of the head to the track, and corresponding signal output. The waveforms in Figure 6.29 correspond to the RF envelope of the Miller^2 channel code signal. Case (a) and case (c) display the same output, although the tracking error has the opposite sense. Simple processing of the RF level only gives the magnitude of the error, not the sense.

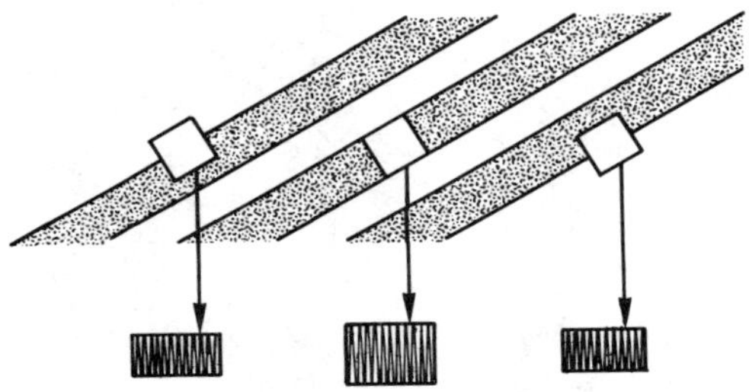

Figure 6.29 Effect of tracking error on playback signal. Signal amplitude in (a) and (c) is identical, despite sense of tracking error. Maximum signal occurs with correct alignment as in (b)

In order to extract the sense of the tracking error, it is necessary to move the tracking head to see if the error becomes greater or smaller. The process of manually tuning an AM radio is very similar.

In the Sony machines, the head elevation is changed by small steps between segments, and if the steps cause the average RF level to fall, the direction of the steps is reversed. This process is known colloquially as 'bump and look' and was originally developed for the Betacam products.

In the Ampex machines, the head is subject to a sinusoidal oscillation or dither, as in C-format machines. One field scan contains many cycles of dither. The effect on the RF envelope, as shown in Figure 6.30, is an amplitude modulation of the carrier, which has little effect on the video, owing to the insensitivity of the digital recording system to amplitude effects. Figure 6.30(a) shows that the effect of dither on a correctly aligned head is a frequency doubling in the RF envelope. Figure 6.30(b) and Figure 6.30(c) show the effect of the head off track. Both cases appear similar, but the phase of the envelope modulation is different and can be used to extract the sense of the tracking error.

Each cantilever carries two heads, one of each azimuth type, and the RF levels from the two heads are averaged together. This gives a 3 dB improvement in signal-to-noise ratio, as well as accommodating manufacturing tolerances in the

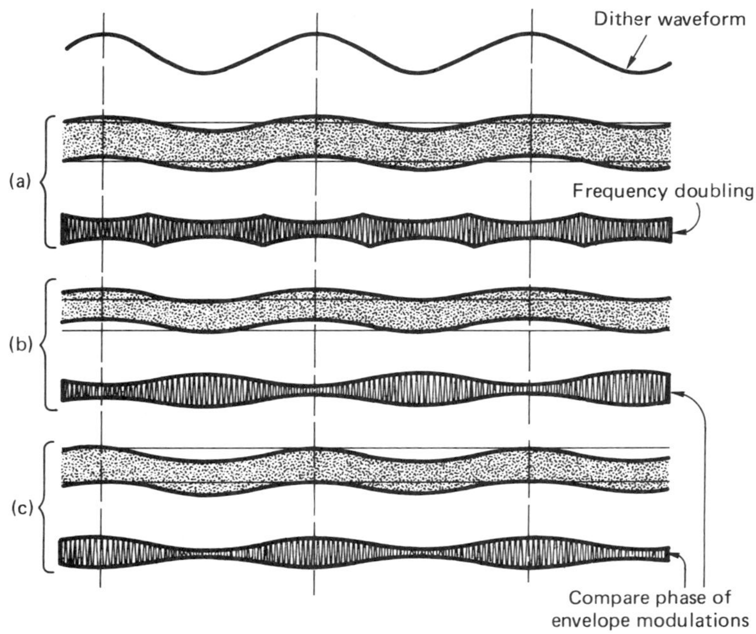

Figure 6.30 Top, dither waveform which causes head to oscillate across track. At (a) is optimum alignment, showing frequency doubling in RF envelope. With head above track centre, as in (b), RF amplitude increases as head reaches lowest point, whereas reverse applies in case (c)

elevation of the video tips. Only one detector or synchronous rectifier is then required per cantilever.

Dither cannot be used with analog VTRs employing azimuth recording because the transverse head motion interacts with the azimuth angle to give the effect of rising and falling head-to-tape speed. In FM recording, this introduces an unwanted signal into the video, which is why Sony developed the bump and look system for Betacam. The effect is still present in a digital recorder, but the result is that the instantaneous offtape data rate rises and falls slightly. This is accommodated by the phase-locked loop in the data separator and has no effect on the data. A harmless dither component will be observed in the VCO control voltage, and the speed variation in the discrete data is completely removed in the timebase correction.

In Figure 6.31 the RF is detected to obtain a level, which is fed to a phase-sensitive rectifier whose reference is the dither drive signal. The output of the phase-sensitive rectifier is a tracking error signal which contains both magnitude and sense and rather a lot of harmonics of dither. Careful choice of the dither frequency allows easier cancelling of the dither harmonics. In the VPR-300, there are five and a half cycles of dither per segment, or 11 cycles per revolution. Both heads are dithered by the same waveform, which means that the dither on odd segments will be phase reversed with respect to even segments. A dither frequency of 1100 Hz results in PAL, whereas in NTSC it is 990/1.001 Hz.

The tracking error is averaged over a pair of segments to cancel the harmonics and to produce an elevation error. The tracking error will also be sampled at

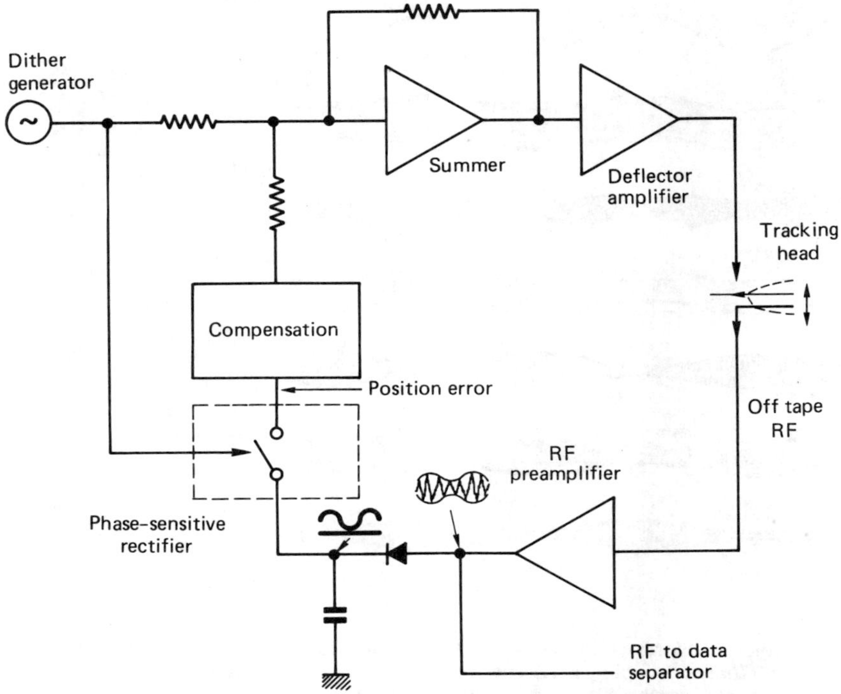

Figure 6.31 The tracking error is extracted from the RF envelope by a phase-sensitive rectifier

several points down the track to see if there is a consistent curvature in the tracking. This can be reduced by adding a correction curve to the deflection waveform which will adjust itself until the best envelope is obtained over the whole track length.

A different approach to measuring track curvature is used in Sony transports. As has been stated, deflecting a head which uses azimuth recording causes timing changes. If a pair of heads of opposite azimuth are tracing the appropriate tracks, tracking error will result in differential timing changes. This can be measured by comparing the time at which sync patterns are detected in the two channels. Figure

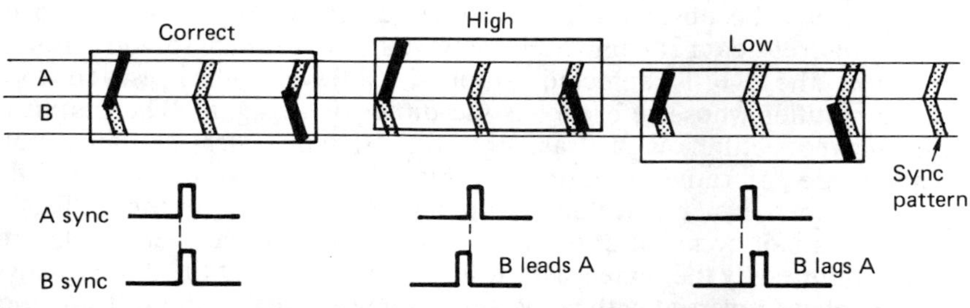

Figure 6.32 In the Sony D-2 DT system the tracking error is detected by comparing the times at which sync patterns are detected by the two heads on each arm. Owing to the use of azimuth recording a component of transverse movement results in a movement of the effective position of the head along the track

6.32 illustrates the principle. This system will detect tracking errors due to track distortion. Unfortunately, manufacturing tolerances in the physical stagger between the heads on both the machine which made the recording and the player will combine to put a permanent offset in the tracking error derived using this principle. It is necessary to calibrate the system each time it is used. The bump and look system maximizes the RF level at the beginning of the track, and the control system then inserts a changing offset into the sync timing comparator until it too gives zero tracking error at the beginning of the track. The offset is then held at that value. This process removes the static deflection and allows the tracking error to be used to deflect the heads within the track.

6.22 The ramp generator

The positional feedback generated by the observation of the dithered RF envelope or the sync pattern phase is not sufficiently accurate to follow the tracks unassisted except at normal speed where it can be used as an interchange aid. In variable speed the deflection of the head is predicted to produce a feedforward signal which

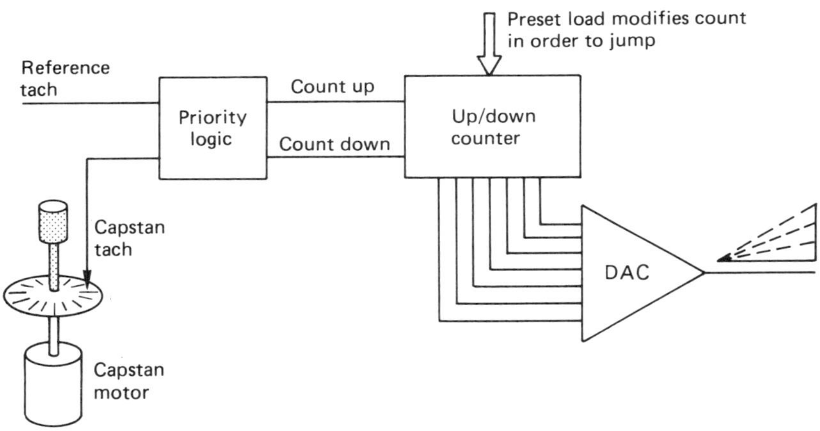

Figure 6.33 An up/down counter is fed with pulses from the capstan tach and from a reference whose frequency is precisely that of capstan tach at normal speed. The counter will thus integrate speed deviation from units, and the DAC will produce a range whose slope is proportional to speed difference. This is the feedforward signal which drives the head deflection system

adds to the head deflection. The feedback then only has to correct for the difference between the feedforward and the actuality. When the tape travels at the wrong speed, the track angle changes, and so the head needs to deflect by a distance proportional to the angle it has rotated in order to follow the track. The deflection signal will be in the form of a ramp, which becomes steeper as the tape speed deviates more from normal. The actual speed of the capstan can be used to generate the slope of the ramp. One possible implementation of this is shown in Figure 6.33. A counter has count-up and count-down inputs. The first of these is driven by the capstan tacho disk, and the second is driven by a clock whose frequency is exactly equal to that generated by the capstan tacho at normal speed. When the capstan runs at normal speed, the counter receives as many up clocks as down clocks, and remains unchanged. However, if the capstan speed is raised, the counter will begin to count up at a rate proportional to the speed difference. The counter drives a DAC which produces the deflection ramp. The RF amplitude detector will generate pulses which are also fed into the counter so that the elevation correction can be made. It is also possible to predict the average or static deflection of the heads from the control track phase. The capstan frequency generator can be used to measure control track phase. The capstan FG wheel will be designed so that a known number of FG pulses is generated when the capstan transports the tape by the distance between control track segment pulses. The FG pulses are fed to a counter which is reset by segment pulses from the control track head. The counter then produces control track phase which can be used to establish head elevation.

The same deflection ramp slope will be fed to both head pairs so that they will read alternate segments for an entire field. At some point it will be necessary to jump the heads to reduce the deflection, and this requires some care. The heads are 180° opposed and contact the tape alternately. The jump takes place while the head is out of contact with the tape. The two head pairs will have to jump at different times, half a scanner revolution apart. Clearly if one head jumps, the second head must also jump, otherwise the resulting picture will have come from two fields. As the jumps are half a revolution apart, it follows that the decision to jump must be made half a scanner revolution *before* the end of the current field. The decision is made by extrapolating the deflection ramp forwards to see what the deflection *will be* when the end of the field is reached. If the deflection will exceed half a field, it can be reduced by jumping one field. If the deflection will exceed one field, it can be reduced by jumping two fields. The jump can be executed by adding or subtracting a number of pulses in the ramp counter.

Tape tension changes can cause the track width to vary minutely. This is normally of no consequence, but when taken over all of the tracks in a segment the error may be significant. It is possible to compare the tracking error before and after a jump, and if the error becomes greater, the jump distance was inappropriate for the tape being played. It is possible to modify the distance jumped simply by changing the number of pulses fed to the ramp counter during the jump command. In this way the jump distance can optimize itself for the tape being played.

Figure 6.34(a) shows the ramping action for the two moving heads in PAL with a tape speed of +2×. It will be seen that when one head jumps, the other one will also jump, so both heads always play the same field. The timing shift due to the heads being separated by half a revolution can be seen. As there are four segments per field in PAL, corresponding to two scanner revolutions, all fields will contain the same sequence. Figure 6.34(b) shows the equivalent sequence of events for NTSC. As there are three segments per field and two moving heads, the action differs between odd fields and even fields. During one field, a given head may jump twice, but in the next field it will not jump at all.

6.23 Track-following block diagram

Figure 6.35 shows an Ampex D-2 track-following system. There is an inner feedback loop which consists of the head actuator, the position feedback sensor and

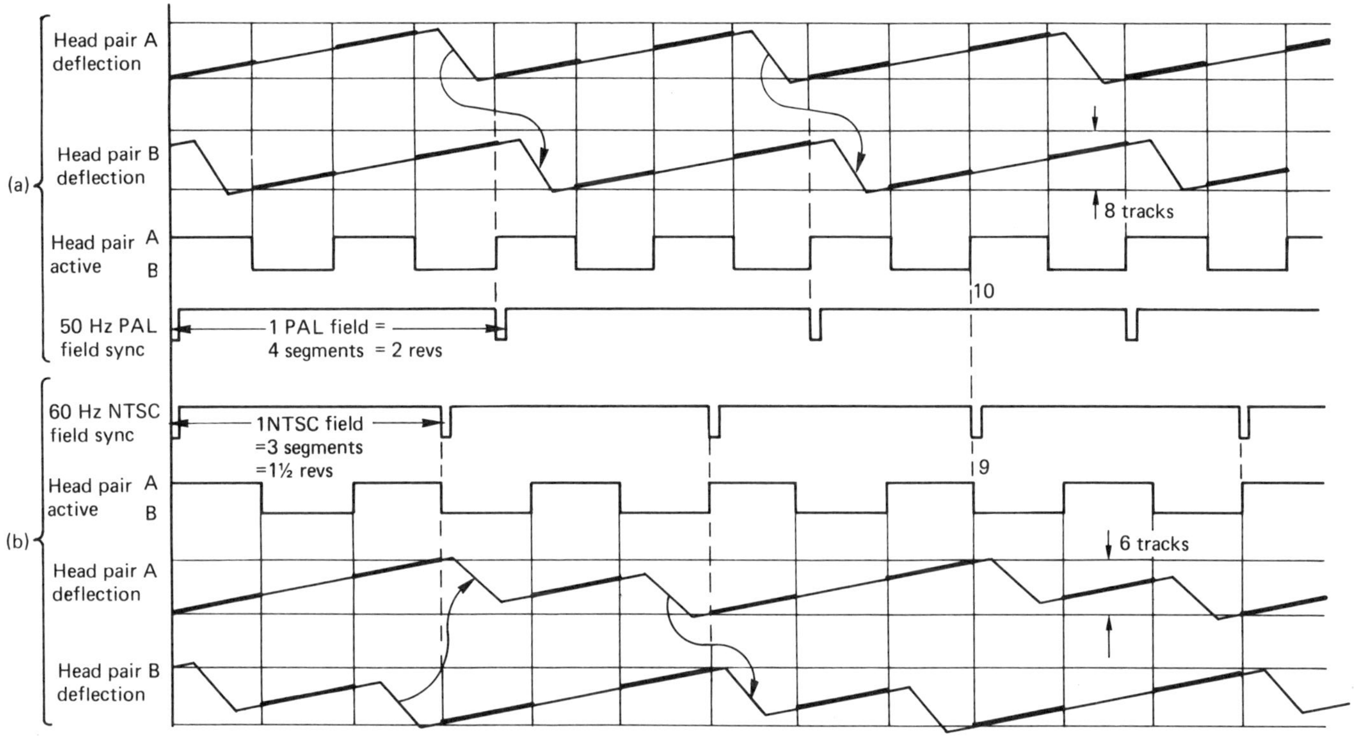

Figure 6.34 The timing and deflections necessary at a speed of +2× in D-2 for (a) PAL and (b) NTSC. Note the segment ratio of 9:10 between the two formats, which is also the ratio of scanner speeds. In order to play entire fields, jumps must take place about the field boundary, so that if one head pair jumps, the other must follow (serpentine arrow). In NTSC there are three segments per field, so the jumping follows a two-field sequence, where the head fails to jump in one field, but jumps twice in the next

an amplifier. This loop is a position servo which makes the head deflection proportional to the input voltage. Like any position servo it is provided with compensation to prevent oscillation and maximize frequency response. The inner loop is driven by the sum of four signals. These are:

(a) The dither signal, which is a sinusoid designed to wobble the head either side of its average elevation.

(b) The elevation correction obtained by averaging the tracking error over the entire track.

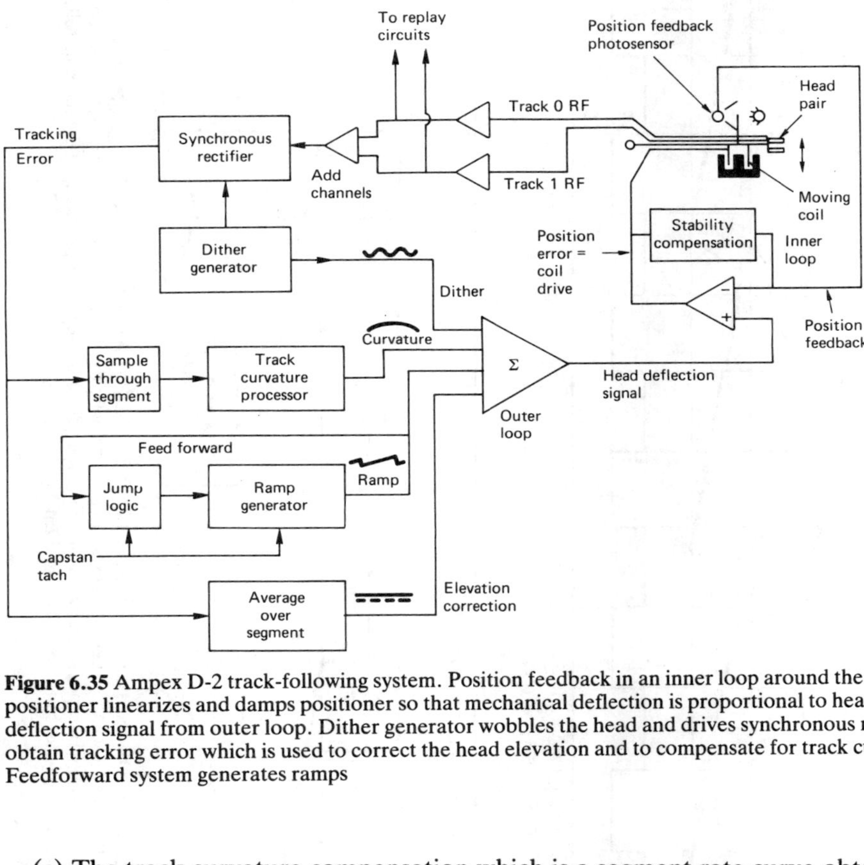

Figure 6.35 Ampex D-2 track-following system. Position feedback in an inner loop around the head positioner linearizes and damps positioner so that mechanical deflection is proportional to head deflection signal from outer loop. Dither generator wobbles the head and drives synchronous receiver to obtain tracking error which is used to correct the head elevation and to compensate for track curvature. Feedforward system generates ramps

(c) The track curvature compensation which is a segment rate curve obtained by sampling the tracking error at several points along the track.

(d) The feedforward ramp and jump signal. The slope is obtained from capstan speed difference and the jumps are initiated by analysing the deflection.

Figure 6.36 shows the Sony D-2 track-following system. The capstan FG and control track are used together for elevation prediction, and capstan FG is used alone for ramp slope prediction. The bump and look system is controlled by the RF detector, which is also used to calibrate the differential sync pattern timing detector.

In both types of machine there are two moving heads, and so two of these systems are necessary. They are largely independent except for a common jump control system which ensures that when one head jumps during the last segment of a field the other head will follow once it has played that segment.

When a head jumps, it will move to a track which began at a different place along the tape, and so the timing in that track will not be the same as in the track which the head left. In a segmented format, the effect is magnified by the number of

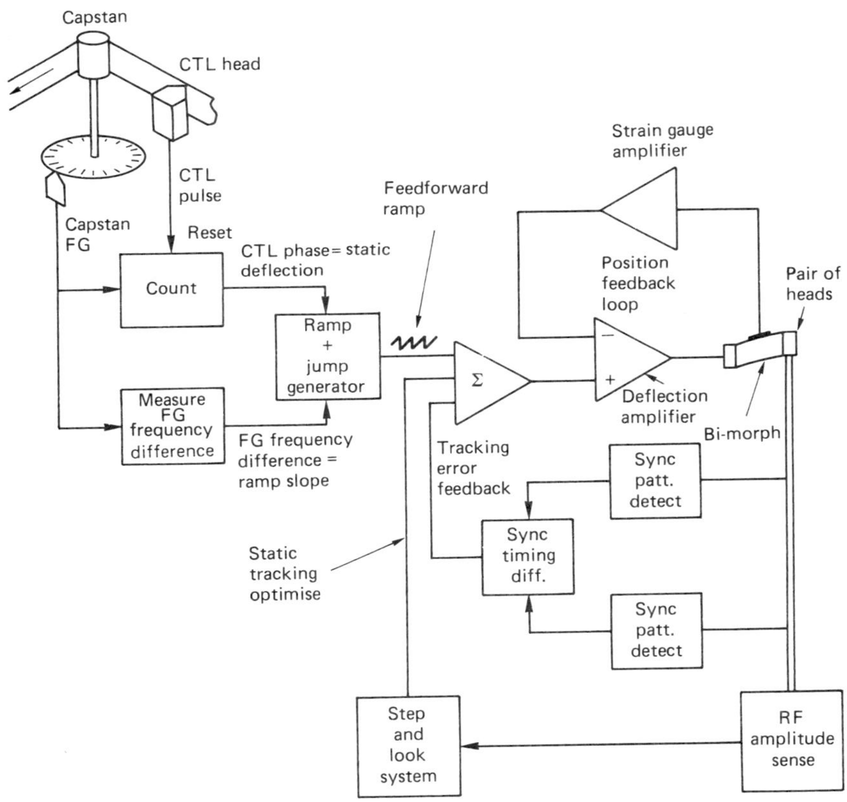

Figure 6.36 The DT system used by Sony D-2 machines combines ramp feedforward from the capstan speed and control track with tracking feedback from the differential sync detection of Figure 6.32. In addition the bump and look system slightly offsets the heads by a different amount each scan and assesses the RF amplitude to find an optimum static deflection. This is really a slow dither process. Note that this is a conceptual diagram. In practice many of these stages are carried out by microprocessor

tracks in a field. To support a speed range from $-1\times$ to $+3\times$, it has been seen that an overall head travel of three fields is necessary. From the start of a given segment to the start of the same segment three fields away is nearly 8 mm in PAL, and over 6.5 mm in NTSC. The replay circuitry must be able to accommodate the timing uncertainty which amounts to about 5% of the segment period. Clearly the timing error can be eliminated by the timebase correction processes within the playback system.

References

1. BRUSH, R. Design considerations for the D-2 PAL composite DVTR. *IERE Conf. Publ.*, no. 79, 141–148 (1988)
2. BRUSH, R. Design considerations for the D-2 NTSC composite DVTR. Presented at SMPTE Technical Conference (Los Angeles, 1987)
3. RYAN, D. M. Mechanical considerations in the design of a digital video cassette recorder. Presented at International Broadcasting Convention (Brighton, 1988), *IEE Conf. Publ.*, no. 293, 387–390 (1988)
4. OLDERSHAW, R. Design of an automatic scan tracking system for a D-2 recorder. *IEE Conf. Publ.*, no. 293, 395–398 (1988)

Chapter 7

The D-2 signal system

The signal system of a D-2 recorder is responsible for producing suitably encoded record waveforms at the correct time so that the format definition is met. On replay, the process will be reversed so that a standard composite waveform which is colour framed to reference results. This chapter examines the circuit techniques necessary to implement the format.

7.1 Introduction

Although the recording technique used in D-2 is digital, it is vital to remember at all times that the data which D-2 records are no more than a numerical representation of a composite video waveform. All of the characteristics of composite video remain, including the requirement for a colour framing system to synchronize the four- or eight-field sequences embedded in the subcarrier.

No description of a D-2 VTR can be complete without a discussion of the techniques necessary to handle colour framing and the intimately related ScH phase. There are considerable differences between the subcarriers of PAL and NTSC, and so there will be corresponding differences between the approaches to colour framing and between the data rates which result from digitizing the two standards. As a result there are really two D-2 formats which leave different tape footprints, but which nevertheless have a lot in common in order to use the same circuitry and mechanical parts as much as possible. There is also a PAL-M version, which as far as the tape format is concerned is essentially NTSC with an eight-field control track. Since colour framing has such a far-reaching effect on composite recording, it will be necessary to treat the subject first so that the implications for subsequent topics become clear.

7.2 Colour framing in PAL

In PAL, the eight-field sequence comes about because of the quarter-cycle offset of subcarrier against line rate. There are 283¾ cycles of subcarrier in one line, which means that four lines must pass before the same relationship occurs between subcarrier and horizontal sync. The odd number of lines in a frame necessary for interlace mean that four frames must elapse before the four-line sequence assumes the same relationship with vertical. Unlike NTSC which has a half-cycle offset,

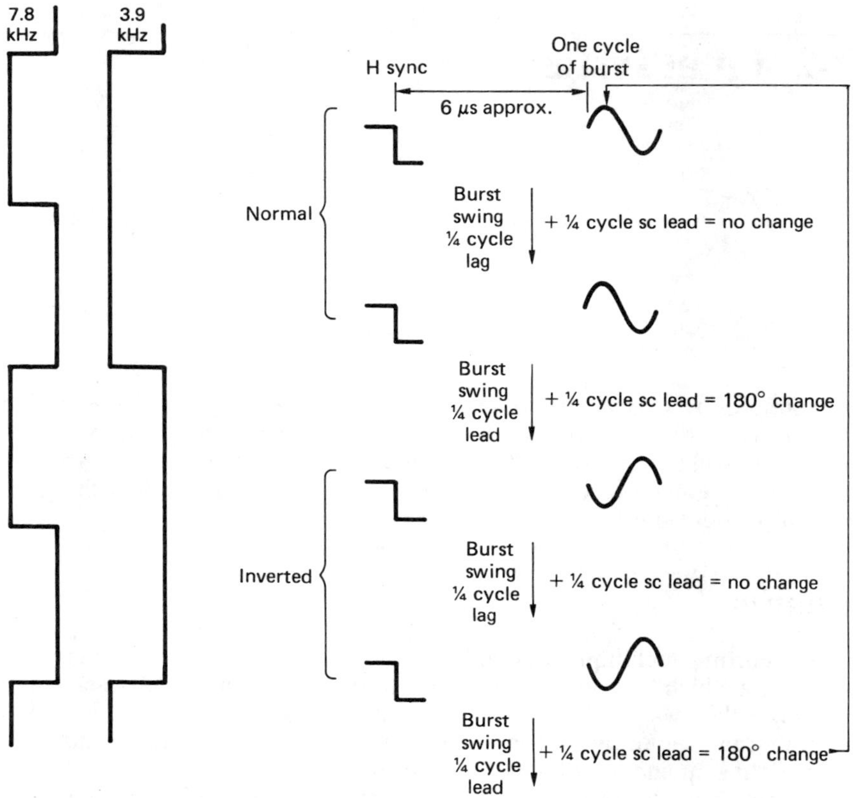

Figure 7.1 Quarter-cycle subcarrier/H relationship and burst swing combine to give four-line sequence

PAL does not naturally have good cancellation of subcarrier on the screen. This is achieved by the addition of 25 Hz to the subcarrier frequency, which causes subcarrier on a given line to be anti-phase with subcarrier on the line in the next field which will be physically adjacent on the interlaced display. Although this 25 Hz component achieves its goal, it also complicates the colour framer because instead of having four different relationships of subcarrier to horizontal there are now 2500!

Since D-2 samples using burst as a phase reference, it is necessary to follow what the burst does with respect to horizontal. Owing to the quarter-cycle offset, subcarrier phase advances by 90° from line to line, but the burst swings forward 90° on one line and back 90° on the next to convey the state of *V*-switch. This means that the burst swing will alternately cancel or double the subcarrier phase advance, giving two lines with the same burst to sync relationship followed by two lines with an inverted burst to sync relationship. Figure 7.1 shows that these four lines repeat at 3.9 kHz and contain two complete cycles of *V*-switch which runs at 7.8 kHz. Once every four frames, line one of the frame will be the first of the four line types, and this fact will be indicated by the colour framing flag on the control track. It is the job of the record colour framer to determine when to produce that colour framing flag. Figure 7.2(a) shows the record colour framing process.

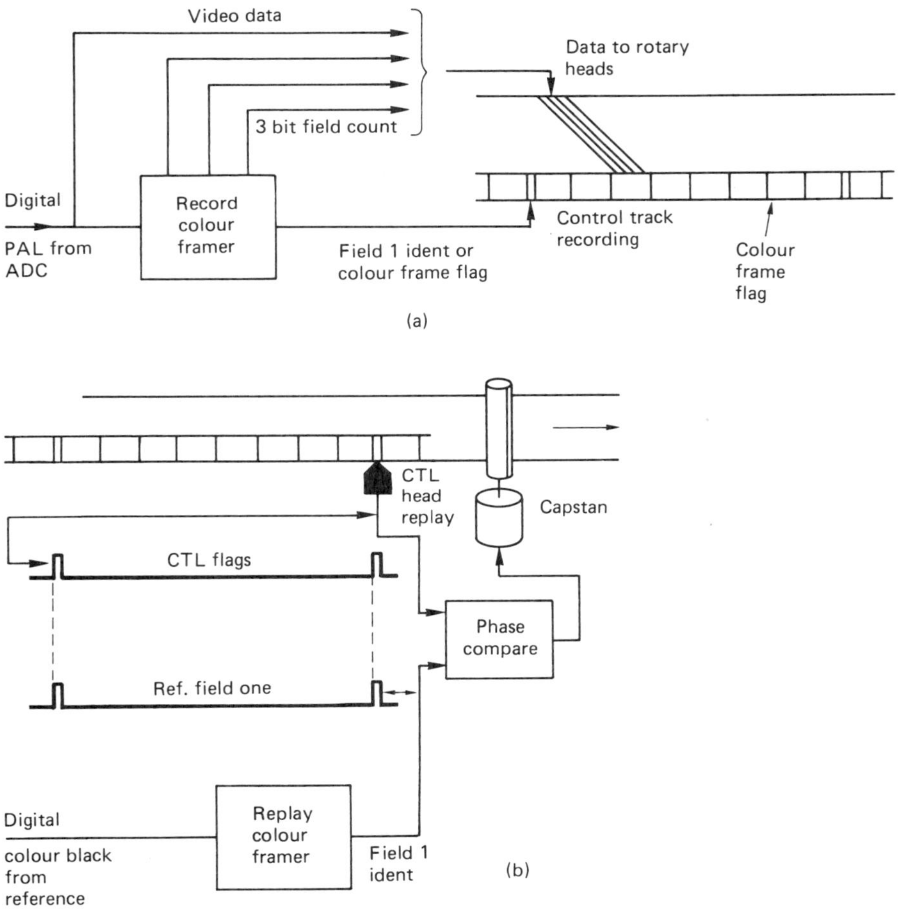

Figure 7.2 At (a) the presence of field 1 on the input is denoted by a colour frame flag on the control track. The field sequence is also written in the rotary tracks.
At (b), on replay the capstan speed will be changed until offtape colour frame flags coincide with reference field 1 identification. Replay video will then be colour framed to reference

On replay, a second colour framer examines the reference input to determine when the start of the colour framing sequence occurs. Figure 7.2(b) shows that the phase difference between the reference colour frame flag and the replay colour frame pulse from the control track controls the capstan speed until the pulse and the flag coincide. When this is achieved, the tape playback is said to be colour framed, and the lines from the tape will be of exactly the same type as the lines in the reference signal. The offtape samples can be fed directly to the machine output. In all other replay modes, a colour processor will be necessary to convert the offtape line type to the reference line type. Since the control track phase is meaningless at anything other than the correct speed, D-2 machines record identification codes on the helical tracks which uniquely locate a segment within the eight-field sequence. In some machines there is only one colour framing circuit, and it is simply switched to the input or the reference according to whether the machine is recording or playing back.

As was seen in Chapters 2 and 3, the sampling clock of PAL D-2 uses burst as a phase reference. The colour framer must use sample values obtained by digitizing the leading edge of sync with a burst phased sampling clock.

The first stage in colour framing is to determine the phase of burst swing. This process is integral with the operation of the sampling clock phase-locked loop. The phase-locked loop includes the video ADC so that chroma phase errors are minimized. The ADC runs constantly and digitizes the entire video signal. Control of the VCO is achieved by examining the numerical value of samples taken during the burst.

Figure 7.3 shows one possible implementation. Owing to PAL burst swing, burst phase can have one of two values, 90° apart, so both of them will have the same phase relationship to $4 \times F_{sc}$. A two-phase divider running from the VCO produces a pair of quadrature square waves at subcarrier frequency. These are used on alternate lines to synchronize the numerical processing of sampled burst. In the first quadrant, A, there should be a positive burst crossing and blanking level, or 64 is subtracted from the sample value to give a phase error measurement. In the next quadrant, B, a positive burst value above 64 should be obtained. In the third quadrant, C, there should be a negative burst crossing, so 64 will be subtracted from the sample value and the result inverted to give a phase error value. The phase error values for several cycles of burst are numerically averaged and fed to a DAC which produces the VCO control voltage. Upon first applying an input, the alternate selection of the quadrature control signals will have a random phase which may be right or wrong. If by chance it is correct, the above sequence of events will hold true for all lines. However, if the selection is in the wrong phase,

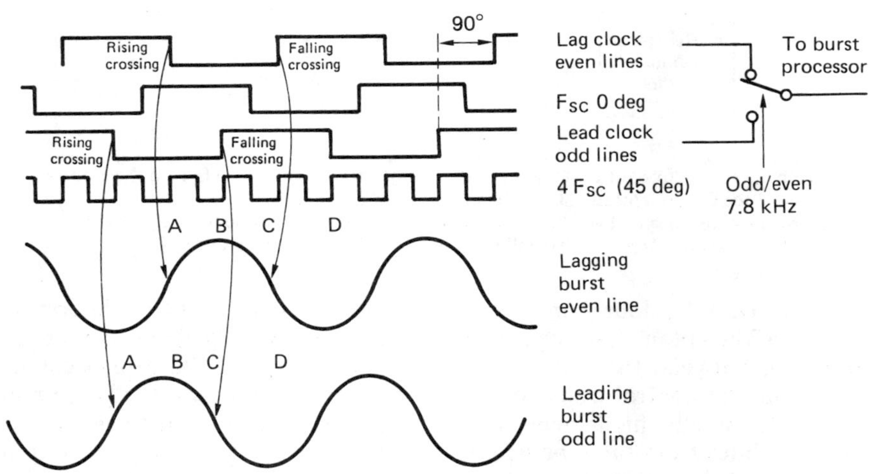

Figure 7.3 Two quadrature clocks are derived from 4 F_{sc} whose crossings are at 45° phase in PAL. These two clocks are used on alternate lines, where they predict the position and direction of a burst crossing. If it is a downward crossing, the phase error needs to be inverted. If the odd/even sequence gets out of step, the clock moves forward a quadrant when an incoming burst moves back a quadrant, so the twist is perceived as inverted. The phase error alternates at half line rate and can be used to reset the 7.8 kHz divider

the timing will lag by a quadrant when burst leads by a quadrant and vice versa. This will make the burst invert with respect to the measurement timing on alternate lines, and a negative peak value below 64 will be obtained instead of a positive peak value above 64 in quadrant B. This condition is easily detected and is used to reset the phase of the line frequency divider. This divider will now be truly reflecting the input burst swing, and an accurate phase comparison can be made between burst and sampling clock on every line. The line frequency divider is now producing 7.8 kHz.

By numerically analysing the ADC output, it is easy to detect the sync pulses, and using programmable counters it is possible to detect both vertical and horizontal syncs from their duration. In this way a gate can be opened during the first line of each frame. During the gate, the state of 7.8 kHz is examined; this will only be true on alternate frames and allows four-field colour framing to be achieved. The burst swing (7.8 kHz) derived colour framing cannot tell whether it has found field 1 or field 5, and this can only be resolved by looking at the 3.9 kHz relationship between burst and sync.

The problems involved in determining the colour frame are summarized below, prior to the somewhat involved solutions.

Since sampling is subcarrier locked, it is necessary to consider how sync pulses move with respect to subcarrier rather than the other way round. The quarter-cycle offset means that the position of a sync edge will move forwards by one quadrant of subcarrier per line.

The presence of the 25 Hz offset causes sync edges to slide back constantly with respect to the sampling clock by four cycles or 225 ns per frame. The absolute position of the sync edge depends on the ScH phase of the input signal which can drift.

Assessment of ScH phase according to the definition can only be made once every eight fields, and reliance on a single measurement is prone to noise. Techniques must be found to allow a colour framer to make a continuous assessment of framing so that a faster, more reliable lock is achieved.

Figure 7.4(a) shows how the phase of sync with respect to subcarrier can be found for any line in the eight-field sequence. The presence of the 25 Hz component is responsible for the diagonal slippage. Since phase only has meaning with respect to a cycle, it is possible to redraw Figure 7.4(a) to produce Figure 7.4(b). The timing of sync with respect to subcarrier now takes on the appearance of a frame-rate sawtooth. The sync edge has a finite slope, and so several samples will be taken during the transition. Figure 7.4(c) shows how the phase of the instant when the 50% level of sync is crossed can be computed to a fraction of a subcarrier cycle. Since this position moves by a fixed amount per line (0.36 ns) it is possible to subtract a variable time which is obtained by multiplying the line number in the frame by 0.36 ns. This process is called horizontal modulation and results in a modulated sync edge that appears to have four phases which are stationary with respect to subcarrier. The four phases are due to the quarter-cycle offset. A further offset of one sample period is subtracted on alternate lines according to the state of 7.8 kHz, giving modulated sync edges which now have two phases. During the colour frame, the sync edge will be in phase with subcarrier, but during frame 3 it will be anti-phase. The colour frame is thus uniquely identified. If the input is not exactly EBU ScH phase, there will be a fixed phase error in the sync timing, and the system must pick the sync edge which is most nearly in phase with subcarrier to identify frame 1.

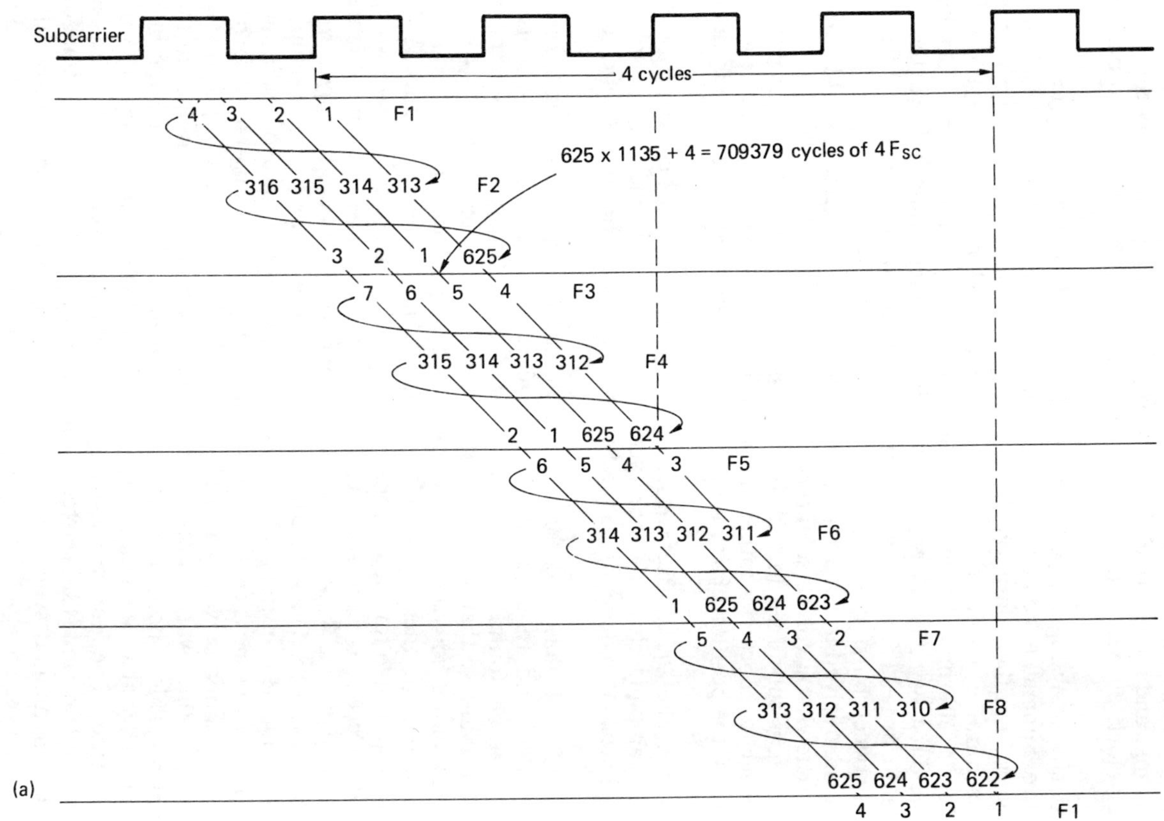

Figure 7.4(a) The position of sync with respect to subcarrier can be found for any line in an eight-field sequence using this chart. The quarter-cycle offset causes sync to move forward one quadrant per line over a four-line sequence, but the addition of 25 Hz causes sync to slip back towards subcarrier by one cycle per frame

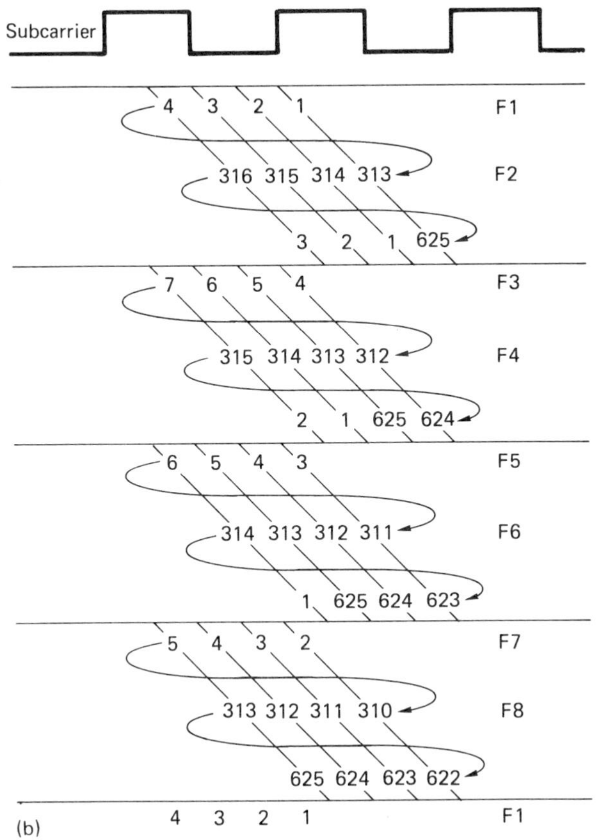

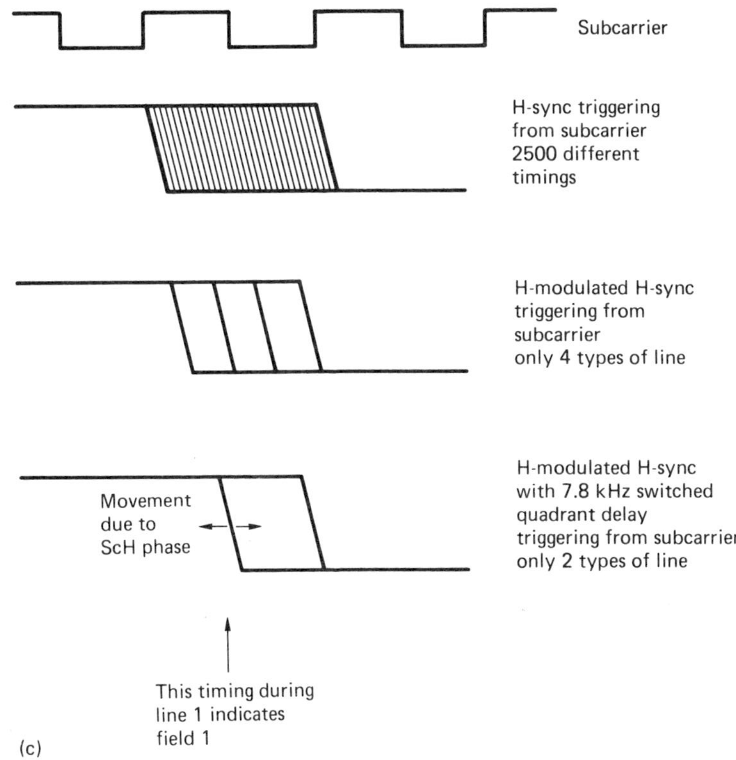

Figure 7.4(b) Since all cycles of subcarrier are identical the drift of sync with respect to subcarrier shown in (a) can be opposed by inserting an extra cycle of subcarrier or four sample periods every frame. The sync timing now slips one cycle per frame and jumps back. There will be $1135 \times 625 + 4 = 709\,379$ sample periods in a frame exactly. By modulating the position of sync by 0.36 ns per line, it can be made to have a constant relationship to subcarrier over a four-line period. (c) Horizontal modulation gives four types of line by cancelling the effect of 25 Hz. Using 7.8 kHz from Figure 7.3 to insert or remove delays of one sample period results in two relationships. Sampling during line 1 will determine whether field 1 or field 5 is present

7.3 Colour framing in NTSC

The four-field sequence of NTSC arises as a result of the half-cycle offset of subcarrier with respect to line frequency. The phase-locked loop which produces the sampling clock synchronizes to burst, but with a 57° offset so that samples are taken on the *I* and *Q* axes. The standard ScH phase relationship will occur on the first line of the four-field sequence and on every other line. On the lines between, the sync edge will be located two sample periods away. These lines are easily detected and so the colour frame flag is readily produced.

7.4 The data to be recorded

It has been seen that, in both line standards, the sampling clock is precisely derived from subcarrier. The very existence of the samples indicates the phase of subcarrier, and so it is not necessary to record the burst. The subcarrier has a mathematical relationship with sync, which can readily be reconstructed on replay, and so it is not necessary to record syncs either.

Accordingly, D-2 only records the active line, along with sufficient positioning information so that syncs and burst can be regenerated on replay. As a result, if a non-standard ScH signal is fed into a D-2 recorder, standard ScH will be automatically generated on playback.

PAL D-2 records 948 samples symmetrically disposed around the unblanked active line, and NTSC records 768 samples similarly. In each case, sample zero is defined as the first sample to be recorded in each line.

Since not all of the line period is recorded, it then becomes important to ensure that sampling commences in a consistent place on each line of each frame and that the first sample has a consistent relationship to subcarrier. Clearly the position of the first sample in a line must be on one cycle or another of the sampling clock, and, as has been seen, this is derived from input burst. The edge of the picture is, however, defined by the sync pulse leading edge, and the input stage has to resolve the conflict of sampling at a fixed relationship to H-sync in order to keep the edge of the picture straight with the need to sample coherently with subcarrier. In NTSC this is relatively easy, owing to the simple relationship between subcarrier and sync. It is only necessary to wait a fixed number of sample periods after sync, and the first sample of the active line will be consistently located.

In the PAL system it becomes more complex because of the 25 Hz component of subcarrier. In the absence of 25 Hz, there would be exactly 283¾ cycles per line and exactly 1135 sample clocks per line. Instead, there will be an additional four sample clocks in one frame period.

The H-modulator in the colour framer allows the ScH phase of the input to be established, and this can be used to ensure that the first sample period of line 1 of the eight-field sequence is identified. From then on, the beginning of subsequent lines is deemed to occur 1135 sample periods later. This means that, at the end of the frame, the virtual beginning of a line will be a whole cycle of subcarrier early. During the vertical interval, the line count of one line is modified to 1139 and the error is removed for the beginning of the next frame. The first sample of the active line is obtained by counting a whole number of cycles of sampling clock from the virtual beginning of the line.

Using the above mechanisms, samples can be put into memory in a consistent way. Because H-sync was displaced to align it with subcarrier, the picture on tape will be slightly rhombic, but this is of no consequence because identical circuitry, including an H-modulator, is necessary on the replay side of the machine to generate a read clock from reference subcarrier and to generate a subcarrier coherent read start pulse from reference sync. The picture distortion caused by H-modulating the input sync is cancelled by H-modulating the output sync.

In either standard, if the ScH phase of the input changes, the sync pulse will move with respect to the sampling structure, but the system will continue to select the same cycle of subcarrier with which to start the sampling by adapting the measured ScH phase value. If subcarrier phase drifts so much that it inverts, the colour framer cannot reframe if the machine is recording, since this would result in a discontinuous control track. It will be necessary to stay with the same subcarrier cycle which results in a drift of the picture position. If drift continues for an entire cycle, the input stage will need to dump four samples to prevent the picture winding across the screen. This will cause a picture shift.

Not all lines of the frame are recorded, and it will be seen from Figure 7.5 that the picture lines recorded differ from field to field. This is necessary to allow picture in shuttle.[1] The recording is composite and contains an embedded chroma signal. Colour processing is necessary in shuttle to change the phase of the offtape signal to that of reference. When tracks are crossed in shuttle, sync blocks are picked up at random, and a framestore is updated by information which has come from several different fields. This is only possible if samples offtape can be colour processed. The necessary interleave and shuffle processes in a DVTR mean that each sync block contains samples which have come from various places within a field segment. When a framestore is updated in shuttle, successive samples in a given line could have come from a different colour line type, and they would no longer represent the modulation of a continuous subcarrier. The colour processor would not be able to operate.

The solution adopted in D-2 is to record more lines than necessary, so that the first line of each field recorded can change over a four- or eight-field sequence. The choice of the first line to be recorded is made by the input colour framer. This ensures that all fields on tape have the same chroma phase at the same line number. At normal speed and in slow motion, the recorded lines are put back into the correct position in the reference fields and the normal chroma sequences result, but in shuttle the colour processor now only needs to know the line number of recovered pixels and they can be correctly processed.

The offset of the line start causes a loss of vertical resolution in shuttle (four lines in PAL, two lines in NTSC), but this is acceptable since resolution falls in shuttle anyway due to movement in the picture.

The raw video data rate which must be sustained can now be derived.

For PAL, 948 samples in 304 lines at 50 Hz results in 14 409 600 samples per second, whereas for NTSC, 768 samples in 255 lines at 59.94 Hz results in 11 738 650 samples per second. Each sample is of 8 bit resolution.

In practice, these figures will need to be increased somewhat to allow for an error-correction overhead of about 15%, the presence of addressing or identification codes and for the double recording of four audio channels. Additionally, a proportion of each track will be used up with the preambles and sync patterns necessary to the operation of the channel coding, and tolerance gaps needed to allow independent editing of audio and video.

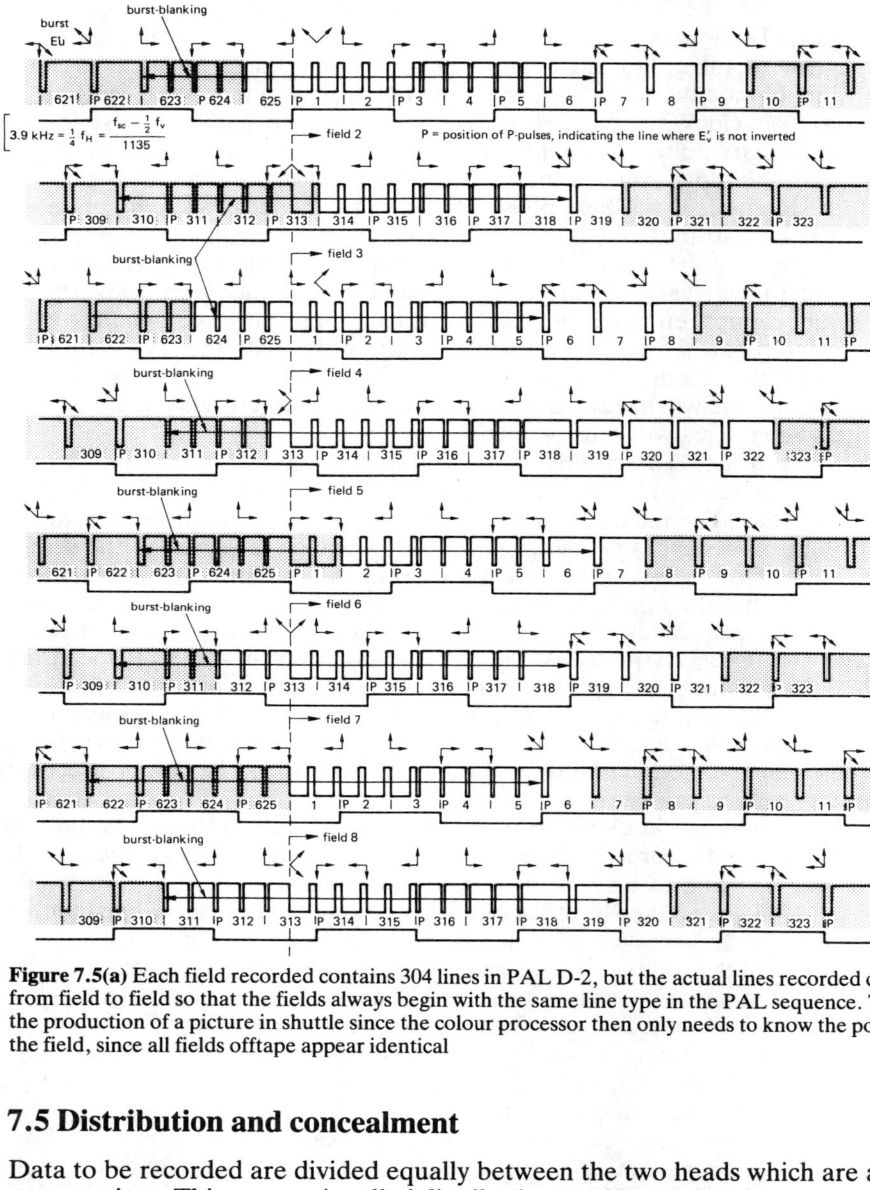

Figure 7.5(a) Each field recorded contains 304 lines in PAL D-2, but the actual lines recorded change from field to field so that the fields always begin with the same line type in the PAL sequence. This eases the production of a picture in shuttle since the colour processor then only needs to know the position in the field, since all fields offtape appear identical

7.5 Distribution and concealment

Data to be recorded are divided equally between the two heads which are active at any one time. This process is called distribution, and is necessary because one head could not sustain the necessary data rate alone.

If the playback process could be guaranteed to be error free, the distribution scheme could be arbitrary, but in a practical machine it is necessary to take into account the fact that the error-correction system may be overwhelmed, albeit infrequently, by corruptions greater than those it was designed to handle. In this case the lost samples can be approximated in a concealment process, the success of which is partly determined by the distribution strategy.[2,3]

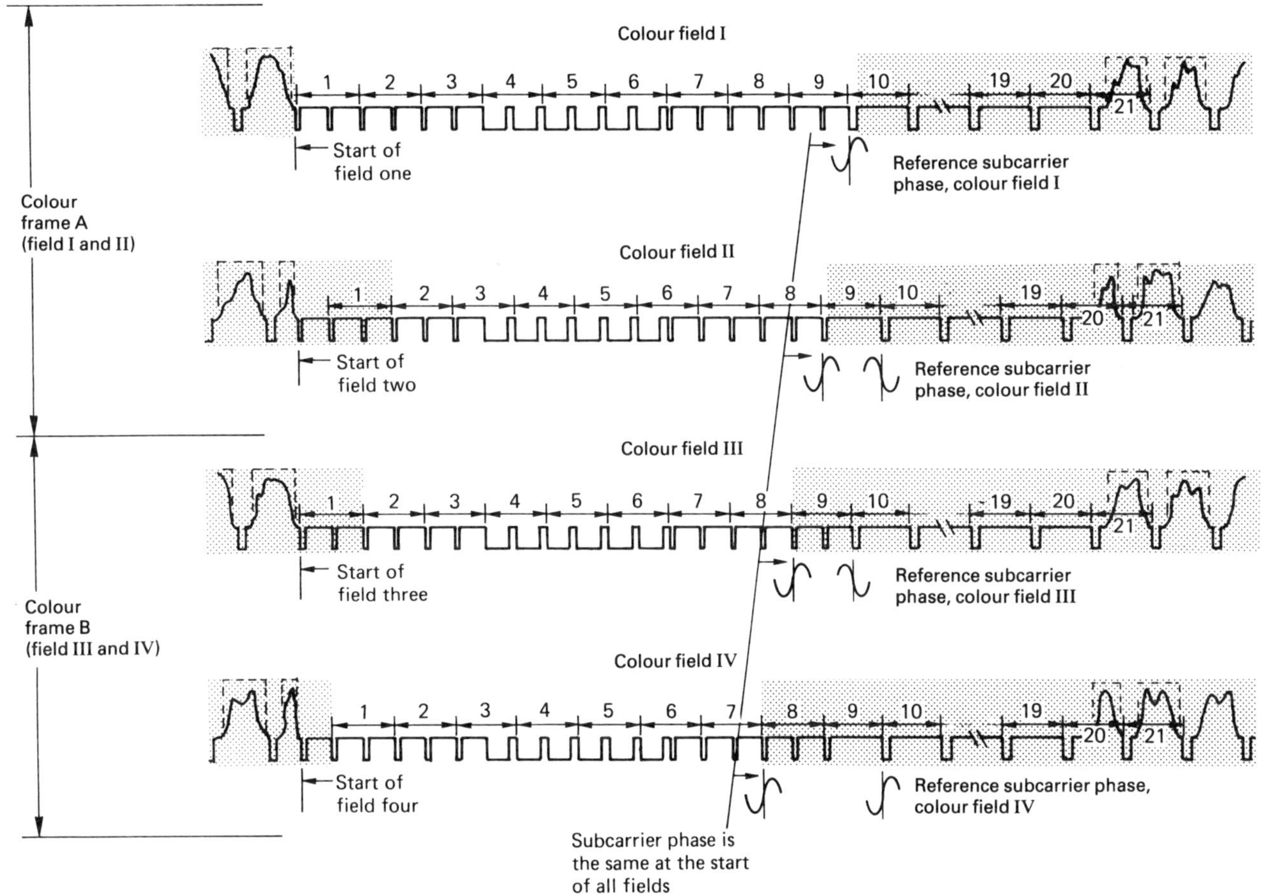

Figure 7.5(b) The recorded lines in NTSC D-2 are shown shaded here. The starting point in each field is chosen so that the subcarrier phase is the same at the beginning of all fields on tape

Figure 7.6 shows the concept of distribution, which ensures that in both rows and columns alternate samples are recorded by a different head. If samples from a given head are lost for any reason, each lost sample is surrounded on four sides by samples from the other head. In the case of complete loss of signal from one head, due perhaps to clogging, it is still possible to produce a picture of reasonable quality from the remaining head because the quality will be as if the picture had originally been sampled at $2\times F_{sc}$. The bandwidth available with one head out of action could, with suitable concealment circuitry, be greater than is available in slow motion in most C-format recorders.

As the recording is segmented, care must be taken to maintain the distribution pattern over segment boundaries. In PAL, there is an even number of lines in a segment, so the distribution in all segments is the same. In NTSC, there is an odd number of lines in a segment, and so it is necessary to reverse the distribution scheme on alternate segments. There are thus two distinct types of segment in NTSC, but this causes no difficulty since there are six segments in a frame, and so the same distribution will always be applied to a given segment in all frames. The distribution reversal is also shown in Figure 7.6.

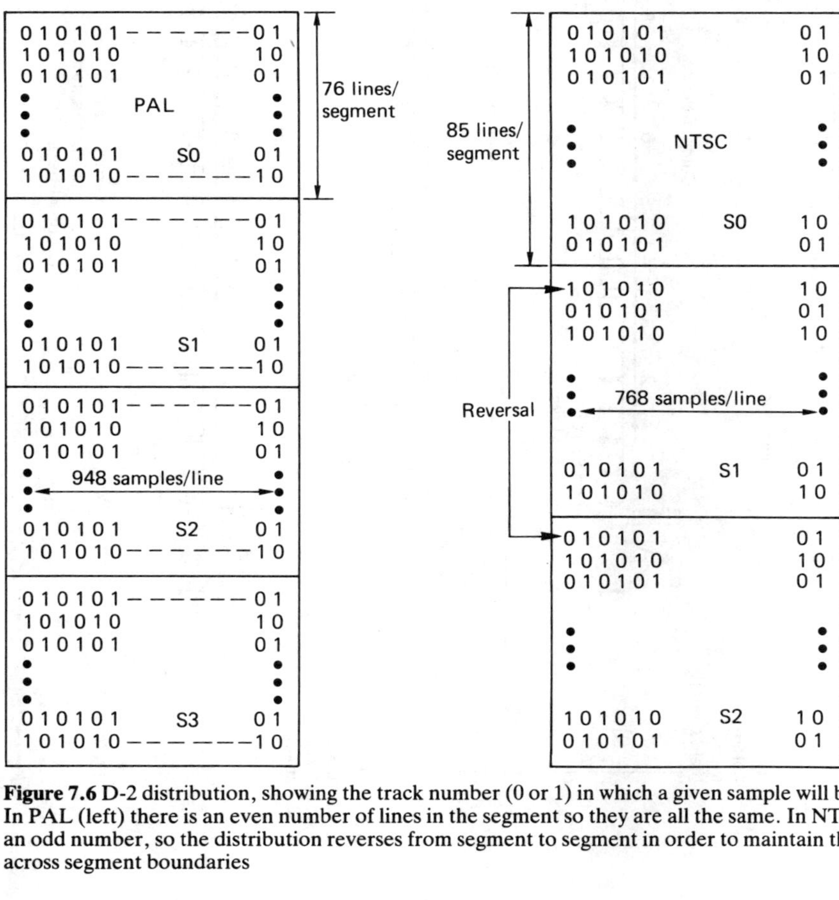

Figure 7.6 D-2 distribution, showing the track number (0 or 1) in which a given sample will be recorded. In PAL (left) there is an even number of lines in the segment so they are all the same. In NTSC there is an odd number, so the distribution reverses from segment to segment in order to maintain the pattern across segment boundaries

7.6 The error-correction strategy

Error correction in DVTRs is more complex than in other forms of recording because of the number of conflicting requirements. Chapter 5 showed that interleave is necessary to break up burst errors, and that a product code gives an efficient and reliable system. Unfortunately the product code produces a regular structure of rows and columns, and if such a system is overwhelmed, regular patterns of uncorrected errors are produced, and the effectiveness of concealment is reduced because the eye can perceive more easily the regular structure.

An irregular interleave, or *shuffle*, can be designed to disperse contiguous errors on the tape track over two dimensions on the screen, and the concealment is then more effective.

Clearly for the best concealment, the shuffle should work over the entire picture, but this is not possible because of the requirement to produce a picture in shuttle.

In shuttle the track-following process breaks down, and the heads cross tracks randomly so that the recovery of entire sectors is impossible. A picture of some sort is needed in shuttle, in order to assist in the location of wanted material, but the quality does not need to be as high as in other modes.

Figure 7.7 shows that, in shuttle, the path of the head crosses tape tracks obliquely. The use of azimuth recording allows the replay head to be about 50% wider than the track, and so a useful replay signal results for a reasonable proportion of the head path, at the times where the head is near the centreline of a track of the same azimuth type, interrupted by noise when the head crosses tracks of the wrong azimuth type. The sectors are broken into short elements called *sync blocks* which are smaller than the length of track which can be recovered at typical shuttle speeds. Each sync block contains two Reed–Solomon codewords, and so it is possible to tell if the sync block was correctly recovered or not, without reference to any other part of the recording.

If a sync block is read properly, it is used to update a framestore which refreshes the picture monitor. Sync blocks which are not read correctly will not update the

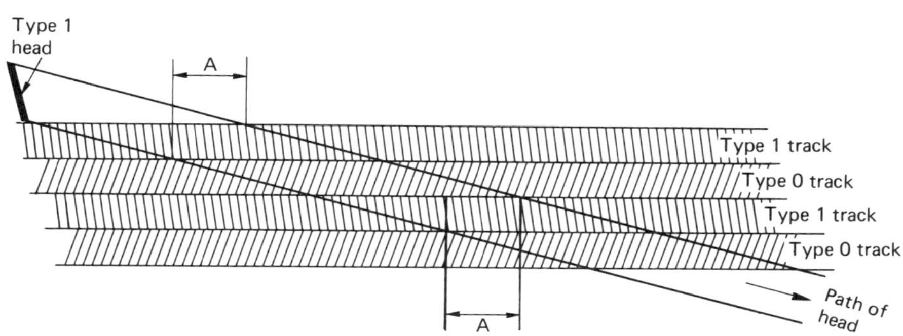

Figure 7.7 During shuttle, the heads cross tracks randomly as shown here (exaggerated). Owing to the use of azimuth recording, a head can only play a track of its own type (0 or 1) but as the head is typically 50% wider than the track (see Figure 6.11(a)) it is possible to recover normal signal for the periods above marked A. If sync blocks are shorter than this period, they can be picked up intact. In fact a slightly longer pickup will be possible because the replay system may tolerate a less-than-perfect signal. In any case the final decision is made by checking that the sync block recovered contains valid or correctable codewords

framestore, and so if concealment cannot be used, data from an earlier frame will be used to refresh the display.

Using this mechanism, the picture in shuttle is composed of data from many tape frames, and a frame-based shuffle would be extremely complex to implement, so that in practice the shuffle is restricted to operate only over one segment of the field.

A further concern is that at certain tape speeds the passage of tracks can beat with the head rotation, such that some parts of the track are consistently not recovered. If this were allowed to happen, parts of the framestore would never be updated, and the picture would be refreshed with stale data which are no longer representative of the recording. This problem is reduced if the shuffle is random, but a random shuffle is less effective for concealment, because it can result in variable density of uncorrected errors, which are harder to conceal than the constant error density resulting from a maximum distance shuffle.

Finally it is necessary to consider the likely cost of implementing shuffling schemes. Essentially the screen is considered to be a two-dimensional memory, and the shuffle is achieved by address mapping. The address generator used should not be unnecessarily complicated.

The product codes used in D-2 have 8 bytes of redundancy in the inner codes and 4 bytes in the outer codes. The inner codes are faced with essentially random data during dropouts and when crossing tracks in shuttle, and so their reliability has to be high. Although four random errors could be corrected, there is a significant probability that noise will produce what appears to be a codeword with a 4 byte error. If the inner code is only used to correct up to three random errors, the remaining 2 bytes of redundancy can be used to double-check, thereby avoiding the potential miscorrection. In the case of more than three errors, the inner code declares the entire block bad, and the outer code must correct by erasure. If the inner code were too large, this would effectively magnify burst errors to the size of the codeword. A further consideration is that if each sector were to be one long data block, loss of synchronization in the data separator due to a dropout would result in the loss of the rest of the sector, again magnifying the error.

Instead, one row of the sector array is made into six inner code blocks, and this reduces error magnification due to small bursts. To identify each inner block individually would result in an excessive overhead, so each sync block contains two codewords which share a common ID.

7.7 The shuffling mechanism

The conflicting requirements mean that compromise is inevitable in the design of the shuffling process, to give a balanced performance in different operating modes.

Figure 7.8(a) shows the sector array of D-2 PAL, and Figure 7.8(b) shows the equivalent in NTSC. Outer codes are vertical columns, and six inner codes are formed along each row. In normal play, an error which is too large to be correctable will have to be concealed. Since the outer codes contain four redundancy symbols, each can correct four symbols by erasure. This corresponds to 4 bytes per column of the sector array. The largest data loss which can be handled will be when this damage is experienced in every column of the array, so the correctable burst error size is given by multiplying the number of columns in the sector array by four. The resulting figures are 1824 bytes for PAL and 2040 bytes for

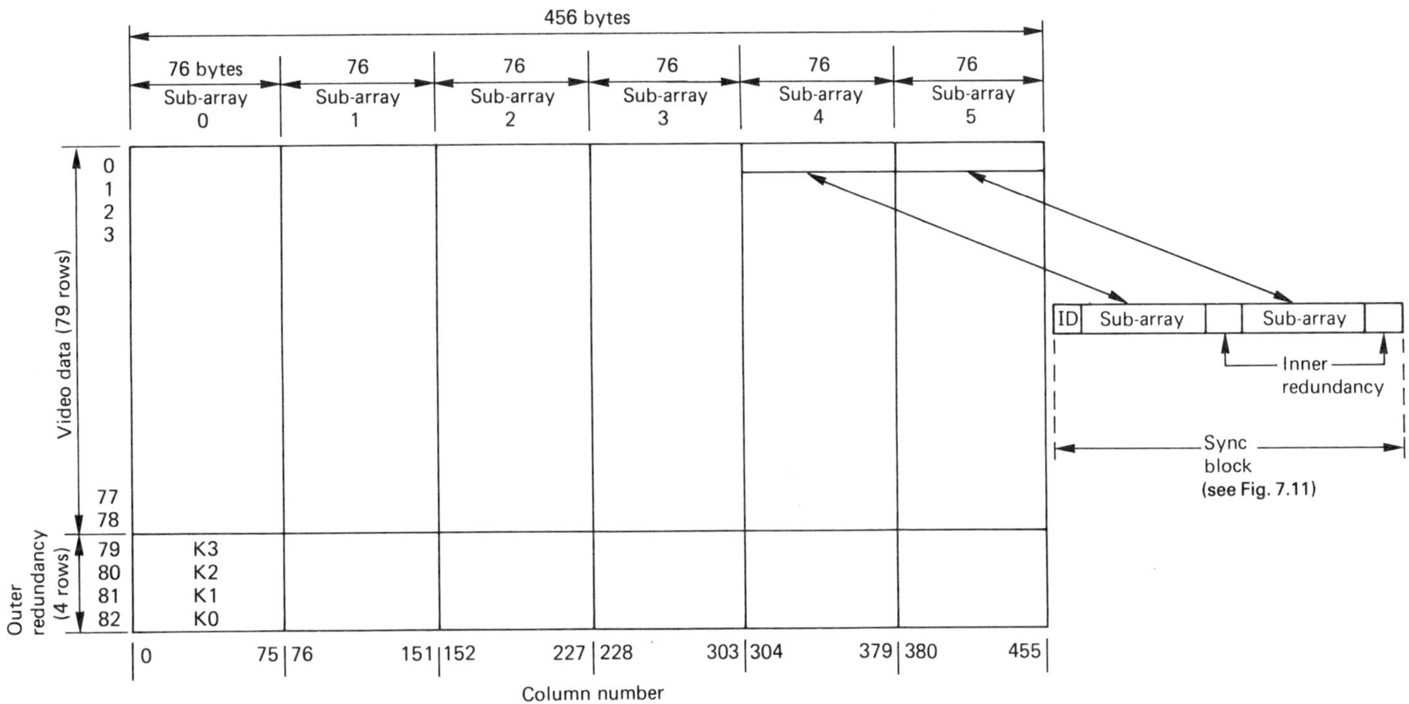

Figure 7.8(a) The sector array of D-2 PAL contains columns which are outer codewords. Rows are read, and two horizontal sub-arrays are used to construct a sync block by adding inner redundancy and ID codes. The video section of the track will thus contain $83 \times 3 = 249$ sync blocks. Array reading begins at the bottom so the outer redundancy appears at the beginning of the track. Data distributed over two such tracks (0 and 1) make a segment. 4 segments make a field

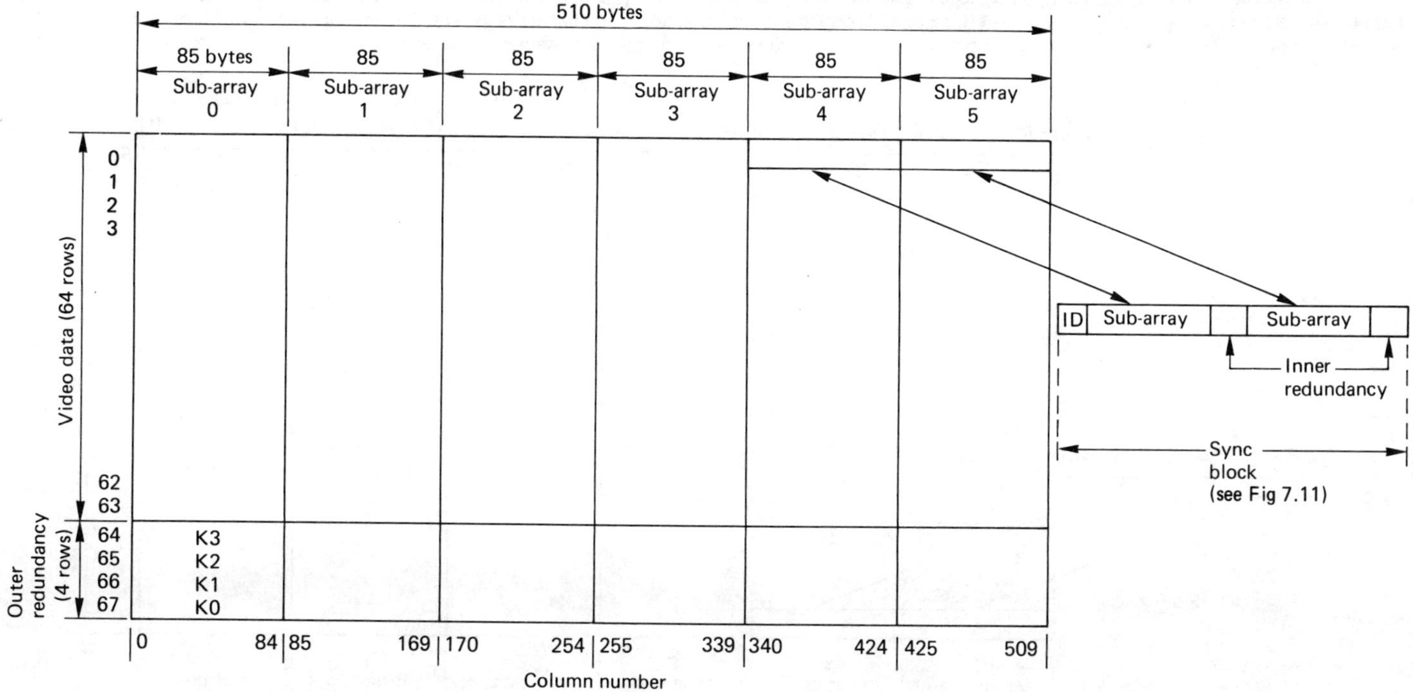

Figure 7.8(b) The sector array of D-2 NTSC contains columns which are outer codewords. Rows are read, and two horizontal sub-arrays are used to construct a sync block by adding inner redundancy and ID codes. The video section of the track will thus contain 68 × 3 = 204 sync blocks. Array reading begins at the bottom so that outer redundancy appears at the beginning of the track. Data distributed over two such tracks (0 and 1) make a segment. 3 segments make a field

NTSC. These figures are for video data only; the actual size of the burst will be larger because the length of the track involved also contains sync and ID patterns and inner redundancy for 12 sync blocks. The maximum correctable burst corresponds to about 13 mm of track length, or a longitudinal tape scratch of about 1.4 mm width.

If this condition is exceeded, total correction is no longer possible, and rows of flagged erroneous data will appear in the sector array. Concealment will become necessary, but the eye is adept at detecting patterns and would perceive the shape of the concealment area if no steps were taken to reduce its visibility.

In fact the concealment patterns are made irregular by using shuffling. When the outer codes are generated, the symbols are assembled in a non-sequential order by address permutation. Successive outer codes are non-adjacent in the sector array owing to permutation of the column address. As a result, each inner code block contains one sample from each television line in the segment, but in a different place on each line. If an entire inner code block becomes uncorrectable, the concealments will then be scattered about the segment and will be much less visible.

In shuttle, sync blocks will be recovered at random, and the shuffle means that when a sync block is successfully read, the pixels it contains are spread widely, so the framestore will be updated all over instead of just in certain areas. The twinkling effect on the picture in shuttle is caused by the shuffle.

For good concealment, the shuffle should disperse samples widely and at even spacing, whereas to prevent patterns appearing in shuttle when the head rotation beats with the passing of tracks, the shuffle should be random.

The solution is to optimize the shuffle in a given track primarily for concealment, but with shuttle in mind, and to increase the randomizing effect for shuttle by changing the shuffle from track to track. In D-2 the two parallel channels feeding the pair of heads of different azimuth type have a different shuffle pattern. The two heads are mounted together, and so in shuttle both will find themselves able to read a section of track at about the same time. At certain shuttle speeds, where the rate at which tracks are passing the scanner is a multiple of the head passing rate, some parts of the track will never be read. By having a different shuffle pattern in each track of the segment, the presence of stale data in the framestore is reduced.

Clearly the shuffle process on recording has to be exactly reversed on replay. The shuffle mechanism is one of address mapping, where samples in each channel of a segment are given addresses which are sequential on the television line and in line number. The samples are selected non-sequentially by generating a suitable address sequence. In principle, the address sequence could be arbitrarily generated from a lookup table, but the sector arrays are large, and this approach would be extremely expensive to implement.

It is desirable that the address generation should be algorithmic to reduce cost, but an algorithm must be found which gives irregular results.

The solution adopted in D-2 is to obtain a good compromise by using address multiplication by a prime number in modulo arithmetic. This is easier than it sounds, as the simple example of Figure 7.9(a) shows. In this example, 16 pixels are to be shuffled. The pixel addresses are multiplied by 11, which is relatively prime to 16, and the result is expressed modulo-16, which is to say that if the product exceeds an integer multiple of 16, that integer multiple is subtracted. It will be seen from Figure 7.9(b) that an effective shuffle is created. Modulo arithmetic has a similar effect to quantizing; in fact a number expressed modulo is the quantizing

Address A	A x 11		A x 11 mod 16
0	0		0
1	11		11
2	22	(−16)	6
3	33	(−32)	1
4	44	(−32)	12
5	55	(−48)	7
6	66	(−64)	2
7	77	(−64)	13
8	88	(−80)	8
9	99	(−96)	3
10	110	(−96)	14
11	121	(−112)	9
12	132	(−128)	4
13	143	(−128)	15
14	154	(−144)	10
15	165	(−160)	5

Figure 7.9(a) The permuted addresses for a shuffle can be obtained by multiplication of the addresses by a number relatively prime to the modulo base

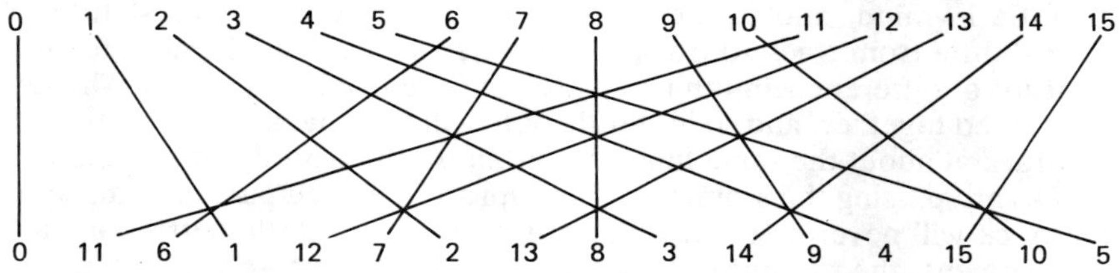

Figure 7.9(b) The shuffle which results from the calculation in Figure 7.9(a)

error due to quantizing in steps equal to the modulo base as Figure 7.9(c) shows. When the input has a large range, the quantizing error becomes random.

Generating a randomizing effect with modulo arithmetic depends upon the terms in the expression being relatively prime so that repeats are not found in the resulting shuffle, and this constrains the dimensions which can be used for the codewords.[2]

7.8 Implementing the shuffle in PAL

The shuffle uses the same principles in both standards, but with different parameters. To avoid confusion, this section contains the PAL parameters, whereas Section 7.9 contains the same text but with the parameters for NTSC. The

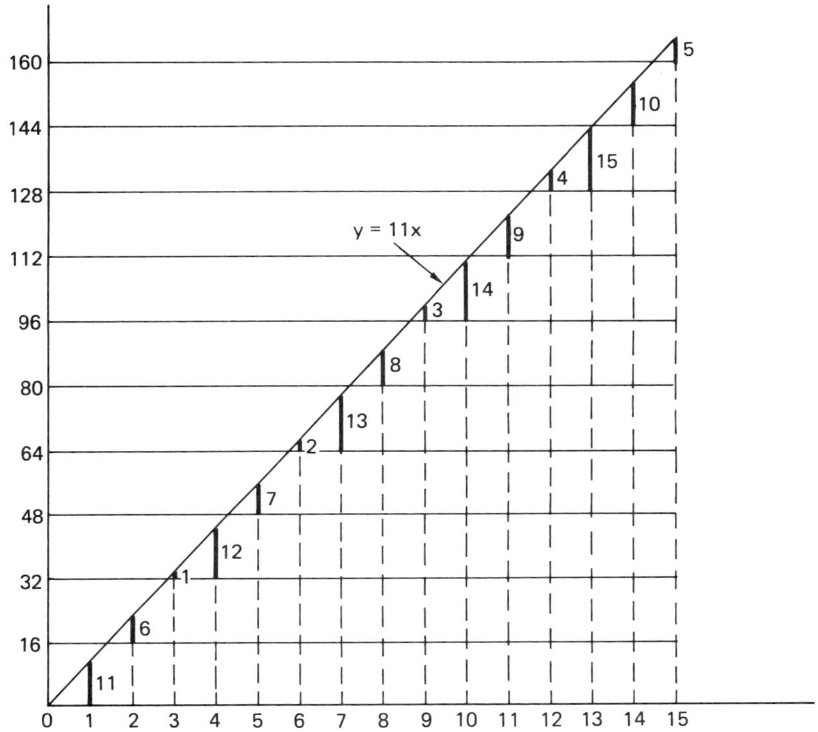

Figure 7.9(c) Modulo arithmetic is rather similar to quantizing. The results of Figure 7.9(a) can be obtained by quantizing the graph of $y = 11x$ with the quantizing intervals of 16 units

shuffle process falls into two main stages: permuting the order of samples within sector array columns, and then permuting the order of the columns. These will be dealt with in turn.

One possible implementation of the first shuffle stage is shown for PAL in Figure 7.10(a).

The source of information is a memory which sequentially stores samples from the distributor. In each line it will thus have 474 samples. Samples from each line are to be distributed over six outer code blocks, but the distribution is permuted and the permutation changes from line to line throughout the 76 lines of the segment.

An incoming byte rate clock controls the system. This clock operates a counter/divider which divides by 79 before resetting. As this is the number of samples in an outer block, it provides the block address, and the reset will increment a block count. The block counter/divider divides by six, the number of blocks in a line, and in turn the reset drives the line counter which counts the 76 lines in a segment.

Operation begins with all of the accumulators cleared. The byte accumulator adds 60 every time it is clocked, and so produces 60× the byte number. In the first block the other accumulators will remain zero, so only this value passes to the modulo-474 circuit which subtracts integer multiples of 474 until the output no longer exceeds 474. Part of the calculation sequence is shown in the diagram, where it will be seen that successive source addresses are six samples apart. The intervening five samples will go into other blocks. The source address is used to obtain a sample from the pixel memory, and the byte count is used to place the sample in the correct place in the block. In this way a permuted block is assembled.

At the end of the first block, the byte counter resets, clearing the byte accumulator and incrementing the block counter. This causes the block accumulator to add 79 to the previously cleared state. The source address now becomes 60× the byte address plus 79× the block address. The process continues until all six blocks of the first line are complete, when the block count will reset, clearing the block accumulator and incrementing the line count. This causes the line accumulator to add 83 to the previously cleared value, so the full source address is given by 60× the byte count plus 79× the block count plus 83× the line count. This process continues throughout the whole segment, until 456 blocks have been assembled.

The second stage of the shuffle is to permute the column number in the sector array. Figure 7.10(a) also shows that the block rate clock also operates a second accumulator which multiplies the block number by 76. To the output of this is added the line number, and the result is taken modulo-456, since 456 is the number of columns in the sector array. This produces the column address into which the blocks produced by the mapping process are written. The row address is equal to the byte address within the block. The shuffle process is then complete. The outer redundancy can be added on the columns, and the inner redundancy and ID codes can be added to the rows. The array contents are then ready to be recorded.

7.9 Implementing the shuffle in NTSC

The shuffle process falls into two main stages: permuting the order of samples within sector array columns, and then permuting the order of the columns. These will be dealt with in turn.

One possible implementation of the first shuffle stage is shown for NTSC in Figure 7.10(b).

The source of information is a memory which sequentially stores samples from the distributor. In each line it will thus have 384 samples. Samples from each line are to be distributed over six outer code blocks, but the distribution is permuted and the permutation changes from line to line throughout the 85 lines of the segment.

An incoming byte rate clock controls the system. This clock operates a counter/divider which divides by 64 before resetting. As this is the number of samples in an outer block, it provides the block address, and the reset will increment a block count. The block counter/divider divides by six, the number of blocks in a line, and in turn the reset drives the line counter which counts the 85 lines in a segment.

Operation begins with all of the accumulators cleared. The byte accumulator adds 258 every time it is clocked, and so produces 258× the byte number. In the first block the other accumulators will remain zero, so only this value passes to the modulo-384 circuit which subtracts integer multiples of 384 until the output no

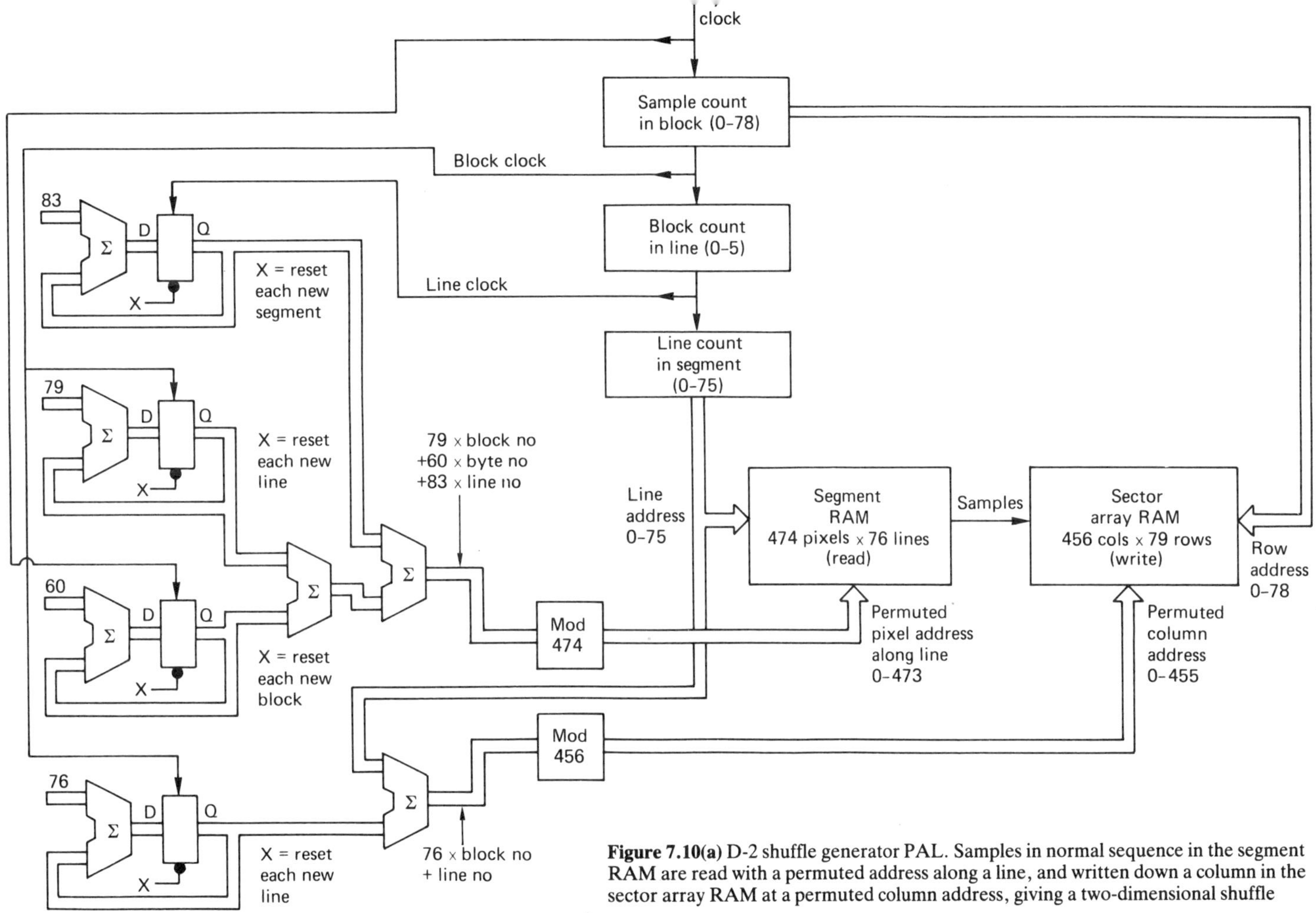

Figure 7.10(a) D-2 shuffle generator PAL. Samples in normal sequence in the segment RAM are read with a permuted address along a line, and written down a column in the sector array RAM at a permuted column address, giving a two-dimensional shuffle

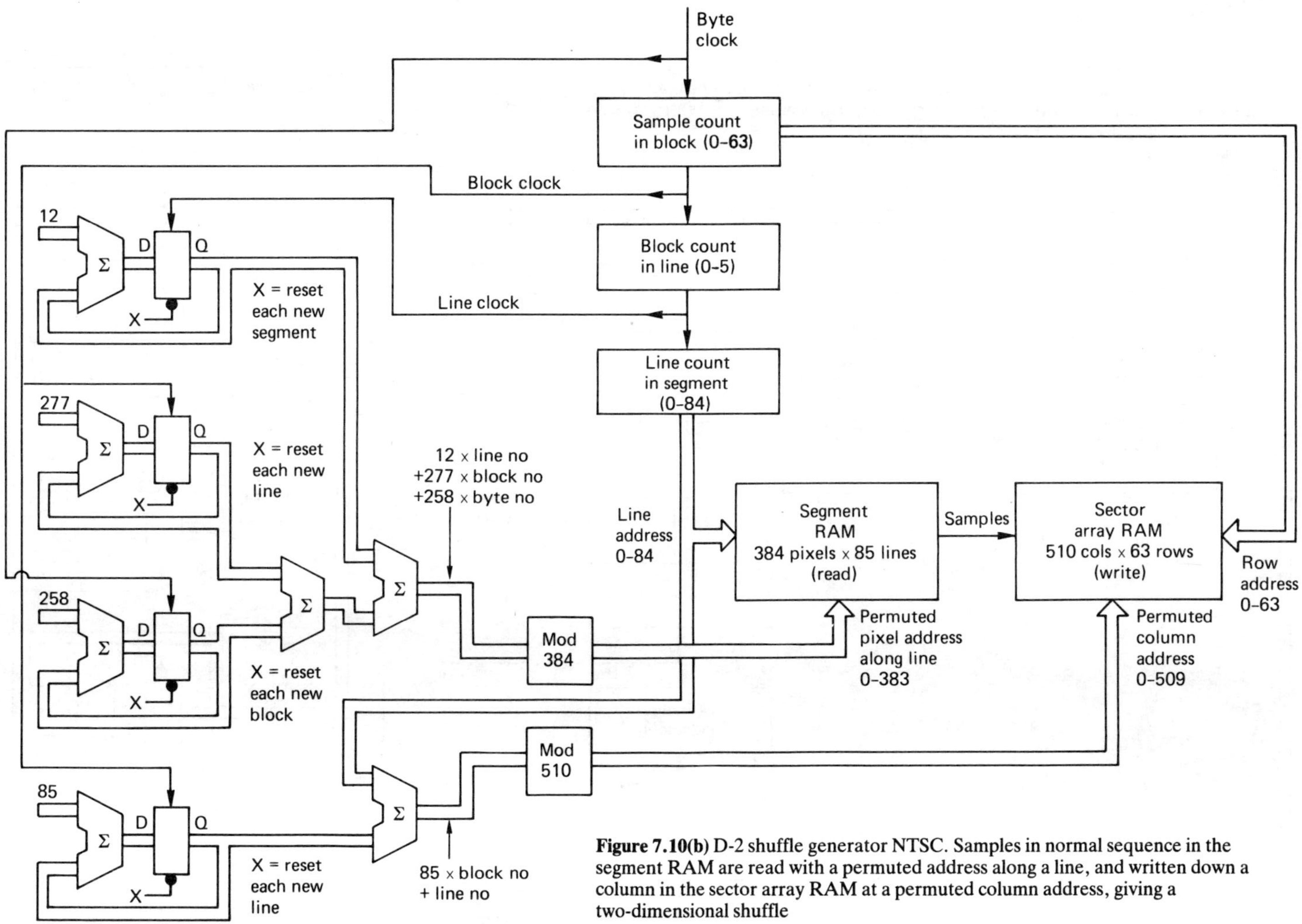

Figure 7.10(b) D-2 shuffle generator NTSC. Samples in normal sequence in the segment RAM are read with a permuted address along a line, and written down a column in the sector array RAM at a permuted column address, giving a two-dimensional shuffle

longer exceeds 384. Part of the calculation sequence is shown in the diagram, where it will be seen that successive source addresses are six samples apart. The intervening five samples will go into other blocks. The source address is used to obtain a sample from the input memory, and the byte count is used to place the sample in the correct place in the block. In this way a permuted block is assembled.

At the end of the first block, the byte counter resets, clearing the byte accumulator and incrementing the block counter. This causes the block accumulator to add 277 to the previously cleared state. The source address now becomes 258× the byte address plus 277× the block address. The process continues until all six blocks of the first line are complete, when the block count will reset, clearing the block accumulator and incrementing the line count. This causes the line accumulator to add 12 to the previously cleared value, so the full source address is given by 258× the byte count plus 277× the block count plus 12× the line count. This process continues throughout the whole segment, until 510 blocks have been assembled.

The second stage of the shuffle is to permute the column number in the sector array. Figure 7.10(b) also shows that the block rate clock also operates a second accumulator which multiplies the block number by 85. To the output of this is added the line number, and the result is taken modulo-510, since 510 is the number of columns in the sector array. This produces the column address into which the blocks produced by the mapping process are written. The row address is equal to the byte address within the block. The shuffle process is then complete. The outer redundancy can be added on the columns, and the inner redundancy and ID codes can be added to the rows. The array contents are then ready to be recorded.

7.10 The D-2 sync block

The inner codes are designed to correct random errors and to detect the presence of burst errors by declaring the whole inner block bad. There is an optimum size for an inner code: too small and the redundancy factor rises; too large and burst errors are magnified, as are errors due to losing sync.

One row of the product block is too large to be one inner code, and so it is split into six inner code blocks. The dimensions of these are chosen to give adequate correction performance. However, individually identifying each inner code would raise the recording overheads. In shuttle, a geometric calculation reveals the shortest length of track which will be recovered at maximum tape speed, and this determines that it is not necessary to identify every inner codeword. Accordingly one sync block contains two inner codewords, but only one ID code. Three sync blocks then form one complete row of the sector array.

Figure 7.11(a) shows a PAL sync block. A 2 byte synchronizing pattern serves to phase the data separator and the deserializer. Following this is a 2 byte ID code which uniquely identifies this sync block in an eight-field sequence. The first data block of 76 bytes then follows, and the 8 bytes of inner redundancy are calculated over the ID code as well as the data, so a random error in the ID can be corrected. The first codeword is thus 86 bytes long. The second inner code block is 2 bytes shorter because it contains no ID. The entire sync block occupies just over half a millimetre of track.

Figure 7.11(b) shows an NTSC sync block. A 2 byte synchronizing pattern serves to phase the data separator and the deserializer. Following this is a 2 byte ID code which uniquely identifies this sync block in a four-field sequence.

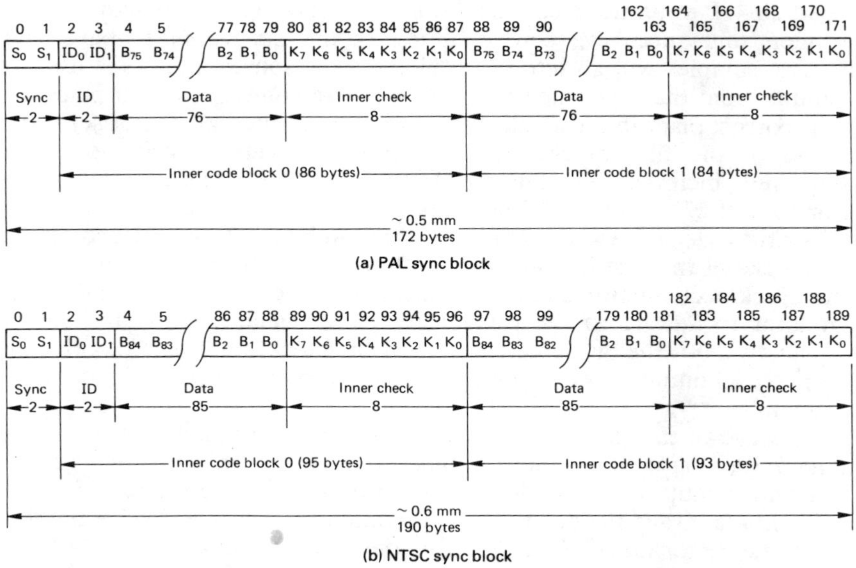

Figure 7.11 The sync blocks of D-2 contain two codewords but only one ID code which forms part of the first codeword. As a result, the first code block is 2 bytes longer than the second

The first data block of 85 bytes then follows, and the 8 bytes of inner redundancy are calculated over the ID code as well as the data, so a random error in the ID can be corrected. The first codeword is thus 95 bytes long. The second inner code block is 2 bytes shorter because it contains no ID. The entire sync block occupies nearly two-thirds of a millimetre of track.

7.11 Miller^2 code in D-2

The need for channel coding has been shown in Chapter 4 and so this section confines itself to the implementation in D-2.

Miller^2 code has the same density ratio but half the jitter window of RNRZ, all other things being equal, and so might be thought inferior. In practice this turns out not to be the case because the narrow spectrum of Miller^2 is less stringent about equalization, the smaller ratio of run lengths causes less peak shift, and the higher clock content eases the task of the data separator. Thus in practice the performance of Miller^2 may be inferior under accurate tracking conditions, but the performance will not fall so rapidly with the onset of tracking error and may actually be better than RNRZ during mistracking.

The main advantage of Miller^2, however, is that it allows overwrite, eliminating erase heads, and makes guard-band-less recording possible, minimizing tape consumption.[4]

The encoding rules of Miller^2 were explained in Chapter 4, but are repeated here briefly. A transition will be recorded in the centre of all data ones and between successive zeros, but in the case of an even number of ones between zeros, the transition of the last one is omitted.

These coding rules are used to generate a variety of patterns used for synchronizing the data separator and identifying the location of the head.

Figure 7.12(a) shows that the run-up pattern or preamble is obtained by feeding alternating ones and zeros to the channel coder. This results in a frequency of half the maximum, since only the ones result in a transition. This frequency will suffer fewer losses, and so will be readily detectable on replay. The transitions all indicate the centre of the bit cell and the PLO can easily synchronize to them. This run-in pattern precedes nearly every sector.

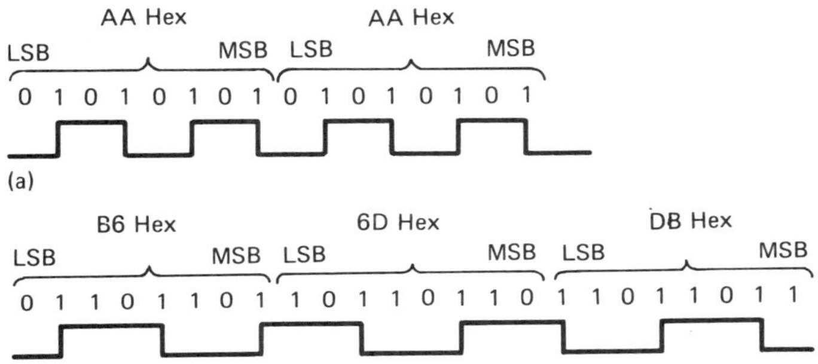

Figure 7.12 (a) The alternation of data ones and zeros produces a frequency of half the maximum which is ideal for bringing the phase-locked loop in the data separator into lock. (b) Under the coding rules of Miller^2, the transition due to the last one in an even number of ones is omitted, so that a repeated sequence of data bits 011 will generate only one transition. This represents the lowest frequency possible in Miller^2, and is used to identify uniquely the first segment in a field

Figure 7.12(b) shows that if 011 repeated is presented to the channel coder, the last coding rule is invoked and the transition due to the second one is omitted, resulting in a frequency of one-third the maximum. This is the lowest frequency which Miller^2 code can generate, and is very easily detectable with a simple filter. This pattern is used only at the beginning of the first segment of a field. It is incorporated in order to simplify the construction of automated systems where several transports share a single signal system. Each transport need only have a simple preamplifier and tuned circuit to detect the field run-in, and so it can, in conjunction with the control track and timecode, synchronize the scanner and colour frame the capstan without the presence of a signal system. A cut between two tapes in two transports can thus be made by switching the input to a single signal system at the transport preamplifiers.

7.12 Synchronization and identification

Each sync block contains two inner codewords and is the smallest quantum of data on the track with which the playback channel can synchronize independently. Synchronization takes place on three levels: to the bit, to the symbol and to the sync block.

The recording density in DVTRs is such that it is not possible to position the heads mechanically with sufficient accuracy to locate an individual bit. The heads are rotated with reasonable accuracy, and the replay signal is accepted as and when it arrives. A phase-locked loop in the data separator must lock to the bit rate and phase of the recording. At the beginning of the track a reference for the phase-locked loop known as a preamble is recorded. This serves to set the frequency of the PLL the same as the offtape bit rate. Once the loop is locked, it can stay in synchronism by phase comparing the data transitions with its own. However, if synchronism is lost due to a dropout, it cannot be regained until the next synchronizing pattern is seen. This is one reason why the tracks are broken into short sync blocks.

Once bit synchronism has been achieved at the preamble, the serial data have to be correctly divided up into 8 bit symbols to restore the original samples. This is symbol synchronization, and it is achieved by the unique pattern which occurs at the beginning of every sync block. Detection of this pattern resets the divide-by-eight counter which deserializes the data. The sync pattern of D-2 is $30F5_{16}$, and is in fact the same pattern as is used for D-1. Figure 4.22(a) showed that this pattern differs from itself shifted by as many bits as possible to reduce the possibility of false sync generation. Within a sector, sync patterns occur at regular spacing, so it is possible for the replay circuitry to predict the arrival of the sync pattern in a time window.

Once symbol synchronizing is achieved, it is then possible to read the inner code

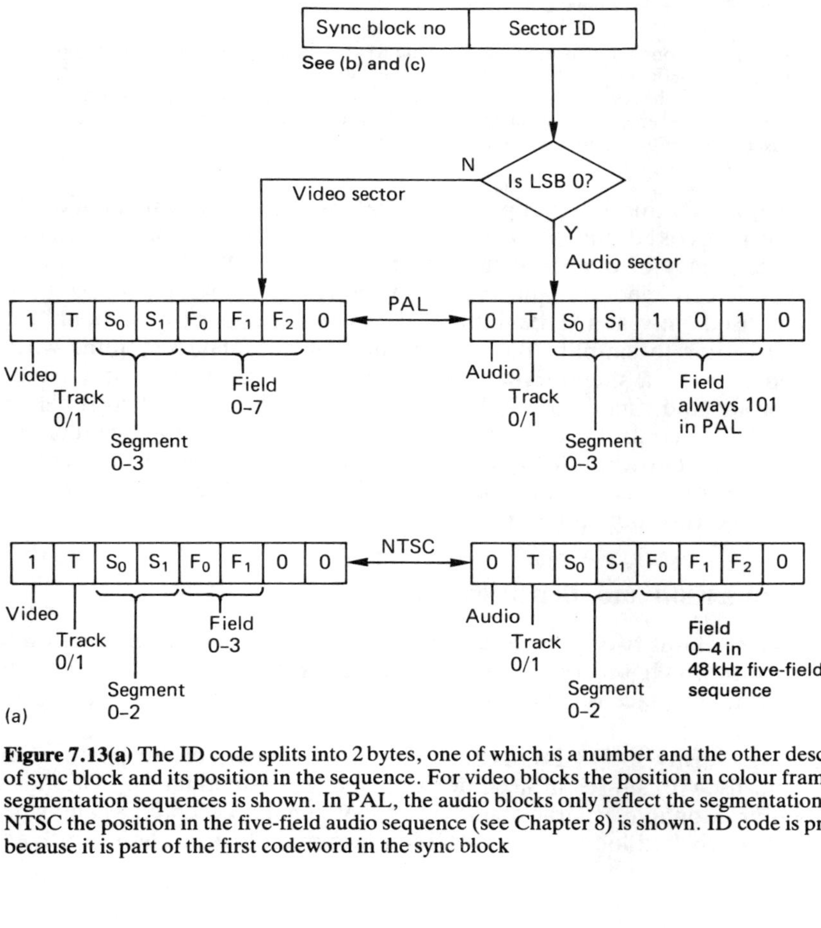

Figure 7.13(a) The ID code splits into 2 bytes, one of which is a number and the other describes the type of sync block and its position in the sequence. For video blocks the position in colour framing and segmentation sequences is shown. In PAL, the audio blocks only reflect the segmentation, whereas in NTSC the position in the five-field audio sequence (see Chapter 8) is shown. ID code is protected because it is part of the first codeword in the sync block

blocks, and the first of these contains the ID code and reveals which sync block has been recovered. In normal play this will be the one after the previous one, but in shuttle the sequence will be unpredictable.

In shuttle, samples from any sync block properly recovered can be put in the correct place in a framestore by reference to the ID code.

Figure 7.13(a) shows that the information recorded in the ID code is split into two parts. The first byte is the sync block number in the track, the second identifies the sector. The V/A bit specifies whether the block is audio or video, and the T bit specifies which track of the parallel pair. The segment bits determine which

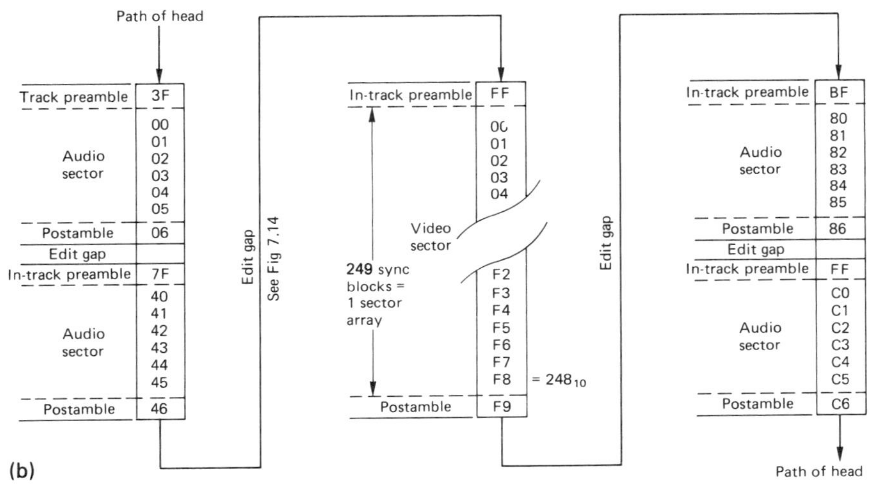

Figure 7.13(b) PAL ID numbering in the first byte of the ID code uniquely identifies every sync block in a sector. Second byte then determines which sector

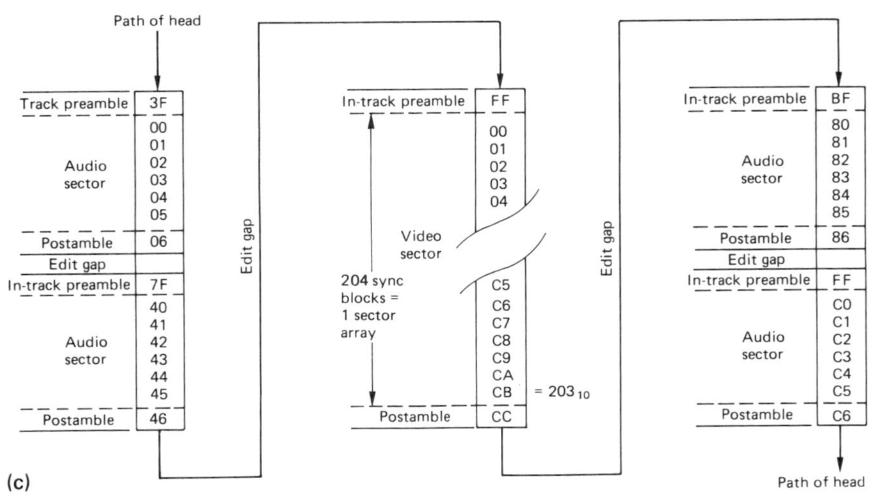

Figure 7.13(c) NTSC ID numbering in the first byte of the ID code uniquely identifies every sync block in a sector. Second byte then determines which sector

segment out of four for PAL and three for NTSC is present, and the field bits specify the position in the eight- or four-field sequences to help colour framing.

Figure 7.13(b) shows the sync block numbering within the segment for PAL, and Figure 7.13(c) shows the equivalent for NTSC.

7.13 Gaps, preambles and postambles

Each slant track in D-2 is subdivided into five sectors: two of audio at each end, and one of video in the centre. It is necessary to be able to edit any or all of the audio channels independently of the video. For this reason short spaces are left between the sectors to allow write current to turn on and off away from wanted data. Since the write current can only turn off at the end of a sector, it follows that the smallest quantum of data which can be written in the track at one time is a whole sector, although the smallest quantum which can be read is a sync block.

A preamble is necessary before the first sync block in a sector to allow the phase-locked loop in the data separator to lock. The preamble also contains a sync pattern and ID code so that the machine can confirm the position of the head before entering the sector proper. At the end of the sector a postamble containing an ID code is written before the write current is turned off. The pre and postamble details can be seen in Figure 7.14.

The space between two adjacent sectors is known as an edit gap and it will be seen in Figure 7.14 to begin with the postamble of the previous sector and end with the preamble of the next sector. Editing may result in a discontinuity nominally in the middle of the gap. There is some latitude, but the new recording must not begin so early that the postamble at the end of the previous sector is damaged, and at least 20 bytes of preamble must be written before the sync pattern at the beginning of the next sector, because it will not be possible to maintain continuity of bit phase at an edit, and the PLL must resynchronize.

In fact the length of the postamble, gap and preamble combined is exactly equal to the length of one sync block, which simplifies the design of the sequencer which controls the track recording. This is half the length of a gap in D-1, and is possible because there is no separate erase head in D-2.

7.14 Block diagram of a D-2 machine

The D-2 format specifies the dimensions and arrangement of the tape record, but provided these are met, there is no compulsion to use a particular mechanical design or circuit architecture.

Figure 7.15 shows a typical block diagram of the record section of a D-2 machine, where the inputs will be composite video and four audio channels. Audio and video can be analog or digital, and most machines incorporate converters which can be bypassed if a digital input is available. Whichever input is used, the input video will always control the timing of the recording circuitry. In the case of a digital video input, the $4 \times F_{sc}$ clock accompanying the data will be used to derive the symbol clock for much of the recorder. In the case of an analog video input, the video burst will be fed to a phase-locked loop which generates the sampling clock locally.

The video data are distributed over two logical channels as was shown in Figure 7.6.

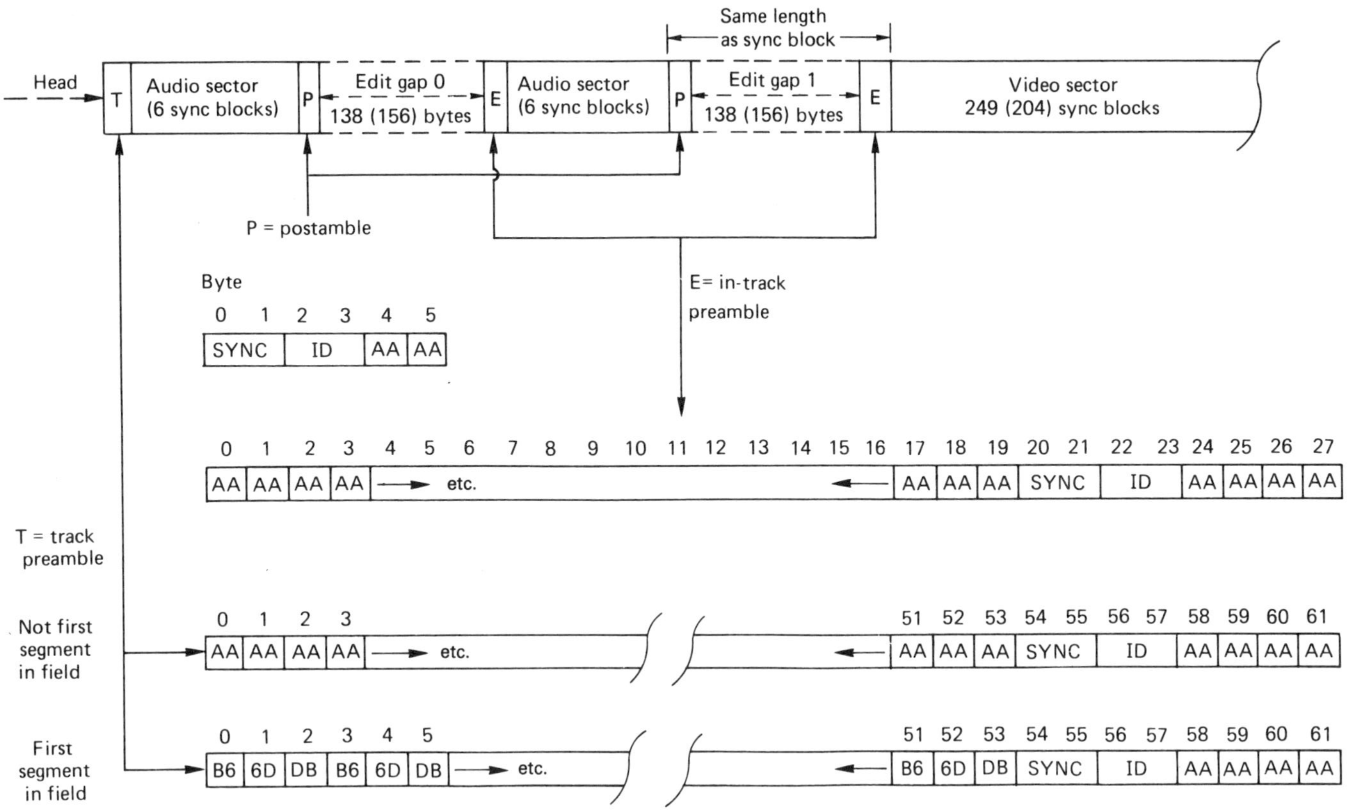

Figure 7.14 Edit gap begins with postamble of previous sector and ends with preamble of next. Remaining space is AA Hex fill which occupies 138 bytes PAL (156 bytes NTSC). Total length of edit gap is then the same as one sync block. Track preamble T is 54 bytes of run-in followed by preamble ID. In the first segment of a field, run-in pattern is changed to B6, 6D, DB sequence (see Figure 7.12(b))

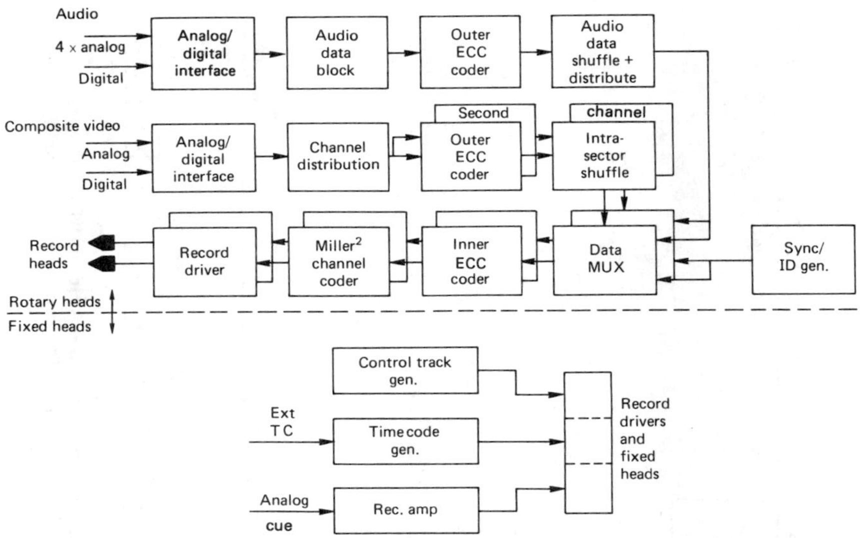

Figure 7.15 D-2 record section showing major stages. Note particularly the distribution of video over two channels, and the merging of the audio and video data into common inner coding circuits

The distribution process results in each logical channel having half of the original number of samples in each line, 474 in PAL and 384 in NTSC. Each of these lines will be interleaved into six outer codewords, such that samples in the same codeword are six samples apart in the channel, and 12 samples apart on the television line. This is done by calculating the sample address along the line as a function of the block number and the byte number within the block. The first stage of the shuffle is obtained by changing the address calculation from line to line so that samples in the same place in an outer block on an adjacent line will have come from a different position along the line. This interleave/shuffle process results in blocks of 79 samples in PAL and 64 samples in NTSC. Four bytes of Reed–Solomon redundancy are calculated on each block to make it into a codeword. The size of the outer codewords is equal to the height of the sector array memory, and outer codewords form columns in the sector array. Adjacent outer blocks do not form adjacent columns owing to the second stage of the shuffle. In this, the column number is a function of the line number as well as the block number.

Eventually the sector array will be filled with columns of outer codewords, and it is then read in rows which are used to form inner codewords.

Each row results in six inner codewords, and two inner codewords are used in each sync block. The first codeword in the sync block is extended to protect the ID code and so it is 2 bytes longer than the second codeword.

There will be 249 video sync blocks in a PAL track and 204 video sync blocks in an NTSC track. In both channels, the reading of the sector array begins at the bottom, so that the four rows of outer code redundancy are placed at the beginning of the track. In the case of channel 0, readout continues in descending row order, but the memory readout sequence is modified for channel 1 so that a linear tape scratch crossing both tracks in the same place does not result in an identical corruption pattern in the two channels and so that a more random shuffle is effected in shuttle.

This additional shuffle is implemented by permuting the sector memory row addresses in channel 1 when the memory is being read to tape. In PAL, 39 is added to the row address and the result is expressed modulo-79, whereas in NTSC, 32 is added to the row address and the result is expressed modulo-64.

The audio processing is shown in parallel with the video processing, and uses a distribution scheme to spread samples in all channels over both heads and double recording to compensate for the fact that the audio blocks are small. This is explained more fully in Chapter 8. Since audio is not two-dimensional, no shuffle is necessary, and a regular array is used to form a product block. Audio samples have longer wordlength than video samples, so the outer coding of the audio is somewhat different to the video, but the dimensions of the product block are such that the audio inner codewords are identical in size to the video inner codewords, so that common processing can be used.

The inner code generation, ID generation, serializing and channel coding are common to both audio and video, which effectively share the same circuitry, heads and tracks in a time multiplex.

At the beginning of a segment, the audio and video to be recorded have been interleaved and shuffled and are held in RAM so that they can be made available at the appropriate time.

Input to the serializer will initially be from the sync generator which produces the preamble at the beginning of the track. The serializer input then receives audio sync blocks until the end of the first audio sector, when the edit gap will be written. Following the preamble at the end of the edit gap, the second audio sector will be written, followed by another edit gap. After this edit gap, video sync blocks will be written, followed by a second edit gap and the second pair of audio sectors.

Figure 7.16 shows the replay system. Following the channel decoder, the sync detector phases the shift register in the deserializer so that the serial bit stream is correctly assembled into a parallel byte stream.

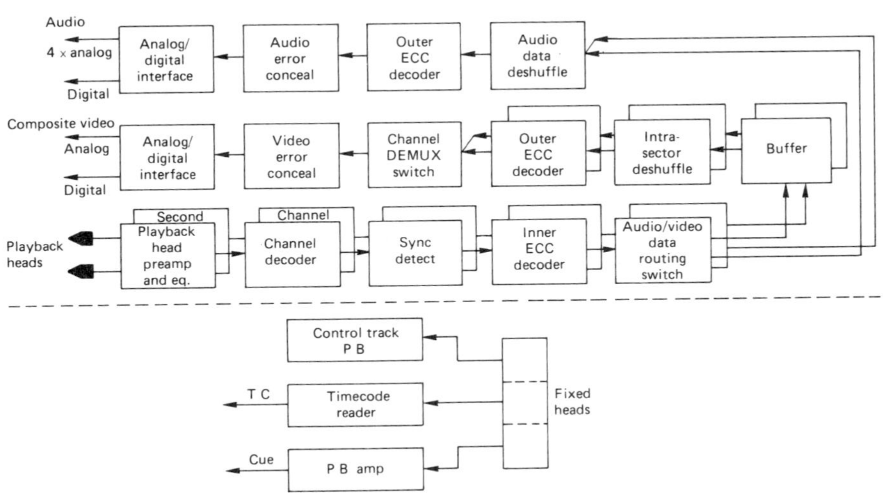

Figure 7.16 D-2 replay section. Two replay channels handle both audio and video as far as inner code error correction, and the data router then splits the audio and video to separate outer code processing

Inner blocks are checked by the inner decoder, which therefore also checks the ID code in the first inner block of a sync block. With 8 bytes of redundancy, the inner decoder could correct up to four random errors, but this would be at the expense of too many miscorrections. Accordingly, the inner code corrects up to three random errors, and a larger number are handled by declaring the entire inner code to be corrupt. An error flag is attached to every symbol in the block. Inner blocks are written into the sector array and de-shuffled, and any burst errors are converted to single byte errors dotted around the array. These are accompanied by the error flags from the inner decoder, and so the 4 bytes of redundancy in the outer codes can correct four errors by erasure. If there are more than four error flags in an outer code, the symbols will be flagged as uncorrectable, and concealment must be used. When the two channels are combined the concealment process takes place.

The samples are now in their original sequence and can be output digitally or fed to a converter.

The audio samples are corrected by the inner and outer codes as for video, but as there are two copies of all audio data, any error-free data will be used to reassemble output samples. This means that audio interpolation to conceal uncorrectable errors is extremely rare.

7.15 Colour processing

When variable speed is used, the transport cannot change its scanner speed, because this determines field rate. If the tape speed is too low, some fields on tape have to be repeated to maintain field rate, whereas if tape speed is too high, some will need to be skipped. The effect of a head jump is that the subcarrier sequence is randomly broken. If variable speed is to be used, then it is necessary to decode the discontinuous incoming subcarrier and re-encode it to continuous reference phase. The unit which does this job is called the colour processor. In NTSC there are only two line types, and the colour processor simply has to invert the subcarrier selectively before adding it back to the luminance. A digital filter is necessary to separate the chroma from the luminance. The operation of such a device has been described in Chapter 3.

In PAL, as usual, the solution is more complex because of the use of *V*-switch. Chroma is filtered from luminance as before, but then the chroma signal will have to be re-encoded not just with the correct phase, but also with the correct sense of *V*-switch.

Figure 7.17(a) shows that, from line to line, the phase of *U* advances 90° as a result of the quarter-cycle offset, and the phase of *V* alternately leads and lags *U*. Figure 7.17(b) shows that if chroma is sampled on the *U* and *V* axes the result of these two effects is that, from one line to the next, the chroma samples represent different sequences of *U* and *V*. The shift to the right of the *U* sample due to the quarter-cycle offset can be seen. It is not immediately obvious how one line of chroma could be made into another with this approach.

The solution adopted in D-2 is to sample at 45° to the *U* and *V* axes, i.e. in phase with the burst. Figure 7.17(c) shows that when this is done, the samples represent $U \times \cos 45 + V \times \cos 45$ etc. If everything is normalized by dividing by cos 45, the samples become $U + V$, $-U + V$, etc.

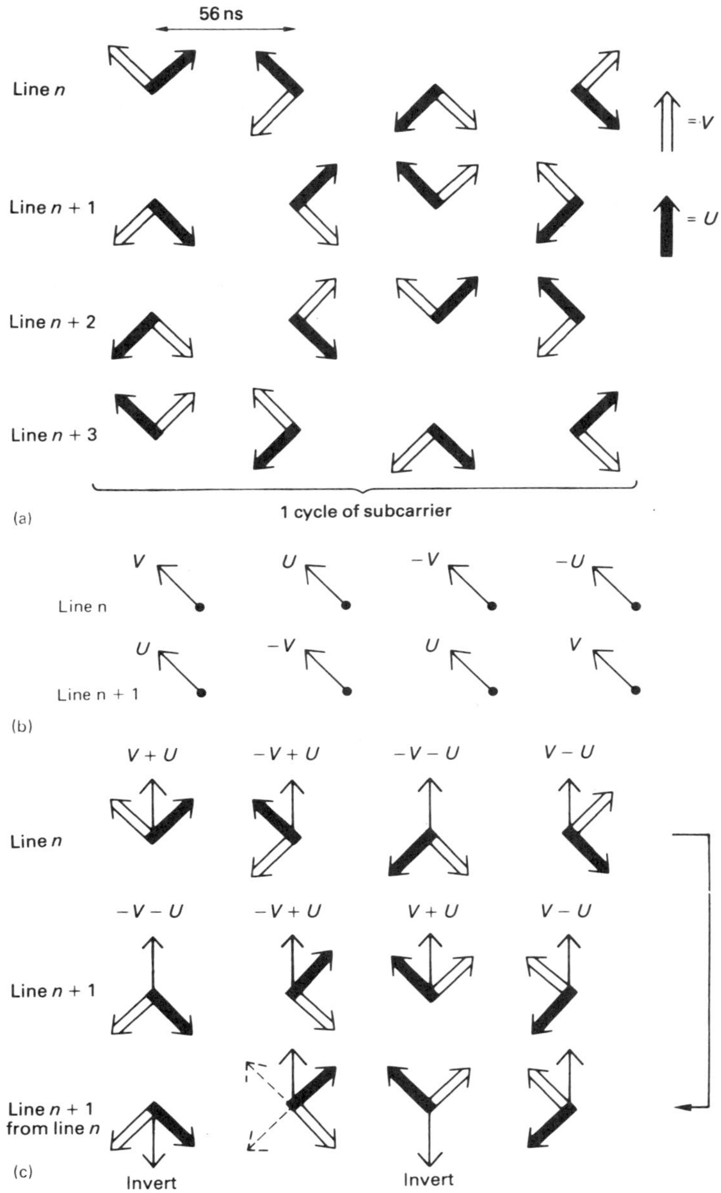

Figure 7.17 (a) The combined effects of *V*-switch and quarter-cycle offset in PAL cause the above relative phases of *U* and *V* from line to line, in a four-line sequence which only repeats after four frames. (b) If a sampling system takes samples on the *U* and *V* axes, it results in the samples having the values shown here. It is not then possible to convert from one line type to the next because, for example, *V* cannot be processed to become *U* and so on. (c) If sampling is performed with sample phase between *U* and *V* axes (thin arrow) the samples represent sums and differences of the two signals (cosine term neglected). It will be seen that line *n* can be made into line $n + 1$ by inverting the samples when *U* and *V* have the same sign. When the sample is made, phase information is lost and only amplitude is known. For this reason, the non-inverted samples phase-mirror (dotted arrow) because the phase is determined in the context of adjacent samples

If it is required to convert a line into the next line type, having the opposite sense of *V*-switch, this can be done by inverting the samples where *U* and *V* have different polarities, but not where they have the same polarity.

If the difference is two lines, the only requirement is to invert the whole chroma waveform, as *V*-switch is the same after two lines, and there have been two quarter-cycle offsets of subcarrier making 180°.

If the difference is three lines, an inversion combined with *V*-switch reversal and a quadrature shift is needed, and this can simply be done by inverting samples where *U* and *V* have the same polarity and not where the polarity is different.

The type of line required on the output is available from the reference colour framer, and the type of line played back is determined from the ID codes in variable speed and from the line address in shuttle. By comparing the two line types, the conversion to be performed by the colour processor can be determined. The chroma must be separated from the luminance signal before the colour processor can operate, and this is done by a digital filter. A combination of chroma bandpass and comb filtering is used. The comb filter gives good separation of sidebands, but fails when there are vertical chroma phase changes in the picture. The chroma phases at the beginning, centre and end of the delays are monitored by a correlator, and when the comb filter fails to mesh, the bandpass signal alone is used. The filtered chroma signal is then subtracted from the composite signal to give digital luminance. Once chroma and luminance are separated, a digital chroma gain control can be incorporated simply by passing the chroma samples through a multiplier.

Owing to the use of interlace, the head jumps may result in an odd field being played when an even field is needed, and vice versa. The vertical picture movement which results from outputting the wrong field can be completely removed by vertical interpolation, as shown in Section 3.20.

References

1. Ampex Corporation, Ampex digital format for video and audio tape recording of composite video signals using 19 mm type D-2 cassette. Redwood City, CA (1988)
2. BRUSH, R. Design considerations for the D-2 NTSC composite DVTR. Presented at SMPTE Technical Conference (Los Angeles, 1987)
3. BRUSH, R. Design considerations for the D-2 PAL composite DVTR. *IERE Conf. Publ.*, no. 79, 141–148 (1988)
4. MORRISON, E. F. Tape format for a PAL DVTR. Presented at 15th Int. Television Symposium (Montreux, 1987)

Chapter 8

Digital audio in D-2

8.1 Introduction

Once audio has been converted to the digital domain, it becomes data which do not differ in principle from any other. A high-quality digital audio channel will require a data rate of only 1 megabit per second, which is considerably less than the rate which is necessary for video. Although the frequency range of audio is much less than that of video, the dynamic range is much greater, and sample wordlengths of 16 bits and greater will be found. The video converter technology described in Chapter 2 is inadequate for these wordlengths, and the special techniques developed for audio will be described here.

In the D-2 format it is necessary to use rotary heads in order to support the substantial data rate. The presence of several channels of digital audio causes an increase in data rate of only a few per cent, and it is logical to record the audio with the same heads as the video by allocating small audio blocks at specified places along the main recorded tracks. This simplifies the construction of the machine tremendously, since there are no separate audio heads, and the same RF and coding circuitry is timeshared between audio and video, and part of the error-correction circuitry can also be common.

Standards exist for the interconnection of audio machines in the digital domain, not the least important being the AES/EBU interface which has found wide acceptance and is fully detailed here along with a four-channel parallel interconnect developed specifically for D-2. Digital audio can also be carried by the serial scrambled video interface which is also treated here.

8.2 Choice of sampling rate for digital audio

In Chapter 2 it was seen that sampling theory demanded a sampling rate of at least twice the baseband frequency to be carried, and then the rate had to be raised a little to allow the use of filters with finite slope. All of this applies equally to audio.

This does not, however, mean that only one sampling rate is necesary for audio; there are a number of rates which have different purposes.

For broadcast, conservation of bandwidth is paramount, and the sampling rate of 32 kHz is the standard for digital landlines to FM transmitters and for the NICAM stereo television sound system. For consumer products, the sampling rate of 44.1 kHz, which is near to twice 20 kHz, came from the PCM adaptors which were used to master Compact Discs on adapted video cassette recorders. For professional products, there is a need to operate at variable speed for pitch correction. When the speed of a digital recorder is reduced, the offtape sampling rate falls, and Figure 8.1 shows that with a minimal sampling rate the first image frequency can become low enough to pass the reconstruction filter. If the sampling frequency is raised without changing the response of the filters, the speed can be reduced without this problem.

The professional sampling rate of 48 kHz was proposed as having a simple relationship to 32 kHz, being far enough above 40 kHz for variable-speed operation, and having a simple relationship with PAL video timing which would allow digital video recorders to store the convenient number of 960 audio samples per video field. This is the sampling rate used by D-2. The field rate offset of NTSC

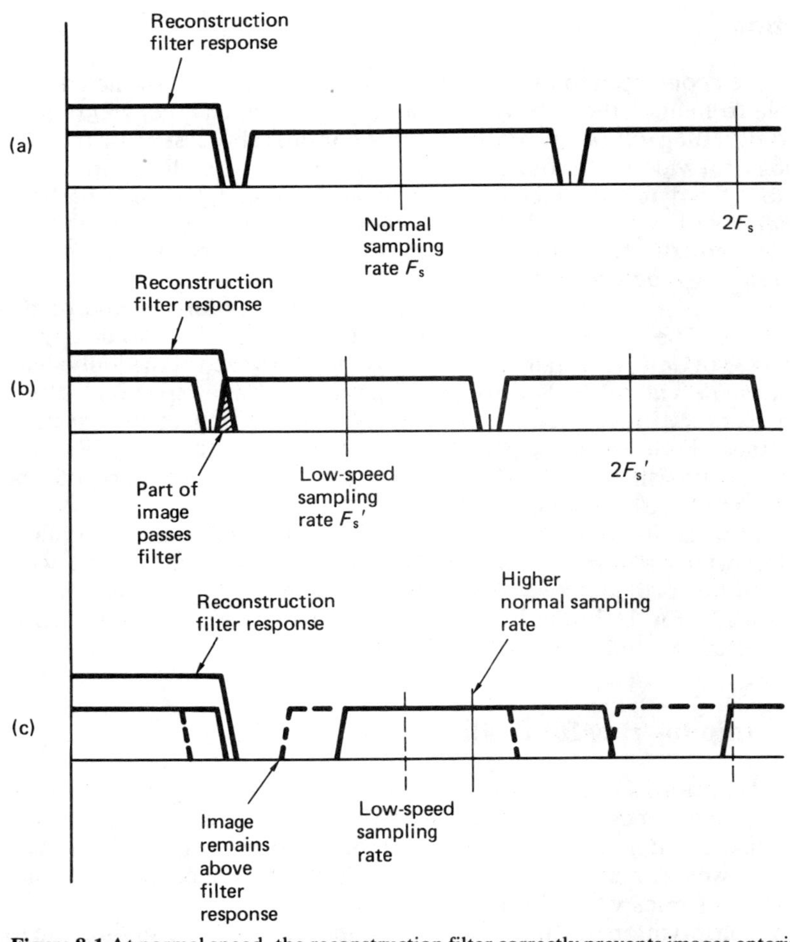

Figure 8.1 At normal speed, the reconstruction filter correctly prevents images entering the baseband, as at (a). When speed is reduced, the sampling rate falls, and a fixed filter will allow part of the lower sideband of the sampling frequency to pass. If the sampling rate of the machine is raised, but the filter characteristic remains the same, the problem can be avoided, as at (c)

does not easily relate to any of the above sampling rates and requires special handling which will be discussed further in this chapter.[1]

The success of the Compact Disc has led to 44.1 kHz becoming dominant in professional audio. Now that oversampling DACs are widely used, the problem of variable speed does not arise, since the response of the oversampling filter tracks the actual sampling rate. CD-video disks use composite video recorded with 44.1 kHz audio. In the circumstances the availability of only a single sampling rate for D-2 is unfortunate, especially as it is quite easy to implement a variety of sampling rates within a common format simply by zero packing at the lower rates.

8.3 Audio quality considerations

The acuity of the human ear is astonishing. It can detect tiny amounts of distortion and will accept an enormous dynamic range. The only criterion for quality that we have is that if the ear cannot detect impairments, we must say that the reproduced sound is perfect. Usually, people's ears are at their most sensitive between about 2 kHz and 5 kHz, and although some people can detect 20 kHz at high level, there is much evidence to suggest that most listeners cannot tell if the upper frequency limit of sound is 20 kHz or 16 kHz.[2,3] For a long time it was thought that frequencies below about 40 Hz were unimportant, but it is becoming clear that reproduction of frequencies down to 20 Hz improves reality and ambience.[4] A digital system can deliver a response down to DC if necessary. The dynamic range of the ear is obtained by a logarithmic response, and certainly exceeds 100 dB. At the extremes of this range, the ear is either straining to hear or suffering pain, neither of which can be described as pleasurable or entertaining, and it is hardly necessary to produce recordings of this dynamic range for consumers since, among other things, they are unlikely to have anywhere to listen to them.

Probably more important than dynamic range is the sensitivity of the ear to distortion. The ear behaves as a kind of spectrum analyser, with frequency bands about 100 Hz wide below 500 Hz and from one-sixth to one-third of an octave wide, proportional to frequency, above this. In the presence of a complex spectrum, it appears to protect itself from information overload by failing to register energy in some bands when there is more energy in a nearby band. This is the phenomenon of auditory masking, defined as the decreased audibility of one sound in the presence of another. The information reduction achieved is considerable; masking can take place even when the masking tone ceases before the masked sound appears. Another example of the slowness of the ear is the Haas effect, in which the source of a sound is attributed to the first arriving wavefront even though a later echo is much louder. Since distortion results in energy moving from one frequency band to another, a knowledge of masking is essential to estimate how audible the effect of distortion will be. Before digital techniques were used for high-quality audio, it was thought that the principles of digitizing were adequately understood, but the disappointing results of some early digital audio machines showed that this was not so. The ear could detect the minute imperfections of filters and converters, which could be neglected in, for example, instrumentation applications. A more rigorous study of digitization was soon applied, and the essentials of it can be found here. The important topic of oversampling which was shown in Chapter 2 to be useful in video is also extremely relevant to audio and will be described in that context.

8.4 Types of digitization

There are several methods of converting an audio waveform to a bit stream, and it is more useful to compare them than to contrast them because they are related; in advanced conversion systems it is possible to move from one system to another to combine the advantages of both. Figure 8.2 introduces the major techniques involved in conversion.

In digital video, the sampling process produces pulses whose amplitude is an analog quantity. When the height of these pulses is quantized and expressed as a number code, the result is known as pulse code modulation, again better known by the abbreviation PCM. The frequencies involved in video virtually preclude any other approach, but this is not the case for audio.

The amplitude of the signal which can be conveyed by PCM only depends on the number range of the quantizer and is independent of the frequency of the input. Similarly, the amplitude of the unwanted signals introduced by the quantizing process is also largely independent of input frequency.

In differential pulse code modulation (DPCM), the parameter which is quantized is the difference between the previous absolute sample value and the current one. This process limits the maximum rate at which the input signal voltage can change, and thus the permissible signal amplitude falls at 6 dB per octave. The unwanted signal amplitude is constant again, so as input frequency rises, ultimately the signal level available will fall down to it.

It is possible to produce a DPCM signal from a PCM signal by simply subtracting successive samples; this is digital differentiation. Similarly the reverse process is possible by using an accumulator to compute sample values from the differences received. The problem with this approach is that it is very easy to lose the baseline of the signal if it commences at some arbitrary time, and a digital high-pass filter is necessary to prevent unwanted offsets.

If DPCM is taken to the extreme case where only a binary output signal is available then the process is described as delta modulation. The meaning of the binary signal is that the analog input is above or below the accumulation of previous bits. The characteristics of the system show the same trends as DPCM, except that there is severe limiting of the rate of change of the input signal. Since the decoder must also accumulate all the difference bits to provide an analog output, this function can be performed by an integrator.

If an integrator is placed in the input to a delta modulator, the integrator's response loss of 6 dB per octave parallels the amplitude limit of 6 dB per octave, and thus the system amplitude limit becomes independent of frequency. This integration is responsible for the term delta sigma modulation, since in mathematics sigma is used to denote summation. The transmitted signal is now the amplitude of the input, not the slope; thus the receiving integrator can be dispensed with, and all that is necessary is an LPF to smooth the bits. Unfortunately the removal of the integration stage now means that the unwanted signal amplitude rises at 6 dB per octave, ultimately meeting the level of the wanted signal.

This principle of using an input integrator can also be applied to a true DPCM system and the result should perhaps be called sigma DPCM. The dynamic range improvement over delta sigma modulation is 6 dB for every extra bit in the code. Because the level of the unwanted signal rises at 6 dB per octave, just as in delta sigma modulation, the system is sometimes referred to as a 'noise-shaping' converter, although it will be seen later in this chapter that the use of the word

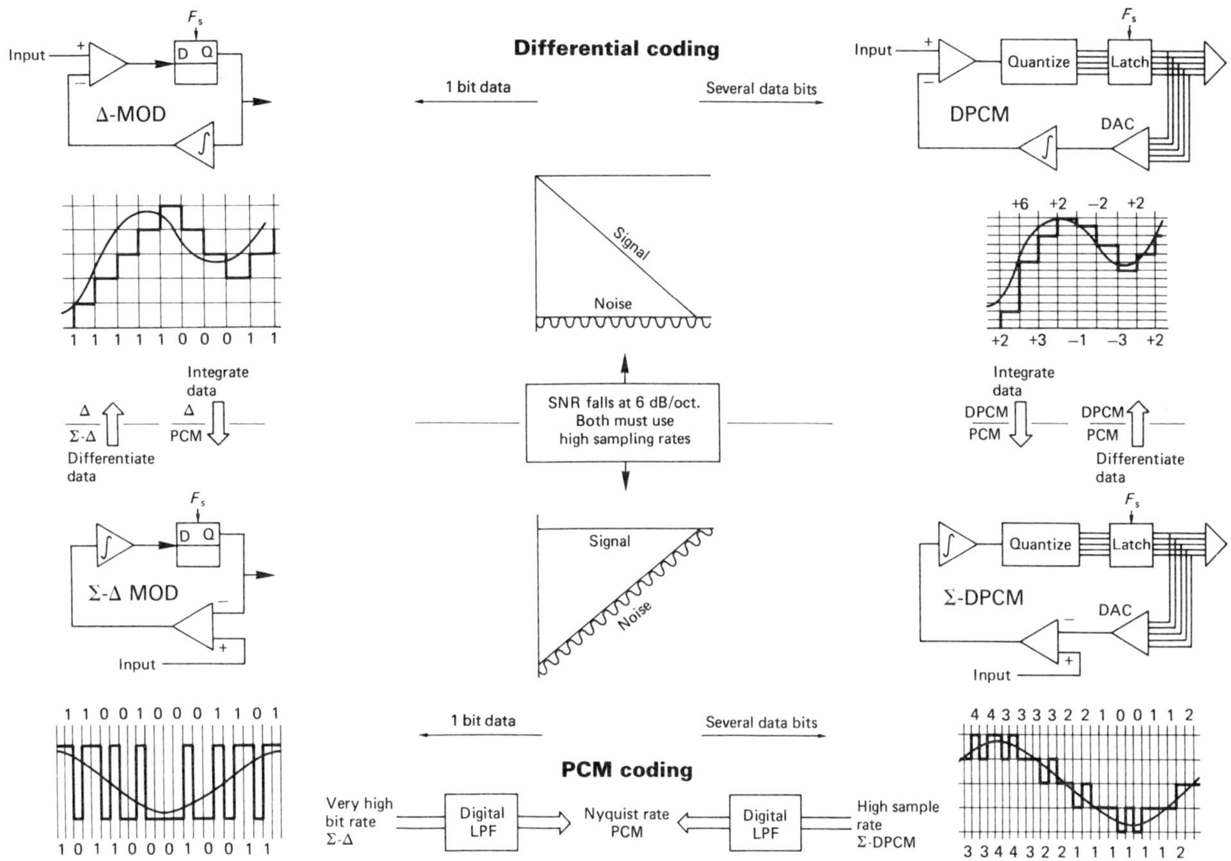

Figure 8.2 The four main alternatives to simple PCM conversion are compared here. Delta modulation is a 1 bit case of differential PCM, and conveys the slope of the signal. The digital output of both can be integrated to give PCM.

Σ–Δ (sigma–delta) is a 1 bit case of Σ-DPCM. The application of integrator before differentiator makes the output true PCM, but tilts the noise floor; hence these can be referred to as 'noise-shaping' converters 'noise' is not wholly appropriate. The output of a sigma DPCM system is again the absolute signal amplitude, and a DAC will be needed to receive it, because it is a binary code.

As the differential group of systems suffer from a wanted signal that converges with the unwanted signal as frequency rises, they all must use high sampling rates.[5] It is possible to convert from sigma DPCM to conventional PCM by reducing the sampling rate digitally. This has some advantages which will become evident when oversampling is discussed (Section 8.7).

8.5 Basic digital-to-analog conversion

The reverse of the quantizing process will be discussed first, since ADCs often use DACs in feedback loops.

There are two main ways of obtaining an analog signal from PCM data. One is to control binary-weighted currents and sum them; the other is to control the length of time a fixed current flows into an integrator. The two methods are contrasted in Figure 8.3. They look simple, but are of no use for audio in these forms because of practical limitations. In Figure 8.3(c), the binary code is about to have a major overflow, and all the low-order currents are flowing. In Figure 8.3(d), the binary input has increased by one, and only the most significant current flows. This current must equal the sum of all the others plus one least significant current to an accuracy of rather better than one least significant current. In this simple 4 bit example, the necessary accuracy is only one part in 16, but for a 16 bit system it would become one part in 65 536, or about 0.0015%. This degree of accuracy is almost impossible to achieve, let alone maintain in the presence of ageing and temperature change.

The integrator-type converter in this 4 bit example is shown in Figure 8.3(e); it requires a clock for the counter which allows it to count up to the maximum in less than one sample period. This will be more than 16 times the sampling rate. However, in a 16 bit system, the clock rate would need to be 65 536 times the sampling rate, or about 3 GHz. Clearly some refinements are necessary to allow either of these converter types to be used in audio applications.

One method of producing highly accurate currents is *dynamic element matching*.[6,7] Figure 8.4 shows a current source feeding a pair of nominally equal resistors. The two will not be the same owing to manufacturing tolerances and drift, and thus the current is only approximately divided between them. A pair of change-over switches places each resistor in series with each output. The average current in each output will then be identical, provided that the duty cycle of the switches is exactly 50%. This is readily achieved in a divide-by-two circuit. Current averaging is by a pair of capacitors which do not need to be of any special quality. By cascading these divide-by-two stages, a binary-weighted series of currents can be obtained, as in Figure 8.5. In practice, a reduction in the number of stages can be obtained by using a more complex switching arrangement. This generates currents of ratio 1:1:2 by dividing the current into four paths and feeding two of them to one output, as shown in Figure 8.6. A major advantage of this approach is that no trimming is needed in manufacture, making it attractive for mass production.

To prevent interaction between the stages in weighted-current converters, the currents must be switched to ground or into the virtual earth by change-over switches. The on resistance of these switches is a source of error, particularly the

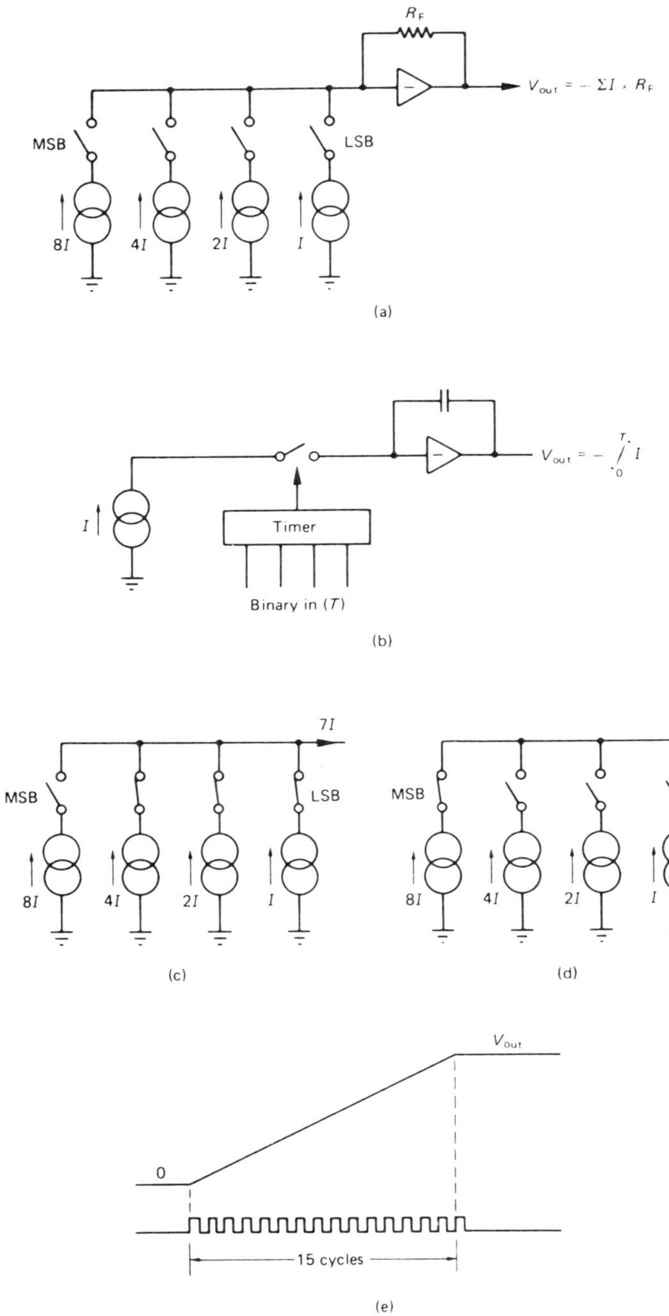

Figure 8.3 Elementary conversion: (a) weighted current DAC; (b) times integrator DAC; (c) current flow with 0111 input; (d) current flow with 1000 input; (e) integrator ramps up for 15 cycles of clock for input 1111

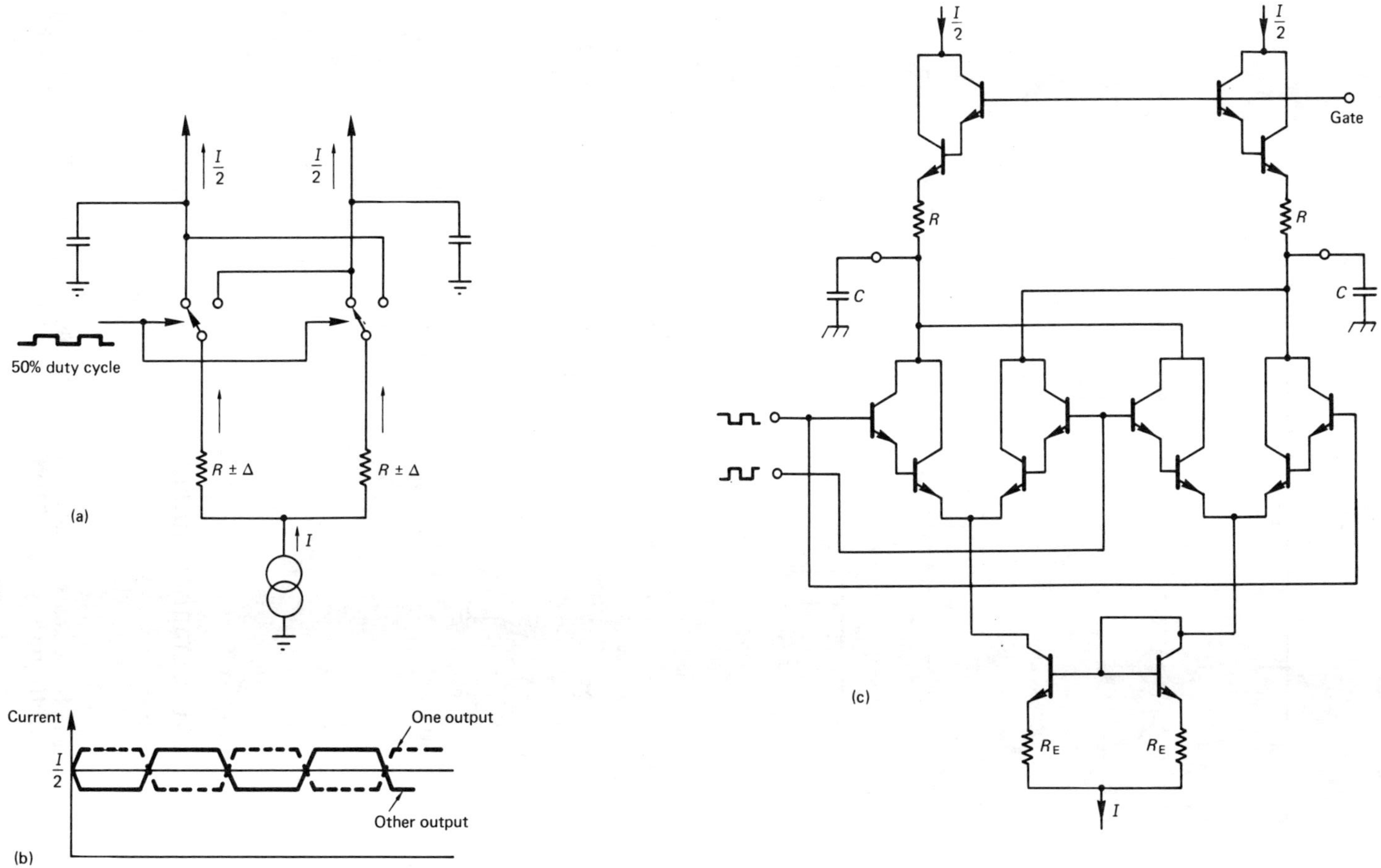

Figure 8.4 Dynamic element matching. (a) Each resistor spends half its time in each current path. (b) Average current of both paths will be identical if duty cycle is accurately 50%. (c) Typical monolithic implementation. Note clock frequency is arbitrary

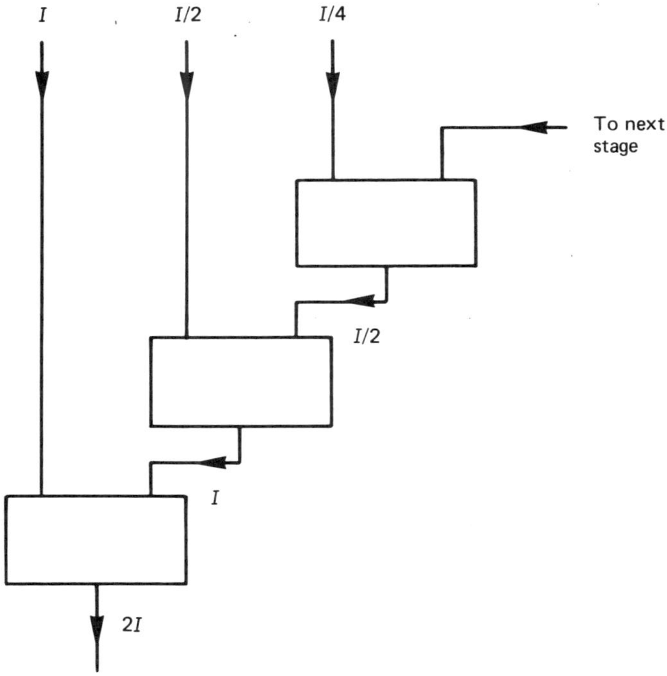

Figure 8.5 Cascading the current dividers of Figure 8.4 produces a binary-weighted series of currents

MSB, which passes most current. A solution in monolithic converters is to fabricate switches whose area is proportional to the weighted current, so that the voltage drops of all the switches are the same. The error can then be removed with a suitable offset. The layout of such a device is dominated by the MSB switch since, by definition, it is as big as all the others put together.

The practical approach to the integrator converter is shown in Figure 8.7 and Figure 8.8 where two current sources whose ratio is 256:1 are used; the larger is timed by the high byte of the sample and the smaller is timed by the low byte. The necessary clock frequency is reduced by a factor of 256. Any inaccuracy in the current ratio will cause errors, but tracking is easier to achieve in a monolithic device. The integrator capacitor must have low dielectric leakage, and the operational amplifier must have high input impedance to prevent non-linearity.

The output of the integrator will remain constant once the current sources are turned off, and an analog switch will be closed during the voltage plateau to produce the resampled output. Clearly this device cannot produce a zero-order hold output without an additional sample–hold stage, so it is naturally complemented by resampling. Once the output pulse has been gated to the reconstruction filter, the capacitor is discharged with a further switch in preparation for the next conversion. The conversion count must take place in rather less than one sample period to permit the resampling and discharge phases. A clock frequency of about 20 MHz is adequate for a 16 bit 48 kHz unit, which permits the ramp to complete in 12.8 μs, leaving 8 μs for resampling and reset.

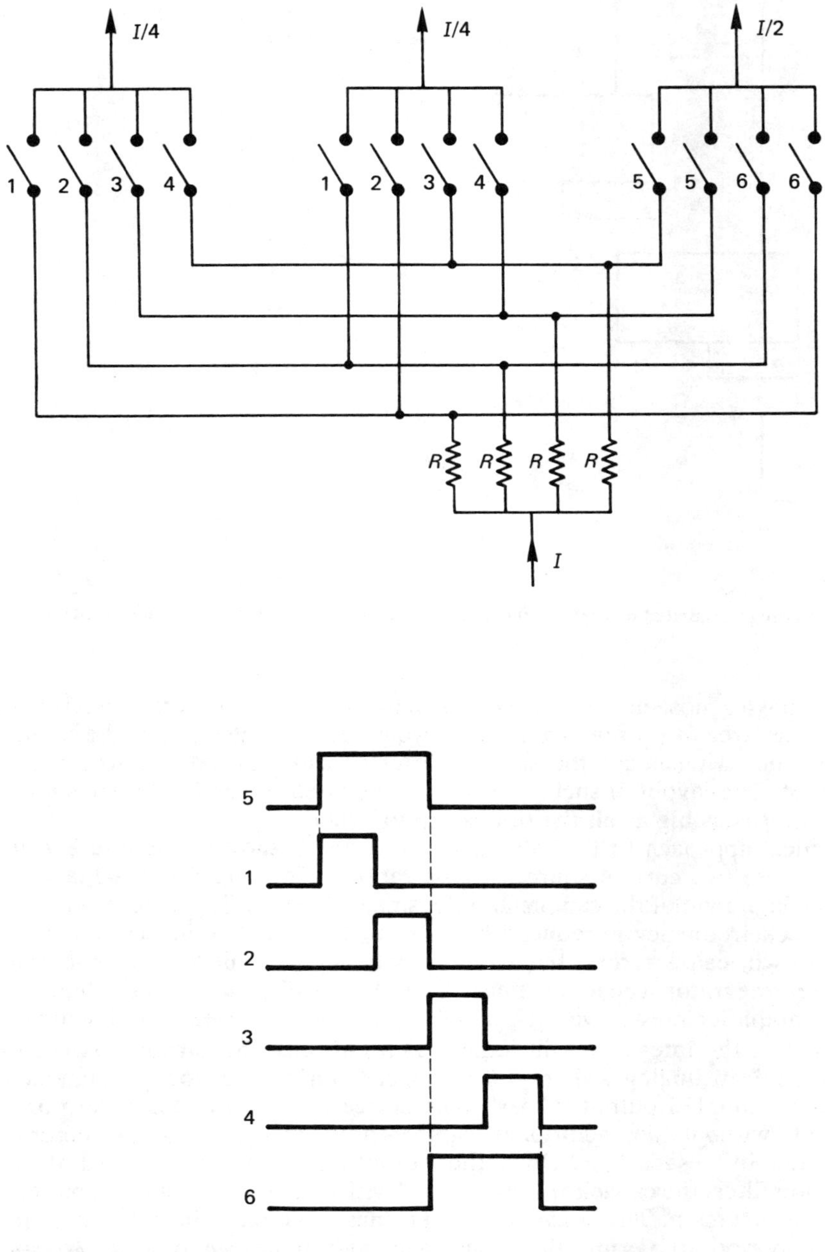

Figure 8.6 More complex dynamic element matching system. Four drive signals (1, 2, 3, 4) of 25% duty cycle close switches of corresponding number. Two signals (5, 6) have 50% duty cycle, resulting in two current shares going to right-hand output. Division is thus into 1:1:2

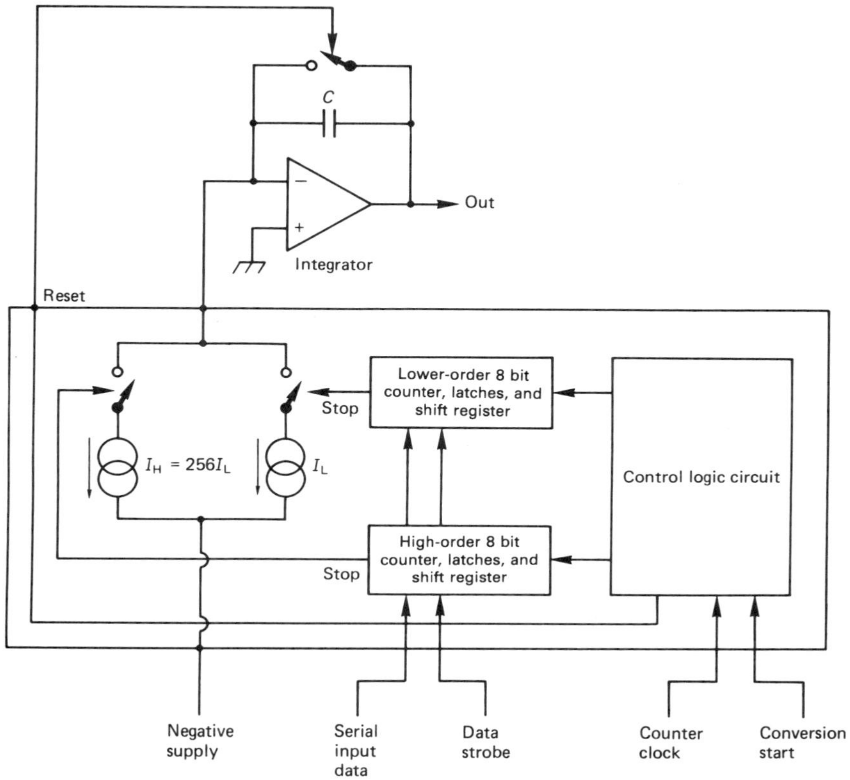

Figure 8.7 Simplified diagram of Sony CX-20017. The high-order and low-order current sources (I_h and I_L) and associated timing circuits can be seen. The necessary integrator is external

8.6 Basic analog-to-digital conversion

Many of the ADCs described here will need a finite time to operate, whereas an instantaneous sample must be taken from the input. The solution is to use a track/hold circuit, which is shown in simple form in Figure 8.9. When the switch is closed, the output will follow the input, and when the switch opens the capacitor holds the voltage of the signal at that instant. In practice, the shortcomings of this simple arrangement are too great for audio use. In particular, the time constant of the capacitor with the on resistance of the switch results in long settling time. The effect can be minimized by putting the switch inside a feedback loop as shown in Figure 8.9(b), since the series resistance of the switch will then be divided by the open-loop gain of the buffer. The specification of the buffer amplifier is stringent, since it requires adequate open-loop gain well beyond the audio band to ensure that the operation is always controlled by feedback. When the switch opens, the slightest change in the input will cause the buffer to saturate, and it must be able to

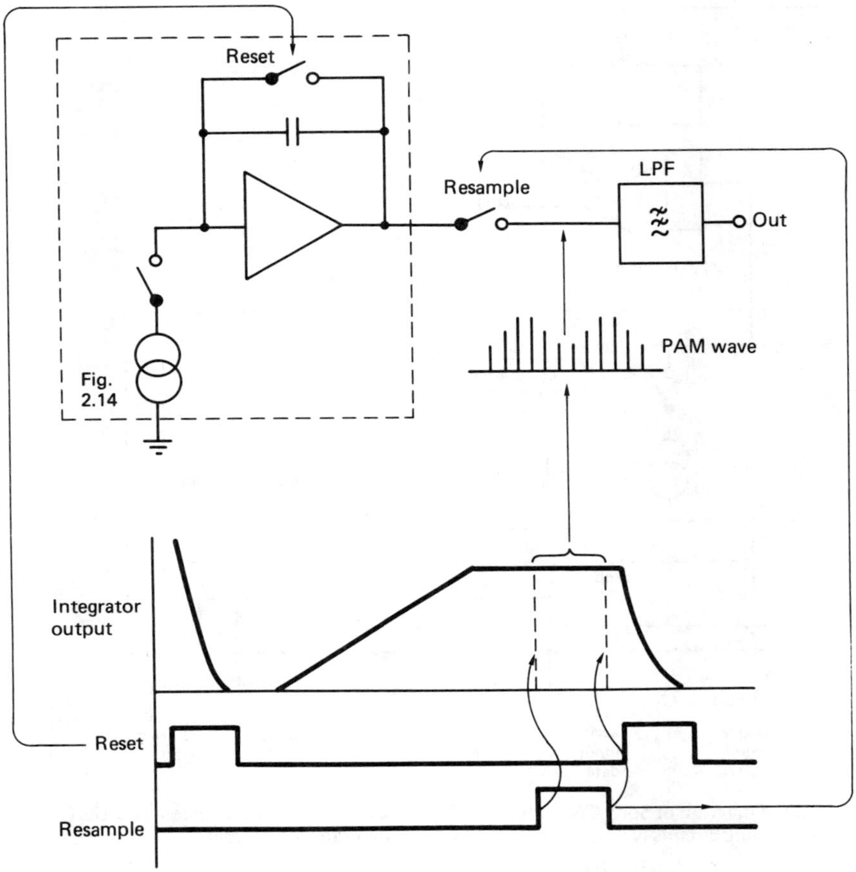

Figure 8.8 In an integrator converter, the output level is only stable when the ramp finishes. An analog switch is necessary to isolate the ramp from subsequent circuits. The switch can also be used to produce a PAM (pulse amplitude modulated) signal which has a flatter frequency response than a zero-order hold (staircase) signal

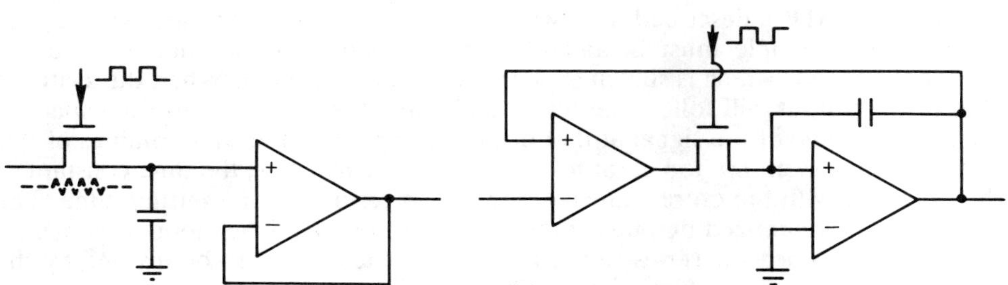

Figure 8.9 (a) The simple track/hold circuit shown has poor frequency response as the resistance of the FET causes a roll-off in conjunction with the capacitor. At (b) the resistance of the FET is now inside a feedback loop and will be eliminated, provided the left-hand op-amp never runs out of gain or swing

recover from this condition rapidly. The feedback eliminates the switch on resistance, but the off resistance must be high enough to prevent the input signal from affecting the held voltage. The impedance seen by the capacitor must be high enough to keep droop of the sample voltage within rather less than one quantizing interval.

The clock which drives the switch must have very low jitter, or noise may be superimposed on the sample values. The mechanism was illustrated in Chapter 2 and is equally relevant to audio. The mistiming due to clock jitter causes the wrong voltage to be sampled. Worst-case assumptions demand an accuracy within about 100 ps!

Figure 8.10 shows the various events during a track/hold sequence and catalogues the various sources of inaccuracy. The track/hold stage is extremely difficult to design because of the accuracy required in audio applications. In particular it is difficult to meet the droop specification for a system of more than 16 bits accuracy. Whenever such performance bottlenecks are found, it is tempting to look for an alternative method which avoids the problems. This alternative will be found in oversampling.

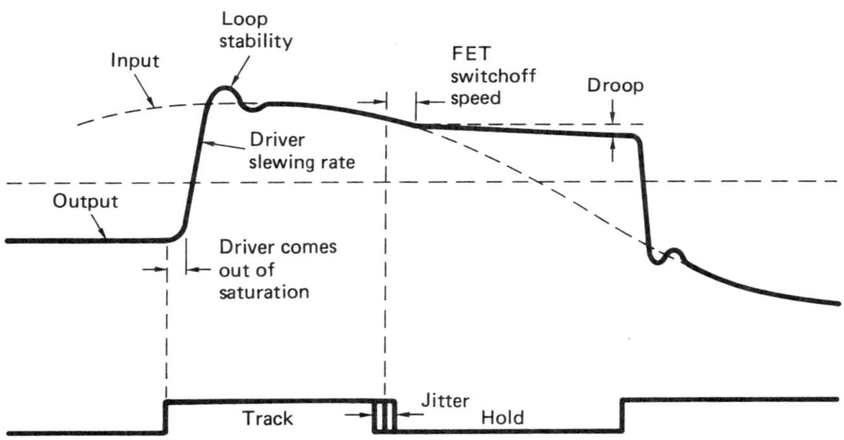

Figure 8.10 Characteristics of the feedback track/hold circuit of Figure 2.16(b) showing major sources of error

The general principle of a quantizer is that different quantized voltages are compared with the unknown analog input until the closest quantized voltage is found. The code corresponding to this becomes the output.

The flash converter is probably the simplest technique available for PCM and DPCM conversion. The principle was shown in Chapter 2. The threshold voltage of every quantizing interval is provided by a resistor chain which is fed by a reference voltage. As one comparator is necessary for each quantizing interval, a 16 bit device would need a ridiculous 65 535 comparators, and thus these converters are not practicable for direct audio conversion, although they will be used in DPCM and in oversampling converters. The analog signal has to drive a lot of inputs, and a low-impedance driver is essential to avoid restricting the slewing rate of the input. The extreme speed of a flash converter which makes it suitable for video use is a distinct advantage in audio where large oversampling factors can then be used.

Because computation of all bits is performed simultaneously, no track/hold circuit is required, and droop is eliminated.

Reduction in component complexity can be achieved by quantizing serially. The most primitive method of generating different quantized voltages is to connect a counter to a DAC as in Figure 8.11. The resulting staircase voltage is compared with the input and used to stop the clock to the counter when the DAC output has just exceeded the input. This method is painfully slow and is not used, as a much faster method exists which is only slightly more complex. Using successive approximation, each bit is tested in turn, starting with the MSB. If the input is

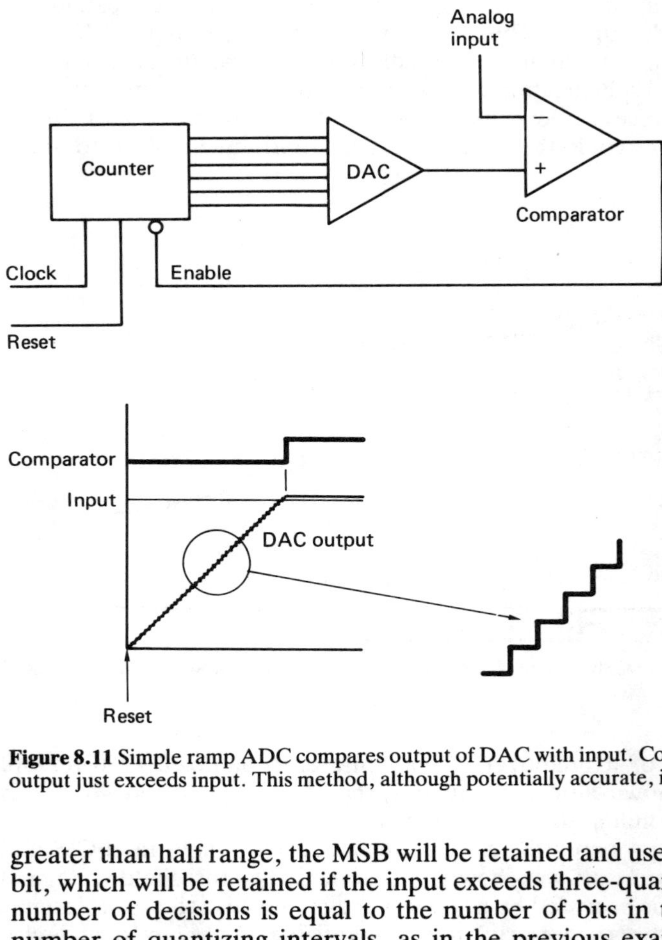

Figure 8.11 Simple ramp ADC compares output of DAC with input. Count is stopped when DAC output just exceeds input. This method, although potentially accurate, is much too slow for digital audio

greater than half range, the MSB will be retained and used as a base to test the next bit, which will be retained if the input exceeds three-quarters range and so on. The number of decisions is equal to the number of bits in the word, rather than the number of quantizing intervals, as in the previous example. A drawback of the successive approximation converter is that the least significant bits are computed last, when droop is at its worst. Figure 8.12 and Figure 8.13 show that droop can cause a successive approximation converter to make a significant error under certain circumstances.

Analog-to-digital conversion can also be performed using the dual-current-source-type DAC in a feedback system; the major difference is that the two current

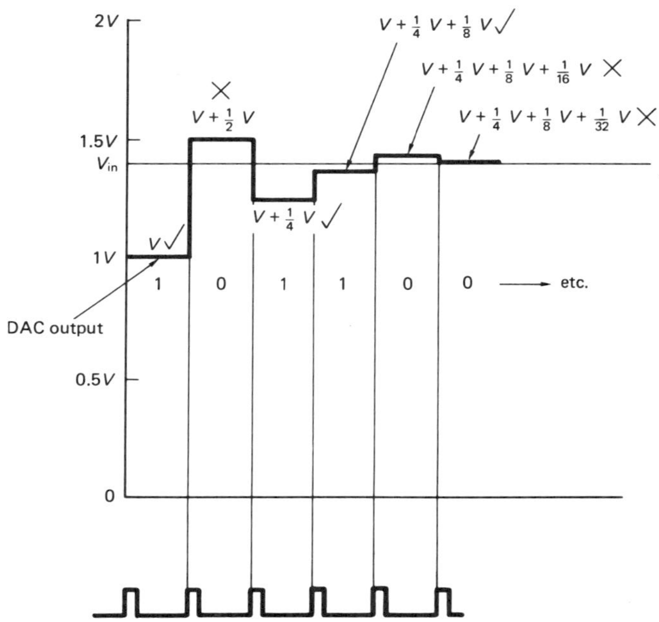

Figure 8.12 Successive approximation tests each bit in turn, starting with the most significant. The DAC output is compared with the input. If the DAC output is below the input (√) the bit is made 1; if the DAC output is above the input (×) the bit is made zero

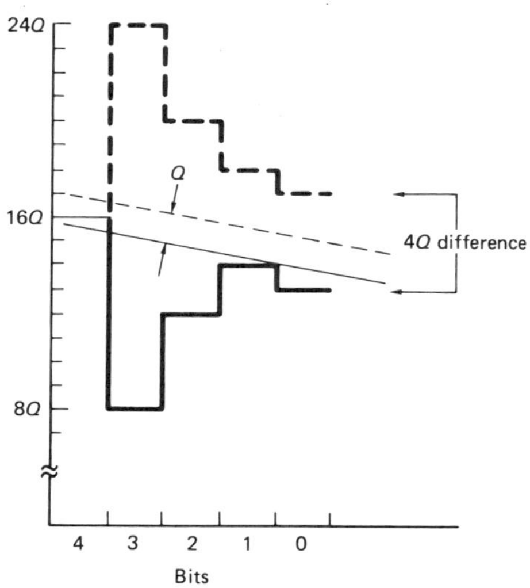

Figure 8.13 Two drooping track/hold signals (solid and dotted lines) which differ by one quantizing interval Q are shown here to result in conversions which are $4Q$ apart. Thus droop can destroy the monotonicity of a converter. Low-level signals (near the midrange of the number system) are especially vulnerable

sources must work sequentially rather than concurrently. Figure 8.14 shows a 16 bit application in which the capacitor of the track/hold circuit is also used as the ramp integrator. The system operates as follows. When the track/hold FET switches off, the capacitor C will be holding the sample voltage. Two currents of ratio 128:1 are capable of discharging the capacitor. Owing to this ratio, the smaller current will be used to determine the seven least significant bits, and the larger current will determine the nine most significant bits. The currents are provided by current sources of ratio 127:1. When both run together, the current produced is 128 times that from the smaller source alone. This approach means that the current can be changed simply by turning off the larger source, rather than by attempting a change-over.

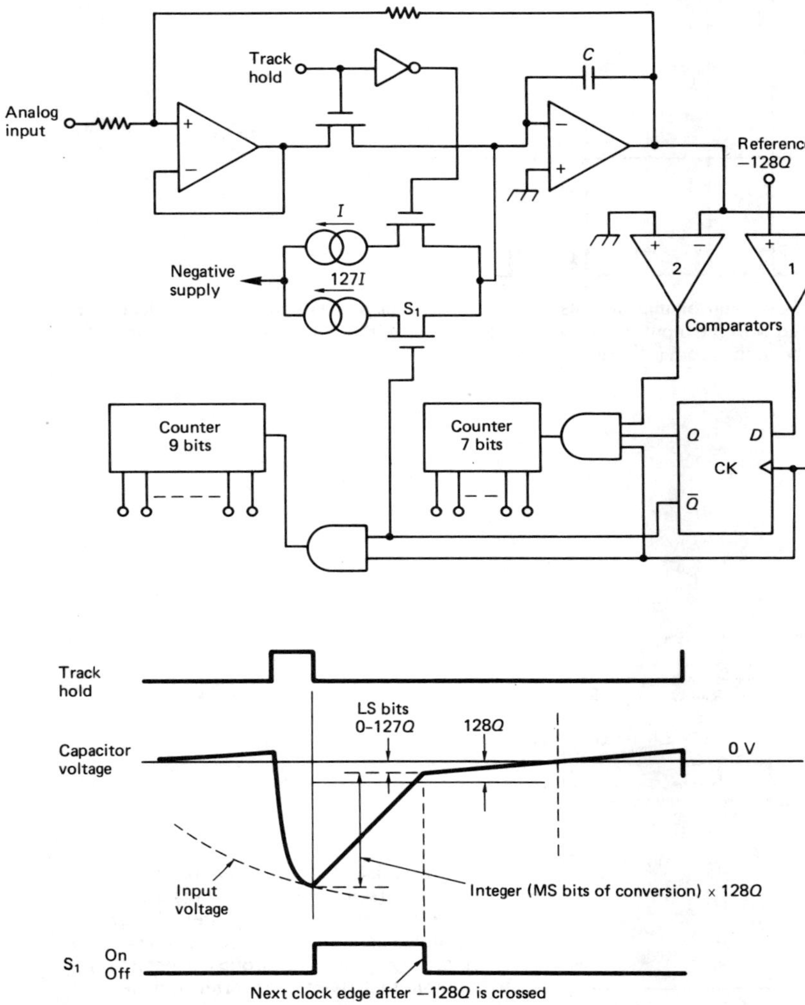

Figure 8.14 Dual-ramp ADC using track/hold capacitor as integrator

With both current sources enabled, the high-order counter counts up until the capacitor voltage has fallen below the reference of $-128Q$ supplied to comparator 1. At the next clock edge, the larger current source is turned off. Waiting for the next clock edge is important, because it ensures that the larger source can only run for entire clock periods, which will discharge the integrator by integer multiples of $128Q$. The integrator voltage will overshoot the $128Q$ reference, and the remaining voltage on the integrator will be less than $128Q$ and will be measured by counting the number of clocks for which the smaller current source runs before the integrator voltage reaches zero. This process is termed residual expansion. The break in the slope of the integrator voltage gives rise to the alternative title of gear-change converter. Following ramping to ground in the conversion process, the track/hold circuit must settle in time for the next conversion. In this 16 bit example, the high-order conversion needs a maximum count of 512, and the low order needs 128, a total of 640. Allowing 25% of the sample period for the track/hold circuit to operate, a 48 kHz converter would need to be clocked at some 40 MHz. This is rather faster than the clock needed for the DAC using the same technology.

An ADC cannot be more accurate than the DAC it contains, and because of the higher operating speed and the imperfections of the track/hold process, ADCs are generally responsible for more signal degradation than DACs. The two devices have the same transfer function, since they are only distinguished by the direction of operation, and therefore the same terminology can be used to classify the shortcomings of both. The major converter errors were shown in Chapter 2 in connection with video conversion and do not differ in audio converters except for the magnitude of allowable error.

Offset error has no effect on sound quality, unless the offset is gross, when the symptom would be premature clipping. DAC offset is of little consequence, but ADC offset is undesirable since it can cause an audible thump if an edit is made between two signals having different offsets. Offset error is sometimes cancelled by digitally averaging the converter output and feeding it back to the analog input as a small control voltage. Alternatively, a digital high-pass filter can be used.

The gain stability is probably the least important factor in a digital audio converter, since ears, meters and gain controls are logarithmic.

For audio the absolute accuracy is rather less important than linearity. For example, if all the current sources in a converter have good thermal tracking, linearity will be maintained, even though the absolute accuracy drifts.

8.7 Oversampling

The information in an analog audio signal is two-dimensional; it was shown earlier that the number of levels which can be resolved unambiguously represents one dimension, and the bandwidth represents the other dimension. Figure 8.15 shows that it is an area which is the product of bandwidth and signal-to-noise ratio expressed linearly. The figure also shows that the same amount of information can be conveyed down a channel with 6 dB less SNR if the bandwidth used is doubled, with 12 dB less SNR if bandwidth is quadrupled, and so on, provided that the modulation scheme used is perfect. This theory predicts that if an audio signal is spread over a much wider bandwidth by, for example, the use of an FM broadcast transmitter, the SNR of the demodulated signal will be higher than that of the channel it passes through, and this is clearly the case. The theory also predicts that

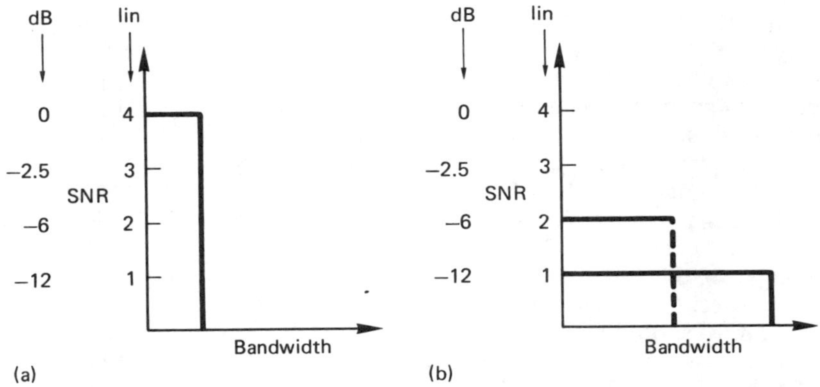

Figure 8.15 The number of levels which can be resolved in a voltage signal determines the amount of information at any instant. The bandwidth determines the rate at which that information can be conveyed; hence the overall information capacity is the product of linear SNR and bandwidth. If a given input signal at (a) is modulated in some perfect way, it may be conveyed down a channel with one-half the signal-to-noise ratio (−6 dB) if the bandwidth is doubled, or with one-quarter the signal-to-noise ratio (−12 dB) if the bandwidth is quadrupled

stereo FM will have more hiss; since two channels of audio are now using the same transmitter bandwidth, each one can have only half the information capacity, so it will lose 6 dB of SNR. In practice things are slightly worse, because the process is not perfect.

The information in an analog signal can be conveyed using some analog modulation scheme in any combination of bandwidth and SNR which yields the appropriate channel capacity. In digital audio, a signal having only two states, known as a binary channel, will be used, needing only a poor SNR but a correspondingly high bandwidth.

It is useful to examine information capacity in the digital domain. Figure 8.16 shows several examples. A single binary digit can only have two states; thus it can only convey two pieces of information, perhaps 'yes' or 'no'. Two binary digits together can have four states and can thus convey four pieces of information, perhaps 'spring summer autumn or winter', which is two pieces of information per bit. Three binary digits grouped together can have eight combinations and convey eight pieces of information, perhaps 'doh re mi fah so lah te or doh', which is nearly three pieces of information per digit. Clearly the further this principle is taken, the greater the benefit. In a 16 bit system, each bit is worth 4096 pieces of information. It is always more efficient, in information-capacity terms, to use the combinations of long binary words than to send single bits for every piece of information. This is one reason why PCM is more popular than delta modulation, despite the simplicity of implementation of the latter. PCM simply makes more efficient use of the capacity of the binary channel.

Information theory is necessary to understand oversampling. The storage or transmission system is usually going to be PCM, where the sampling rate is a little more than twice the audio bandwidth. In the converters, the sampling rate will be higher, because of the advantages stated earlier in the construction of the analog filters. There are other advantages because, when the sampling rate is raised, information theory suggests that the wordlength of the samples can be reduced.

	0 = No 1 = Yes	00 = Spring 01 = Summer 10 = Autumn 11 = Winter	000 do 001 re 010 mi 011 fah 100 so 101 lah 110 te 111 do	0000 0 0001 1 0010 2 0011 3 0100 4 0101 5 0110 6 0111 7 1000 8 1001 9 1010 A 1011 B 1100 C 1101 D 1110 E 1111 F		0000 ⋮ FFFF (Digital audio sample values)
No of bits	1	2	3	4		16
Information per word	2	4	8	16		65536
Information per bit	2	2	≈3	4		4096

Figure 8.16 The amount of information per bit increases disproportionately as wordlength increases. It is always more efficient to use the longest words possible at the lowest word rate. It will be evident that 16 bit PCM is 2048 times as efficient as delta modulation. Oversampled data are also inefficient for storage

Figure 8.17 shows the reverse case of Figure 8.16. The information rate is held constant, and as the sampling rate doubles, one bit can be removed from the wordlength. Thus by using oversampling, not only can the problems of analog filters be overcome, but also the wordlength of the converters is reduced, making them easier to construct. Note that the theory only predicts what is possible; it is not a guarantee of success. Oversampling simply means raising the sampling rate. Further mechanisms are needed to take advantage of the wordlength reduction.

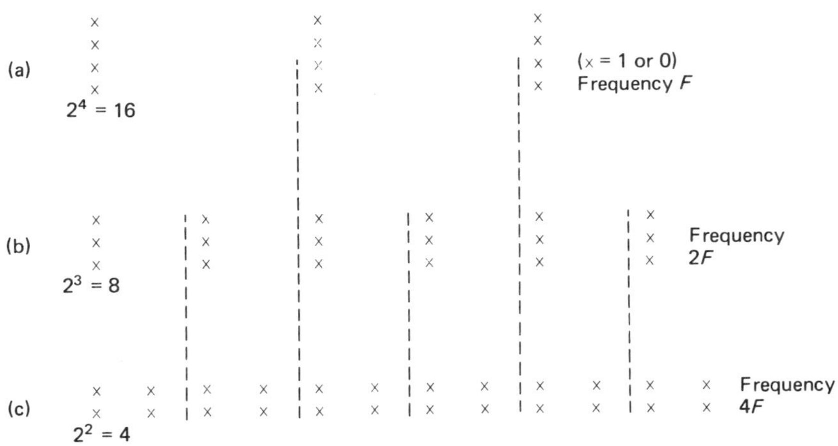

Figure 8.17 Information rate can be held constant when frequency doubles by removing 1 bit from each word. In all cases here it is $16F$. Note bit rate of (c) is double that of (a). Data storage in oversampled form is inefficient

8.8 An oversampling DAC

Consider a system which oversamples by a factor of four. Starting with 16 bit PCM, the 4× oversampling will permit the use of a 14 bit converter, but only if the wordlength is reduced optimally. Simple truncation of wordlength gives the same result as if the original audio had been quantized into fewer levels in the first place. For every bit lost, the same amount of distortion will be obtained with a level 6.02 dB higher. Simple truncation, then, does not allow the results predicted by information theory.

The roundoff mechanism used in oversampling spreads the distortion products due to truncation over the entire oversampling spectrum; thus distortion power within the baseband is only a fraction of the total. The fraction is the reciprocal of the oversampling factor. Thus in our 4× example, removing 2 bits raises the distortion by 12 dB, but this is spread over a spectrum four times as great, thus reducing the distortion by the same 12 dB. The wordlength is reduced by an extension of the technique of rounding up. The error caused by the previous truncation is carried over to the next, so that the average error of the two is smaller. As the sampling rate is much higher than normal, the averaging process will have taken place by the time the signal has returned to baseband audio. Figure 8.18 shows that the accumulated error is controlled by using the bits which were neglected in the truncation, and adding them to the next sample. In this example, with a steady input, the roundoff mechanism will produce an output of 01110111.... If this is low-pass filtered, the three ones and one zero result in a level of three-quarters of a bit, which is precisely the level which would have been

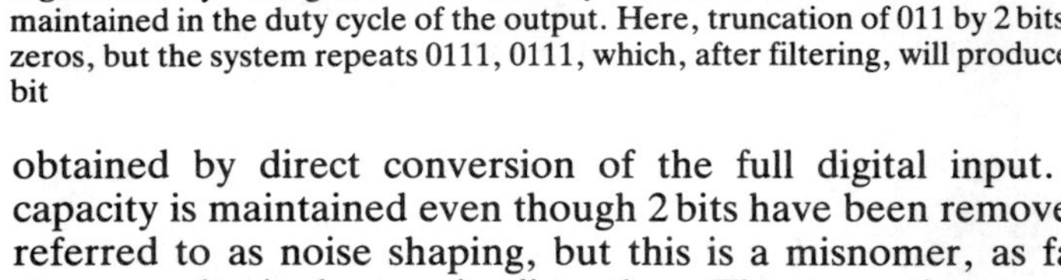

Figure 8.18 By adding the error caused by truncation to the next value, the resolution of the lost bits is maintained in the duty cycle of the output. Here, truncation of 011 by 2 bits would give continuous zeros, but the system repeats 0111, 0111, which, after filtering, will produce a level of three-quarters of a bit

obtained by direct conversion of the full digital input. Thus the information capacity is maintained even though 2 bits have been removed. This process is often referred to as noise shaping, but this is a misnomer, as failure to perform these steps results in harmonic distortion. The term time averaging is also used to describe this kind of resolution extension.

In the oversampling system used in Philips Compact Disc players[8] the aperture effect in the DAC is used as part of the reconstruction filter response, in conjunction with a third-order Bessel filter. Equalization of the frequency response is by the digital filter which produces the oversampled data. The operation of digital filters is described in Chapter 3, where it will be seen that their frequency response is proportional to the sampling rate. If a digital recorder is played at a reduced speed, the response of the digital filter will reduce automatically and prevent images passing the reconstruction process. If oversampling were to become universal, there would then be no need for the 48 kHz sampling rate.

8.9 Oversampling ADCs

The use of oversampling to extend the resolution of ADCs is subject to the same limits set by information theory, but it is harder to approach the limit in ADCs than in DACs.

If a perfect quantizer is used, no amount of oversampling will increase the resolution of the system, since a perfect quantizer is blind to all changes of input within one quantizing interval, and looking more often is of no help. It was shown earlier that the use of dither would linearize a quantizer, so that input changes much smaller than the quantizing interval would be reflected in the output. Dither must be employed to minimize distortion in conventional quantizers. Resolution cannot possibly be extended in ADCs unless some signal is added to the analog input.

Figure 8.19 shows the example of a white-noise-dithered quantizer, oversampled by a factor of four. Since dither is correctly employed, it is valid to speak of the unwanted signal as noise. The noise power extends over the whole baseband up to the Nyquist limit. If the baseband width is reduced by the oversampling factor of four back to the bandwidth of the original analog input, the noise bandwidth will also be reduced by a factor of four, and the noise power will be one-quarter of that produced at the quantizer. One-quarter noise power implies one-half the noise voltage, so the SNR of this example has been increased by 6 dB, the equivalent of one extra bit in the quantizer. Information theory predicts that an oversampling factor of four would allow an extension by 2 bits, so it can be concluded that this method is suboptimal.

The division of the noise by a larger factor is the only route left open, since all the other parameters are fixed by the signal bandwidth required. The reduction of

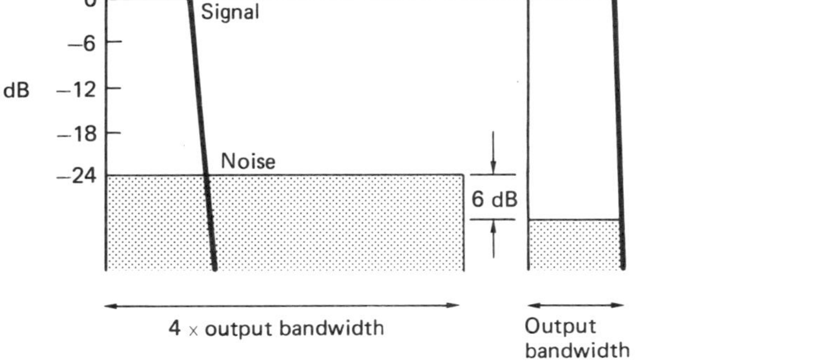

Figure 8.19 In this simple oversampled converter, 4× oversampling is used. When the converter output is low-pass filtered, the noise power is reduced to one-quarter, which in voltage terms is 6 dB. This is a suboptimal method and is not used

noise power resulting from a reduction in bandwidth is only proportional if the noise is white, i.e. it has uniform power spectral density (PSD). If the noise from the quantizer is made spectrally non-uniform, the oversampling factor will no longer be the factor by which the noise power is reduced. The goal is to concentrate noise power at high frequencies, so that after low-pass filtering in the digital domain to the audio input bandwidth, the noise power will be reduced by more than the oversampling factor. The sigma DPCM converter has a natural application here; as seen earlier its noise spectrum rises at 6 dB per octave. Figure 8.20 shows an oversampling system using a sigma DPCM converter and an oversampling factor of four. The sampling spectrum shows that the noise is concentrated at frequencies

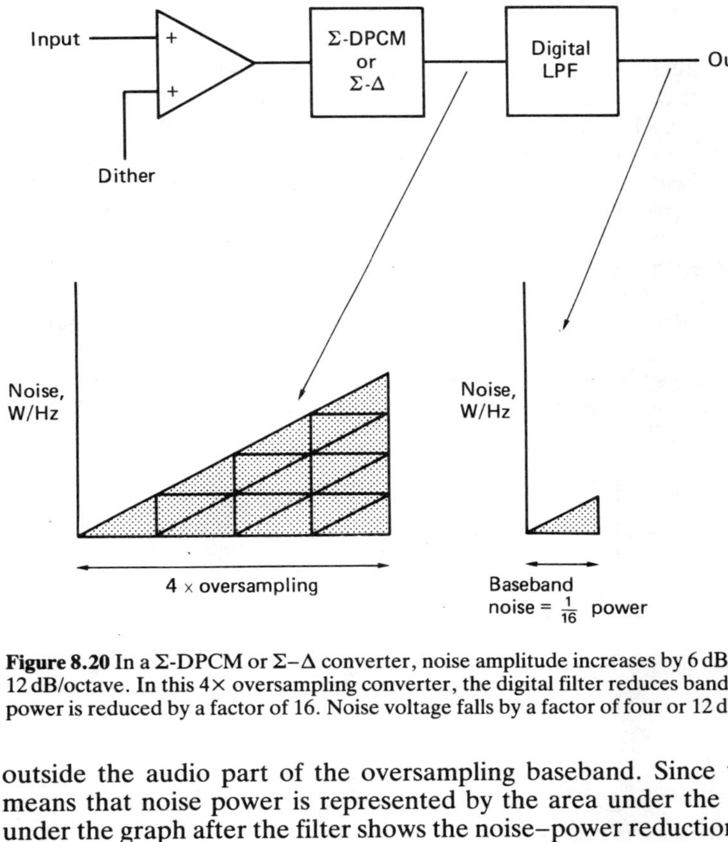

Figure 8.20 In a Σ-DPCM or Σ–Δ converter, noise amplitude increases by 6 dB/octave, noise power by 12 dB/octave. In this 4× oversampling converter, the digital filter reduces bandwidth by four, but noise power is reduced by a factor of 16. Noise voltage falls by a factor of four or 12 dB

outside the audio part of the oversampling baseband. Since the scale used here means that noise power is represented by the area under the graph, the area left under the graph after the filter shows the noise–power reduction. Using the relative areas of similar triangles shows that the reduction has been by a factor of 16. The corresponding noise–voltage reduction would be a factor of four, or 12 dB, in a white-noise system, being exactly that permitted by four times oversampling, but because of the rise of 6 dB per octave in the PSD of the noise, the SNR will be 3 dB worse at the edge of the audio band. This 3 dB loss remains the same whatever the oversampling factor; therefore the greater the oversampling factor, the less significant the loss. The noise injected by the vital dither process will be of the same order anyway, and the digital low-pass filter will introduce some noise, so that the slope of 6 dB per octave will not continue back to the origin; it will level out. The performance of such a system is thus not far short of the information theory limit for a dithered converter. A 16 times oversampling system using this technique allows a resolution extension of 21 dB. The number of bits needed to express this greater resolution will be four; thus the effective quantizing intervals of the system have been made one-sixteenth the size of the quantizer steps. The amount of white-noise dither on the analog input must be reduced accordingly. Adams[9] has described a system based on these principles using an oversampling factor of 128 to achieve a true 18 bit accurate conversion with exceptional phase linearity. At the time of writing this probably represents the state of the art in audio converters.

8.10 Digital audio interconnection

Although digital audio recorders and the audio sections of digital video recorders can be connected to an analog world by the use of converters, many of the advantages of the digital domain are lost if transfer of audio between digital machines has to be via analog signals. The importance of direct digital interconnection was realized early, and numerous incompatible methods were developed by various manufacturers until standardization was reached in the shape of the AES/EBU digital audio interface.

8.11 AES/EBU interconnect

In all of the digital interconnects which were independently devised by various manufacturers there is enough similarity due to the common purpose to make interfacing possible with a little extra hardware, but enough difference to be irritating.

The AES/EBU digital audio interface[10] was proposed to embrace all the functions of existing formats in one standard which would ensure interconnection independent of the manufacture of equipment at either end. For consumer use, a corresponding standard has been arrived at which offers different facilities, yet retains compatibility with the professional interface so that, for many purposes, consumer and professional machines can be connected together.[11]

Many of the older interconnects for professional use have separate lines for bit clocks and sampling-rate clocks, which is acceptable for the short distances required for simple dubbing, but causes problems in the broadcast environment where long lines might be needed in a studio complex. It was desired to use the existing analog audio cabling in such installations, which would be 600 ohm balanced line screened, with one cable per audio channel, or in some cases one twisted pair per channel with a common screen.

If a single channel is to be used, the interconnect has to be self-clocking and self-synchronizing, i.e. the single signal must carry enough information to allow the boundaries between individual bits, words and blocks to be detected reliably. To fulfil these requirements, the AES/EBU interconnect and the consumer equivalent both use FM channel code (see Chapter 4) which is DC free and strongly self-clocking. Synchronization is achieved by violating the usual encoding rules.

Using FM means that the channel frequency is the same as the bit rate when sending data ones. Tests showed that in typical analog audio-cabling installations, sufficient bandwidth was available to convey two digital audio channels in one twisted pair. The standard driver and receiver chips for RS-422A[12] data communication (or the equivalent CCITT-V.11) are employed for professional use, but work by the BBC[13] suggests that equalization and transformer coupling are desirable for longer cable runs, particularly if several twisted pairs occupy a common shield. Successful transmission up to 350 m has been achieved with these techniques. The use of the transformer is mandatory in the equivalent EBU specification.[14] Figure 8.21(a) shows a typical configuration. The output impedance of the drivers will be about 110 ohms, and the impedance of the cable used should be similar at the frequencies of interest. The driver should produce between 3 and 10 V p–p into such an impedance.

The receiver impedance is high at 250 ohms, which allows up to four receivers to be driven from one source. The number of loads may need to be reduced if long cables are used. In Figure 8.21(b), the specification of the receiver is shown in terms of the minimum eye pattern (see Chapter 4) which can be detected without error.

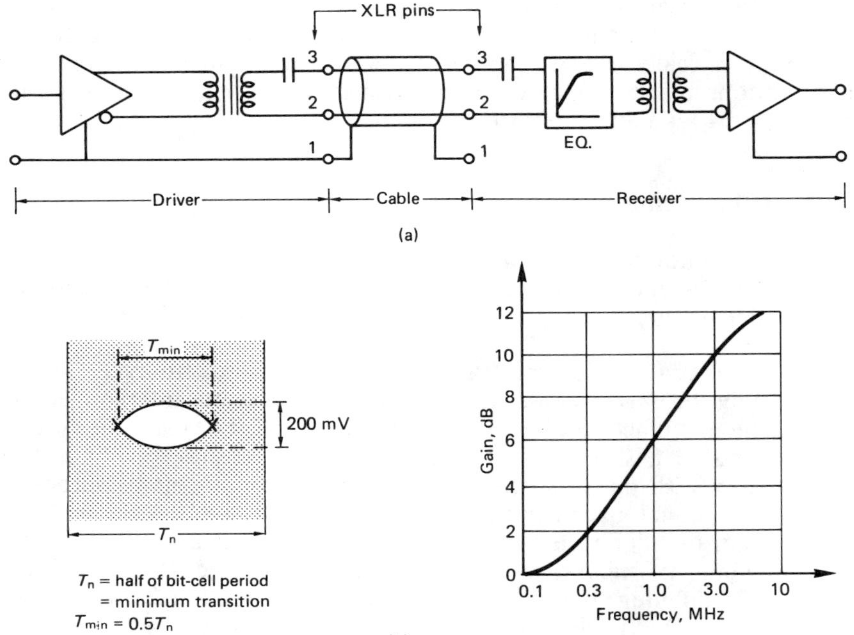

Figure 8.21 The configuration of the AES/EBU interface which is suggested for long cables. XLR pin connections are the same as balanced-line audio cables, which can be used. At (b) the minimum eye pattern which a receiver must be able to detect, and a suggested equalization response

The purpose of the standard is to allow the use of existing analog cabling, and as an adequate connector in the shape of the XLR is already in wide service, the recommendations of IEC 268 Part 12 have been adopted for digital audio use. Effectively, existing analog audio cables having XLR connectors can be used without alteration for digital connections. The standard does, however, require that suitable labelling should be used so that it is clear that the connections on a particular unit are digital.

The need to drive long cables does not generally arise in the domestic environment, and so a low-impedance balanced signal is not necessary. The electrical interface of the consumer format uses a 0.5 V peak single-ended signal, which can be conveyed down conventional audio-grade coaxial cable connected with phono plugs.

In Figure 8.22 the basic structure of the professional and consumer formats can be seen. One subframe consists of 32 bit cells, of which four will be used by a synchronizing pattern. Up to 24 bit sample wordlength can be used, which should cater for all conceivable future developments, but normally 20 bit samples will be available with four auxiliary data bits, which might be used for a voice-grade channel in a professional application. In a consumer RDAT machine, subcode can be transmitted in bits 4–11, and the 16 bit audio in bits 12–27.

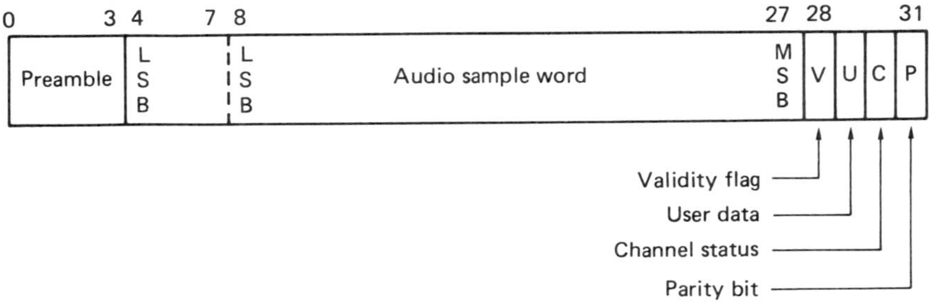

Figure 8.22 The basic subframe structure of the AES/EBU format. Sample can be 20 bits with four auxiliary bits, or 24 bits. LSB is transmitted first

In contrast with preceding formats, this format sends the least significant bit first. One advantage of this approach is that simple serial arithmetic is then possible on the samples because the carries produced by the operation on a given bit can be delayed by one bit period and then included in the operation on the next higher-order bit.

Four status bits accompany each subframe. The validity flag will be reset if the associated sample is reliable. The parity bit produces even parity over the subframe, such that the total number of ones in the subframe is even. This allows for simple detection of an odd number of bits in error and improves the probability of detection of sync. The user and channel-status bits are discussed later.

Two of the subframes described above make one frame, which repeats at the sampling rate in use. The first subframe will contain the sample from channel A, or from the left channel in stereo working. The second subframe will contain the sample from channel B, or the right channel. At 48 kHz, the bit rate will be 3.072 MHz.

In order to separate the audio channels on receipt the synchronizing patterns for the two subframes are different.

The channel-status and user bits in each subframe form serial data streams with 1 bit of each per audio channel per frame. It is in the use of these bits that the differences between the professional and consumer interface standards are most pronounced. The channel-status bits are given a block structure and synchronized every 192 frames, which at 48 kHz gives a block rate of 250 Hz, corresponding to a period of 4 milliseconds.

In professional applications, users are free to assign any desired format to the user bits, but it is recommended that the same block size is used, so that the user-block boundaries align with the channel-status block boundaries. This would minimize problems if the data structure was upset during synchronizing operations between systems with non-phased sampling clocks. It has been suggested that the user data are employed for labelling.[15] Labelling information is assembled into one or more 48 bit words which begin with a uniquely identifiable synchronizing pattern. Suggested uses of the labelling structure include:

Time and date of program origin
Program and identification number
Take number
Programme duration
Cue information
Act or scene number
Timecode-related information
Network information
Radio data information
Generation number (for recording)
Copyright ownership code
Editing information
Signal source information (e.g. originating microphone)

In order to synchronize the channel-status blocks, the channel A sync pattern is replaced for one frame only by a third sync pattern. These three sync patterns are shown in Figure 8.23. As stated, there is a parity bit in each subframe, which means

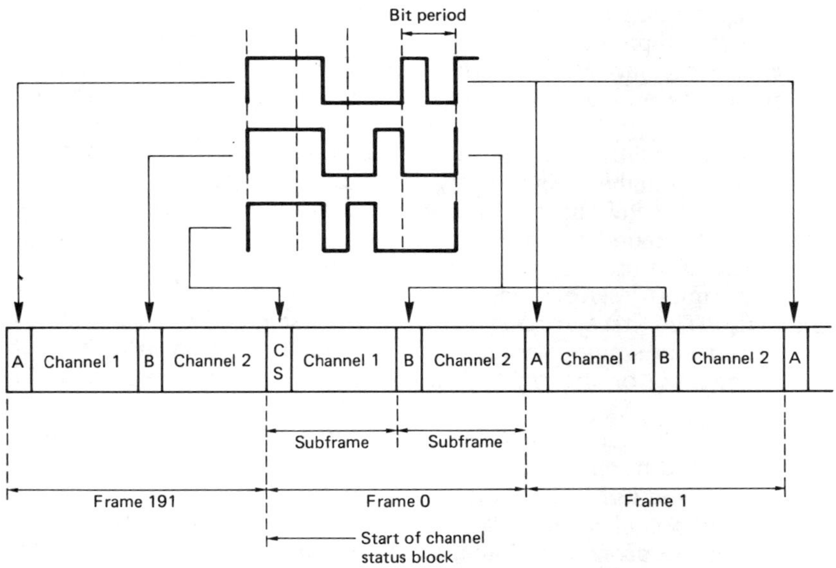

Figure 8.23 The A and B channels are distinguished by different sync patterns. Once every 192 blocks, the A sync is replaced by channel-status block sync. Inverse of sync patterns is also valid; system is not polarity conscious

that the binary level at the end of a subframe will always be the same as at the beginning. Since the sync patterns have the same characteristic, the effect is that sync patterns always have the same polarity and the receiver can use that information to reject noise. The polarity of transmission is not specified, and indeed an accidental inversion in a twisted pair is of no consequence, since it is only the transition that is of importance, not the direction.

In the professional format, the sequence of channel-status bits over 192 subframes builds up a 24 byte channel-status block, shown in Figure 8.24. The first byte determines the use of emphasis and the sampling rate, with details in Figure 8.25. The second byte determines the channel usage, i.e. whether the data transmitted are a stereo pair, two unrelated mono signals or a single mono signal. Figure 8.26 gives details. The third byte determines wordlength as in Figure 8.27.

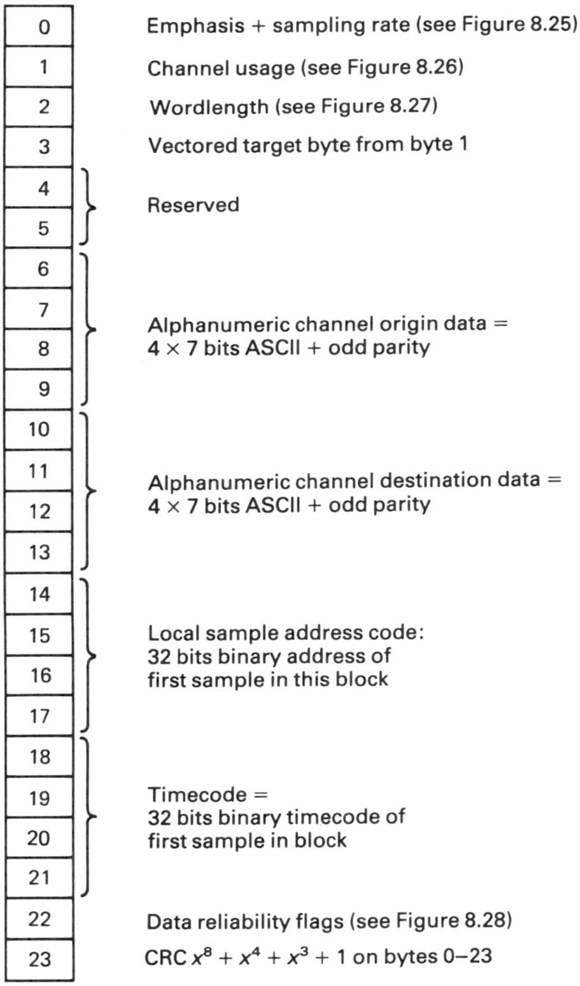

Figure 8.24 The content of the 24 byte sequence of channel-status data in the AES/EBU format

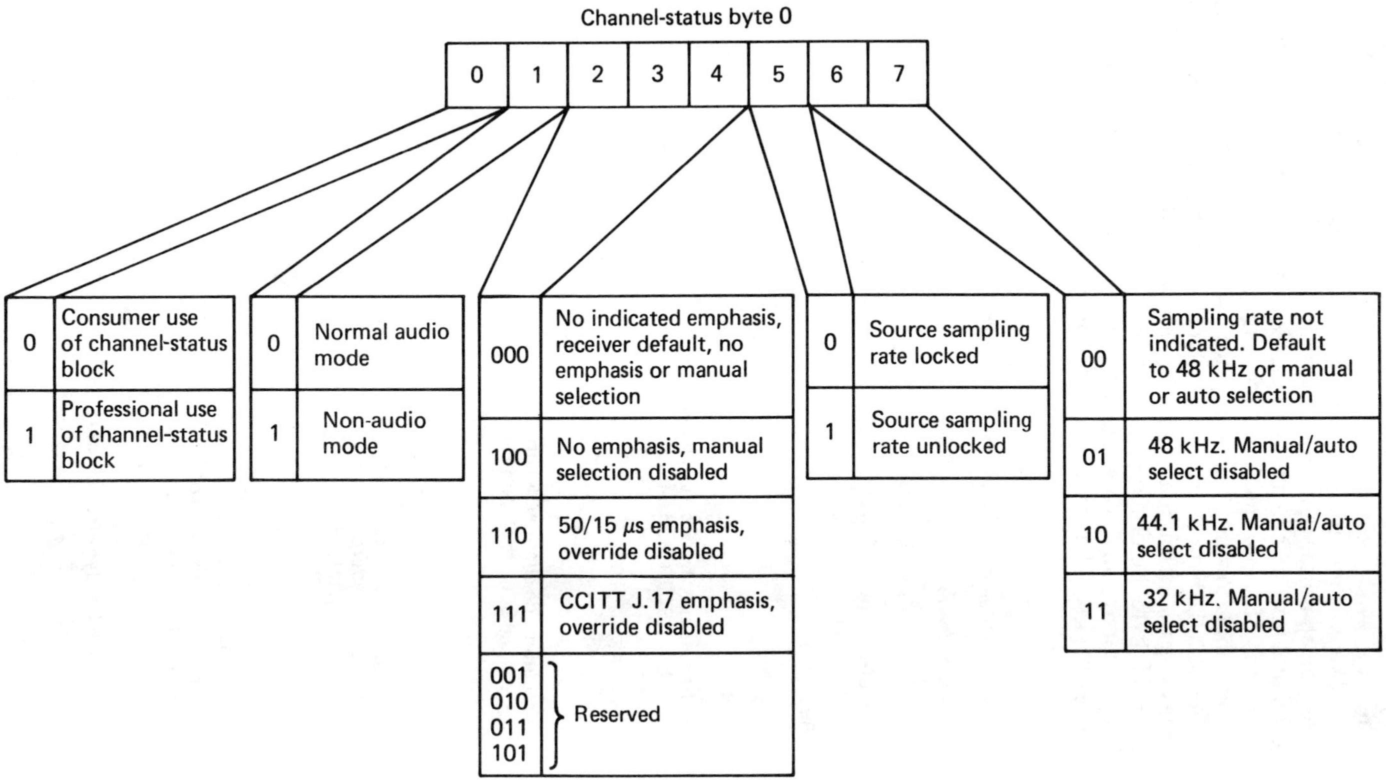

Figure 8.25 The first byte of the channel-status information in the AES/EBU standard deals primarily with emphasis and sampling rate control

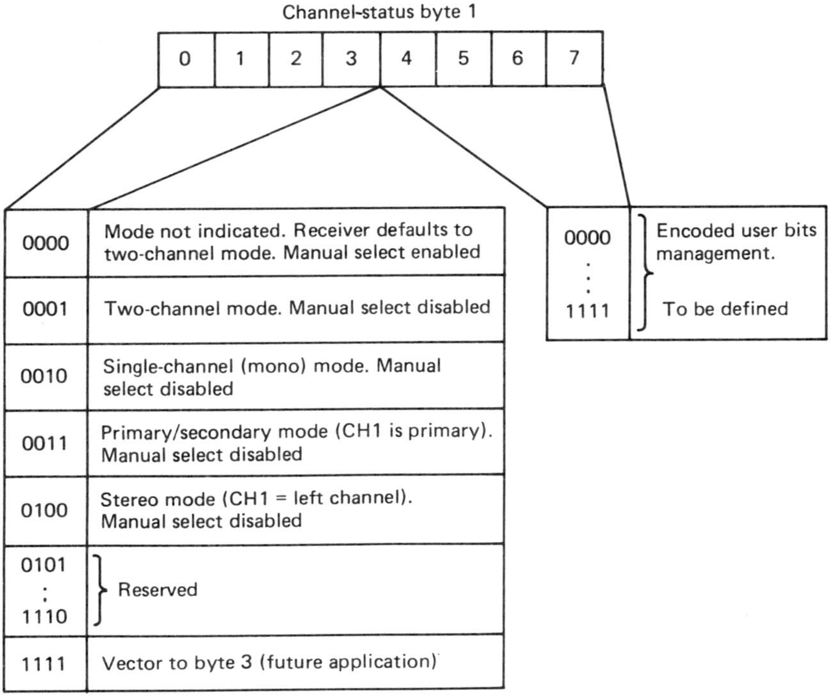

Figure 8.26 The secondary byte of the channel-status information currently deals with audio channel usage, but will be extended in the future to manage user bits

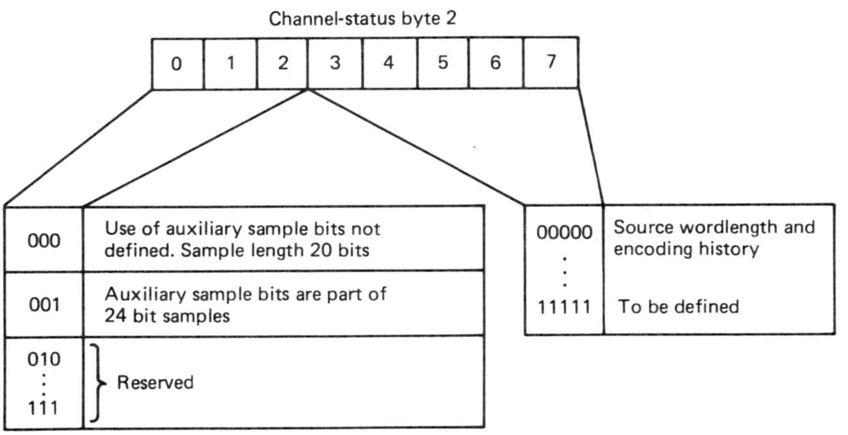

Figure 8.27 Byte 2 of channel status determines whether all 24 bits of the word slot are used for audio samples, or just 20 maximum

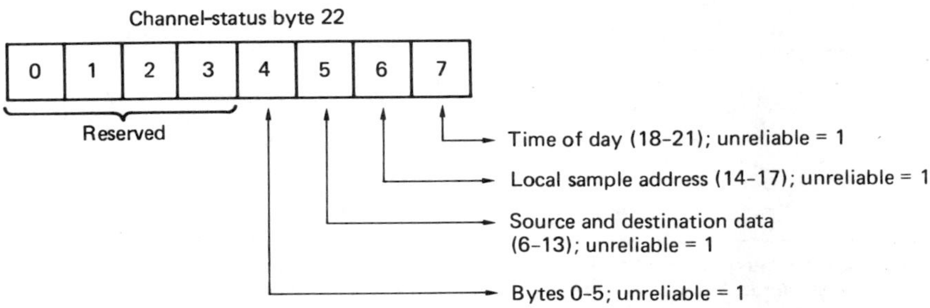

Figure 8.28 Byte 22 of channel status indicates if some of the information in the block is unreliable

There are two slots of 4 bytes each which are used for alphanumeric source and destination codes. These can be used for routing.

Bytes 14–17 convey a 32 bit sample address which increments every channel-status frame. It effectively numbers the samples in a relative manner. Bytes 18–21 convey a similar number, but this is a time-of-day count, which starts from zero at midnight. With a sampling rate of 48 kHz, the binary count represents the number of 4 millisecond intervals from midnight and can easily be converted into, for example, EBU timecode by dividing by ten to obtain a count of 40 ms video frames. At other sampling frequencies, the division ratio would need to be modified appropriately.

The penultimate byte contains four flags which indicate that certain sections of the channel-status information are unreliable (see Figure 8.28). This allows the transmission of an incomplete channel-status block where the entire structure is not needed or where the information is not available. For example, setting bit 5 to a logical one would mean that no origin or destination data would be interpreted by the receiver, and so it need not be sent.

The final byte in the message is a CRCC which converts the entire channel-status block into a codeword (see Chapter 5).

8.12 Ampex parallel interconnect

The AES/EBU interface and the consumer derivative are ideal for a small number of audio channels. In an application, such as D-2, which has four audio channels, it is more convenient to convey them together. The use of a parallel interface has some advantages for this application, since the distances involved are usually short and the cost of the cable is not an issue.

In the parallel digital audio format devised by Ampex for the D-2 format,[16] the audio data have been packed into a format where they can be conveyed by the same electrical and mechanical interface as the parallel digital video interconnect. Figure 8.29(a) shows that one of the spare signals of the 8 bit parallel interface has become a square wave at 48 kHz. Between falling edges of this sync signal, audio data for the four channels are multiplexed. Each audio channel is of 32 bits, so that the whole of the AES/EBU information is conveyed, and this is split into 4 bytes, such that 16 timeslots are necessary in one audio sample period. Each timeslot is clocked by the main 768 kHz clock as shown in Figure 8.29(b). The byte contents are shown in Figure 8.29(c), where it will be seen that the beginning of the AES/EBU channel-status block is specified by the SM bit.

Contact	Signal line	Contact	Signal line
1	Clock		
2	System ground	14	Clock ret.
3	Data 7 (MSB)	15	System ground
4	Data 6	16	Data 7 ret.
5	Data 5	17	Data 6 ret.
6	Data 4	18	Data 5 ret.
7	Data 3	19	Data 4 ret.
8	Data 2	20	Data 3 ret.
9	Data 1	21	Data 2 ret.
10	Data 0	22	Data 1 ret.
11	Spare	23	Data 0 ret.
12	Sync	24	Spare ret.
13	Chassis shield	25	Sync ret.

(a)

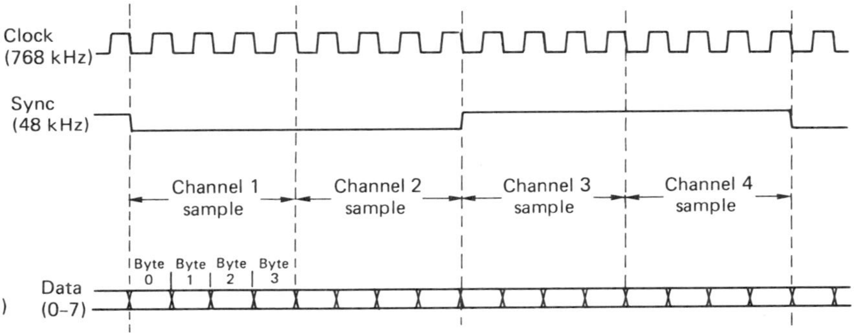

(b)

Figure 8.29 In the Ampex four-channel parallel digital audio interconnect, 4 byte channel samples are sent serially in one sample period. The same connections are used as for the video interconnect except that an additional signal, Sync, shown at (a) is needed to specify which samples correspond to which channel as in (b)

Bit number in (a)

Byte order in (b)	Data 0	1	2	3	4	5	6	Data 7
0	A	A	A	A	A	A	A	A MSB audio
1	A	A	A	A	A	A	A	A
2	WM_1	WM_2	WM_3	SM	A_0 / C	A_1 / U	A_2 / V	A_3 / R
3	EM_0	EM_1	EM_2	V_D	V_J	CH_1	CH_2	CH_3

A = Part of audio sample.
SM = Sync mark. This sample is the first of an AES block.
WM = Word mode
CH = Channel usage.
EM = Pre emphasis mode.
C, U, V, R = Channel status bits,
V_D = 1 if sample was interpolated, else 0..
V_J = 1if any of C, U, V, R bits unreliable.

Figure 8.29(c) The 4 bytes corresponding to each audio channel are assembled as shown here

8.13 Audio in the scrambled serial video interconnect

In this interconnect, which was described in Section 3.12, there is provision for auxiliary data to be sent during vertical and horizontal sync pulses. Figure 8.30(a) shows that following the usual TRS-ID, there will be an auxiliary data flag whose value must be $3FC_{16}$. Following this is a code which identifies the kind of data which are being sent. In the case of digital audio, it will be $2FF_{16}$. Next, a word carrying the data block number is sent. This is an 8 bit binary incrementing count with two parity bits. The ID is completed by a symbol count parameter which specifies how many symbols of user data are being sent in this packet. The audio data proper follow, and at the end of the packet a checksum is calculated.

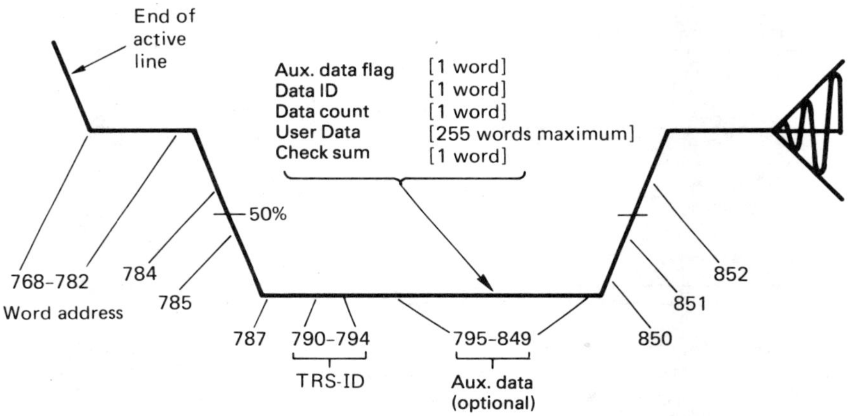

Figure 8.30(a) In the scrambled serial digital video interface, auxiliary data, including digital audio, can be sent during sync pulses (shown) and during equalizing pulses. The auxiliary data must be preceded by the auxiliary data flag and followed by a checksum as shown

Address / Bit	x3	x3 + 1	x3 + 2
B9	$\overline{B8}$	$\overline{B8}$	$\overline{B8}$
B8	A (2^5)	A (2^{14})	P
B7	A (2^4)	A (2^{13})	C
B6	A (2^3)	A (2^{12})	U
B5	A (2^2)	A (2^{11})	V
B4	A (2^1)	A (2^{10})	A MSB (2^{19})
B3	A LSB (2^0)	A (2^9)	A (2^{18})
B2	CH (MSB)	A (2^8)	A (2^{17})
B1	CH (LSB)	A (2^7)	A (2^{16})
B0	Z	A (2^6)	A (2^{15})

Figure 8.30(b) AES/EBU data for one audio sample is sent as three 9 bit symbols. A = audio sample. Bit Z = AES/EBU channel status block start bit

Figure 8.30(b) shows that the system wordlength is 9 bits, and that three 9 bit symbols are used to convey all of the essential bits of the 32 used in the AES/EBU subframe. Since DVTRs have four audio channels, there are two channel bits which specify the channel number to which the subframe belongs. A further bit, Z, specifies the beginning of the 192 sample channel-status message. V, U and C have the same significance as in the normal AES/EBU standard, but the P bit reflects even parity on the three 9 bit symbols.

The exact positions of the packets in the video structure are not specified, but it is only necessary to provide a little RAM buffering at both ends of the link, and this becomes unimportant, as the receiver can determine the sampling rate from the video timing and demultiplex the channels according to the symbol labelling.

8.14 Digital audio in D-2

The audio performance of video recorders has traditionally lagged behind that of audio-only recorders. In video recorders, the use of rotary heads to obtain sufficient bandwidth results in a wide tape whose longitudinal audio tracks travel relatively slowly by professional audio standards. In all rotary-head recorders, the intermittent head contact causes shock wave patterns to propagate down the tape, making low-flutter figures difficult to achieve. This is compounded by the action of the capstan servo which has to change tape speed to maintain control-track phase if the video heads are to track properly.

The requirements of lip-sync dictate that the same head must be used for both recording and playback, when the optimum head design for these two functions is different. When dubbing from one track to the next, one head gap will be recording the signal played back by the adjacent magnetic circuit in the head, and mutual inductance can cause an oscillatory loop if extensive anti-phase crosstalk cancelling is not employed. Placing the tracks on opposite sides of the tape would help this problem, but phase errors between the channels can then be introduced by tape weave. This can mean the difference between a two-channel recorder and a stereo recorder. Crosstalk between the timecode and audio tracks can also restrict performance.

The adoption of digital techniques essentially removes these problems for the audio in D-2. Once the audio is in numerical form, wow, flutter and channel-phase errors can be eliminated by timebase correction; crosstalk ceases to occur and, provided a suitable error-correction strategy is employed, the only degradation of the signal will be due to quantizing. The most significant advantages of digital recording are that there is essentially no restriction on the number of generations of re-recording which can be used and that proper crossfades can be made in the audio at edit points, following a rehearsal if necessary.

Digital audio recording with video is rather more difficult than in an audio-only environment. The special requirements of professional video recording have determined many of the parameters of the format and have resulted in techniques not found in audio-only recorders. Professional video formats must permit different degrees of flexibility in audio editing, allowing at one extreme simple circuits for portable recorders and at the other extreme complex post-production recorders. These will perform track bouncing, synchronous recording and split audio edits with variable crossfade times.

8.15 Track sectoring

The audio samples in D-2 are binary numbers just like the video samples, and although there is an obvious difference in sampling rate and wordlength, this only affects the relative areas of tape devoted to the audio and video samples. The most important difference between audio and video samples is the tolerance to errors. The acuity of the ear means that uncorrected audio samples must not occur more than once every few hours. There is little redundancy in sound, and concealment of errors is not desirable on a routine basis. In video, the samples are highly redundant, and concealment can be effected using samples from previous or subsequent lines or, with care, from the previous frame. Major differences can be expected between the ways that audio and video samples are handled in a D-2 recorder. One such difference is that the audio samples have 100% redundancy: every one is recorded twice with a physical separation.

The audio samples are carried by the same channel as the video samples. The audio could have used separate stationary heads, but this would have increased tape consumption and machine complexity. In order to permit independent audio and video editing, the tape tracks are given a block structure. Editing will require the heads momentarily to go into record as the appropriate audio block is reached. Accurate synchronization is necessary if the other parts of the recording are to remain uncorrupted.

Figure 8.31 shows the structure of the audio portions of the tape tracks in D-2. As there are four independent audio channels, there are four audio sectors, since for reasons which will become clear, it is only possible to edit complete blocks. In

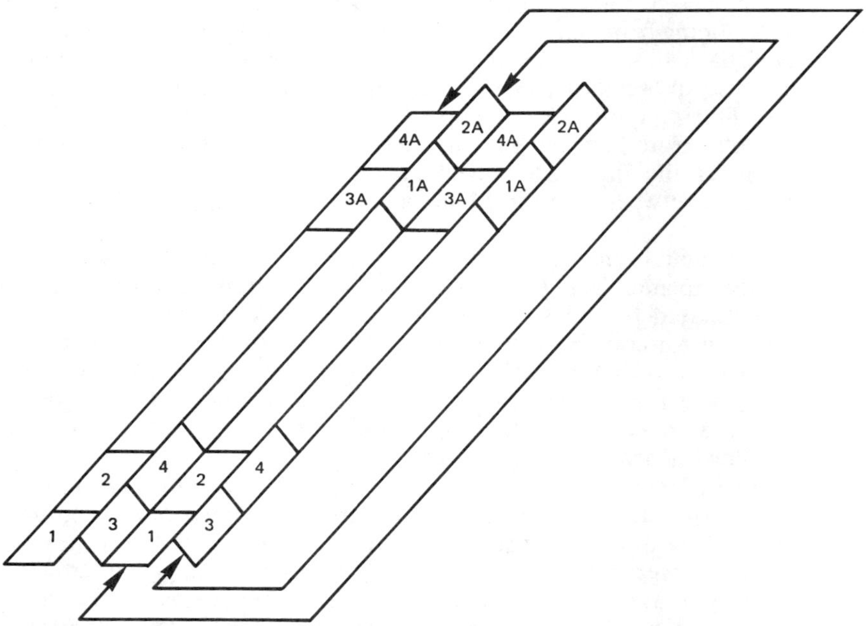

Figure 8.31 In the composite digital format, data from each audio channel are recorded twice, so that the copy is at the opposite edge of the tape, and read by a different head. This gives immunity to head clogs and linear tape scratches

D-2 there are 204 (NTSC) or 249 (PAL) sync blocks in a video sector and six (PAL and NTSC) sync blocks in an audio sector. The sync block structure of the video sector continues in the audio sectors because the same read/write circuitry is used for audio and video data. Clearly the ID code structure must also continue through the audio. To prevent audio samples from arriving in the framestore in shuttle, the audio addresses are different from the video addresses.

The arrangement of the audio blocks is designed to maximize data integrity in the presence of tape defects and head clogs. The audio blocks are at the ends of the head sweeps; the double recording ensures that the second copy of an audio block is at the opposite edge of the tape and that it will be played with a different head, and so if one of the two active heads clogs the audio is still fully recovered. The D-2 format differs considerably between PAL and NTSC versions owing to the difference between the subcarrier frequencies of the two standards. In PAL one field requires eight tracks, so the track rate will be 400 Hz, and since there are four audio blocks in each track, the audio block rate will be 1600 Hz. In NTSC one field requires six tracks, so the track rate will be 360 Hz (minus 0.1%) and the audio block rate will be 1440 Hz (minus 0.1%). Since the audio sampling rate is unchanged by the video format, the number of samples in each audio block will be different in the two versions of D-2.

8.16 Writing audio sectors

Each audio write block commences with a preamble to synchronize the phase-locked loop in the data separator on replay. Each of the sync blocks begins, as the name suggests, with a synchronizing pattern which allows the read sequencer to deserialize the block correctly. At the end of a write block, it is not possible simply to turn off the write current after the last bit, as the turnoff transient would cause data corruption. It is necessary to provide a postamble such that current can be turned off away from the data. It should now be evident that any editing has to take place a sector at a time. Any attempt to rewrite one sync block would result in damage to the previous block owing to the physical inaccuracy of replacement, damage to the next block due to the turnoff transient, and inability to synchronize to the replaced block because of the random phase jump at the point where it began. Owing to the difficulty of writing in exactly the same place as a previous recording, it is necessary to leave tolerance gaps between sectors where the write current can turn on and off to edit individual write blocks. For convenience, the tolerance gaps are made the same length as a sync block. The first half of the tolerance gap is the postamble of the previous block, and the second half of the tolerance gap acts as the preamble for the next block. The tolerance gap following editing will contain, somewhere in the centre, an arbitrary jump in bit phase and a certain amount of corruption due to turnoff transients. Provided that the postamble and preamble remain intact, this is of no consequence.

8.17 D-2 audio error correction

To facilitate editing, each audio channel is recorded independently. A product code is used, and the interleave is generated by the usual memory array, as Figure 8.32 illustrates for NTSC. The 20 bit audio samples are written into the array in rows,

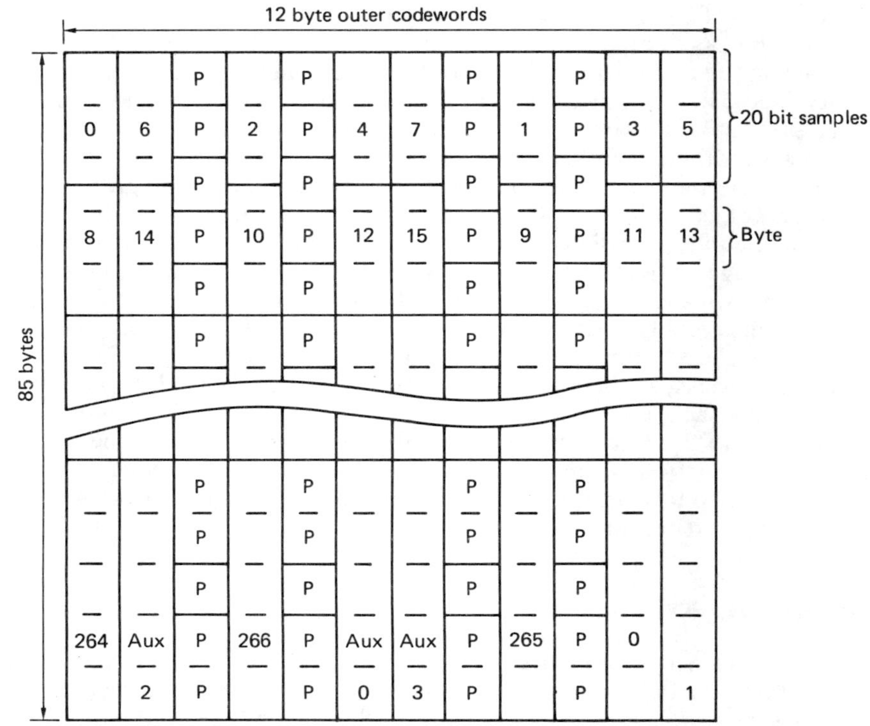

Figure 8.32 In the composite digital VCR, the audio samples are written into the interleave array in rows to produce outer codewords, and read out in 85 byte columns to form sync blocks as shown in Figure 8.33

where 2½ rows of bytes are needed for one row of samples. Each row of the array is loaded with eight samples, and then 4 bytes of Reed–Solomon redundancy are added to form a 12 byte outer code. Reading the array in columns and adding 8 bytes of R–S redundancy produces inner codewords. Two of these inner codewords can be accommodated in one sync block as shown in Figure 8.33. Accordingly six sync blocks are necessary to record one audio block.

In PAL, the audio block rate is higher, so each block contains fewer samples. The interleave block remains the same except that the columns are only 76 bytes tall instead of 85 bytes.

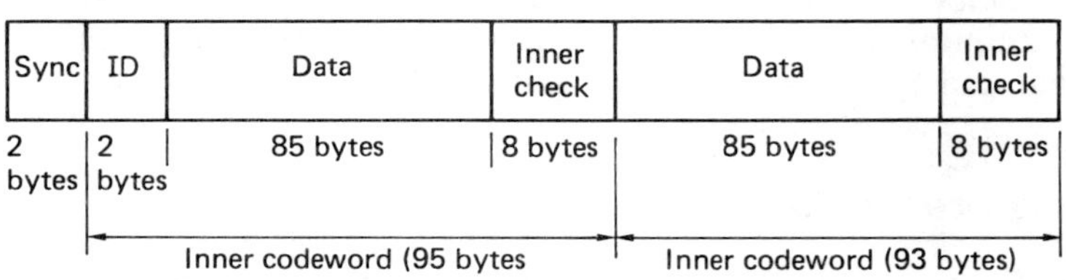

Figure 8.33 One sync block of the composite digital format contains two inner codewords formed from columns of Figure 8.32. Six sync blocks are necessary to hold the entire interleave block and form one audio sector (NTSC)

8.18 Audio functions in D-2

The contents of the audio blocks of the D-2 format can now be detailed in relation to the various audio functions.

8.18.1 AES/EBU compatibility

In order to comply with the AES/EBU digital audio interconnect, wordlengths between 16 and 20 bits can be supported, but it is necessary to record a code in the sync block to specify the wordlength in use. Pre-emphasis may have been used prior to conversion, and this status is also to be conveyed, along with the four channel-use bits. As was seen in Section 8.11, the AES/EBU digital interconnect uses a block-sync pattern which repeats after 192 sample periods corresponding to 4 ms at 48 kHz. Since the block size is different to that of the D-2 audio sector, there can be any phase relationship between sector boundaries and the AES/EBU block-sync pattern. In order to recreate the same phase relationship between block sync and sample data on replay, it is necessary to record the position of block sync within the sector. It is the function of the interface control word in the audio data to convey these parameters. There is no guarantee that the 192 sample block-sync sequence will remain intact after audio editing; most likely there will be an arbitrary jump in block-sync phase. Strictly speaking a machine playing back an edited tape would have to ignore the block-sync positions on the tape and create new block sync at the standard 192 sample spacing.

8.18.2 Synchronization between audio sampling rate and video field rate

Clearly the number of audio sync blocks in a given time is determined by the number of video fields in that time. It is only possible to have a fixed tape structure if the audio sampling rate is locked to video. This causes no difficulty in PAL since there is a simple relationship between audio sampling rate and field rate. In NTSC, it must be recalled that the nominal 60 Hz is actually 59.94 Hz. As this is slightly slow, fields are very slightly longer, and it will be found that in 60 fields exactly 48 048 audio samples will be necessary. Unfortunately 60 will not divide into 48 048 without a remainder. The largest number which will divide 60 and 48 048 is 12; thus in $60/12 = 5$ fields there will be $48\,048/12 = 4004$ samples.

There are six tracks in a D-2 NTSC field, and so in a given audio channel there will be 30 audio blocks in a five-field period. Since double recording is implemented, there are 15 audio blocks repeated twice.

In 14 of these blocks there are 267 audio samples and the remaining block contains 266 samples. The D-2 format does not indicate the audio five-field sequence in the control track, only in the status words of the helical tracks, but nevertheless the sequence must remain unbroken after editing.

Figure 8.34 shows the status word which controls this operation.

8.18.3 Variable-speed replay

Variable-speed operation really falls into three categories: speeds close to normal, speeds close to zero (jogging) and more extreme speeds. At extreme speeds, the heads cross tracks too frequently, and the digital audio system is overwhelmed by errors and lack of data and gives up. If audio is required, the analog linear (cue) track at the edge of the tape must be used.

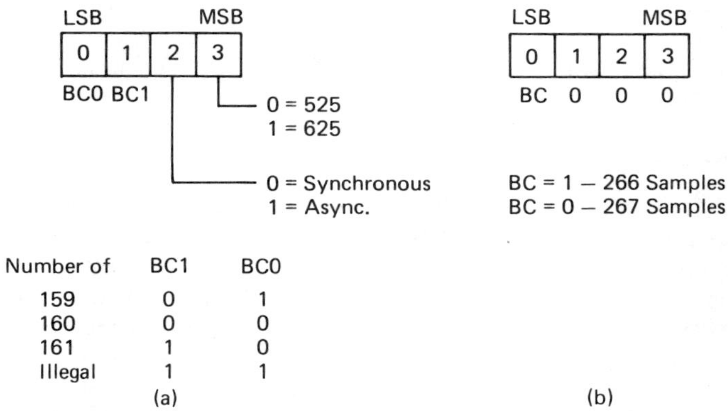

Figure 8.34 At (a) the block count (BCNT) parameter of D-1 allows for slip between 48 kHz and 59.94 Hz, and also for asynchronous operation in 525 or 625 lines. At (b) the block count in D-2 only appears in NTSC. Asynchronous mode is not supported

At speeds close to normal in TSO (Tape Speed Override), used for example to squeeze the length of a programme to insert an additional item, the highest possible video and audio quality are required. One way in which this can be obtained is by changing capstan and scanner speed by the same amount, to allow the heads to follow the tape tracks correctly. The incorrect video field rate will be corrected by a suitable framestore. Clearly if the head-to-tape speed is changed, the audio sampling rate will change also. If the audio samples are simply being fed to a DAC for immediate conversion to the analog domain, this is of no consequence except for a pitch change, but if the samples have to be fed at standard rate into a fully digital studio, a sampling-rate converter will be necessary to avoid the degradation due to reverting to analog.

In jog mode, all tape tracks are played, usually many times because the track-following system maintains field rate with the tape nearly stopped. Since all of the audio data are available, audio can be recovered in jog. The audio samples are placed in a memory, and the memory addressing is controlled from the speed of the capstan, so that samples are obtained at a rate proportional to tape speed. These are fed to a digital interpolator which produces a standard output sampling rate. In this way the digital audio in jog is made to simulate the effect of moving linear analog audio tracks, and so the digital tracks can be used for cueing and locating edit points.

8.18.4 Editing

It has been explained that the audio sectors can be individually written with the use of tolerance gaps to protect adjacent data. There is, however, rather more to editing than that. The D-2 format supports two levels of audio editing, both of which leave a standard formatted tape, but which deliver different performance according to the complexity of the recorder. The reader should take care to distinguish between that which the format permits and that which is implemented on a given model of recorder. Assemble editing, insert editing and split audio/video editing are all supported to achieve maximum flexibility; however, a machine which can only record audio and video together is still capable of meeting the format. In other words a machine can still achieve interchange without supporting every possible edit feature. These features will be described in turn.

(1) Split audio/video editing is achieved in the overall control system of the DVTR. As video and audio data occupy different sync block addresses on tape, they can be written quite independently, and it is only a matter of supplying different timecodes to the video and audio record enables and the split edit will be performed automatically.

(2) Assemble and insert recording of audio are a little more complex because the tape tracks are shared between audio and video. The block structure of the tracks must always be maintained; therefore it is not possible to record only video or only audio on a blank tape. When an assemble is made, the tape beyond the assemble point is assumed to be blank, and a new control track will be written. If only video is supplied to the machine after an assemble, the audio blocks will still be written, filled, most likely, with zero samples, and including the content count which allows the sampling rate to be established. If only audio is supplied after an assemble point, the video blocks will still be written, filled with black. The major difference between assemble and insert is that the control track is read in insert, such that at the end of the insert the original recording carries on without loss of synchronism. During an insert any or all of the audio channels may be changed, by having the heads record during audio sectors, and/or video may be changed, by having the heads record during the video sectors. In practice, once a tape has been recorded, all future work on that tape can be in insert mode as, in digital recording, the format of the tape has a separate existence to the programme material. Many D-2 machines can write the control track and the timecode track without the tape touching the rotary heads, in some cases at a speed higher than normal so that a tape which has been striped in this way can be used entirely in insert mode.

(3) The actual mechanism of video editing is simply to start recording the new data at a field boundary, the result appearing as a cut in the picture. This is also the most basic way of editing the audio; at an interleave block boundary the new recording is substituted for the old. If no action were taken to prevent it, this will result in step discontinuities of the audio waveform, audible as clicks. When the edit is made, the recorder inserts a code in the Edit Flag (EFLG) control word, and on replay this code is detected and used to initiate a short fadeout prior to the end of the old recording, followed by a similar fadein at the beginning of the new recording. No attempt is made to perform a crossfade in this mode, and the position of the cut cannot be changed at less than a block boundary. These restrictions have raised some eyebrows in the audio industry, but do not apply to the second edit mode. In the case of an insert edit EFLG will be set to B_{16} during the first segment of the inserted material and to E_{16} on the last segment. EFLG is otherwise set to zero. Figure 8.35 shows an example of the status of EFLG changing during an edit in one channel only.

(4) For post-production, audio editing needs to be carried out to the same standards of accuracy and fidelity as in the best digital audio-only recorders, albeit with some hardware complexity. The use of interleaving is a powerful tool against dropouts, but it makes editing to sample accuracy much more difficult. The principles involved are exactly those of a digital audio-only machine. Because of the structure of the codewords, only entire blocks can be recorded without error, whereas crossfades need to be performed on de-interleaved data. In order to perform a true crossfade edit with sample accuracy and 100% redundancy, the

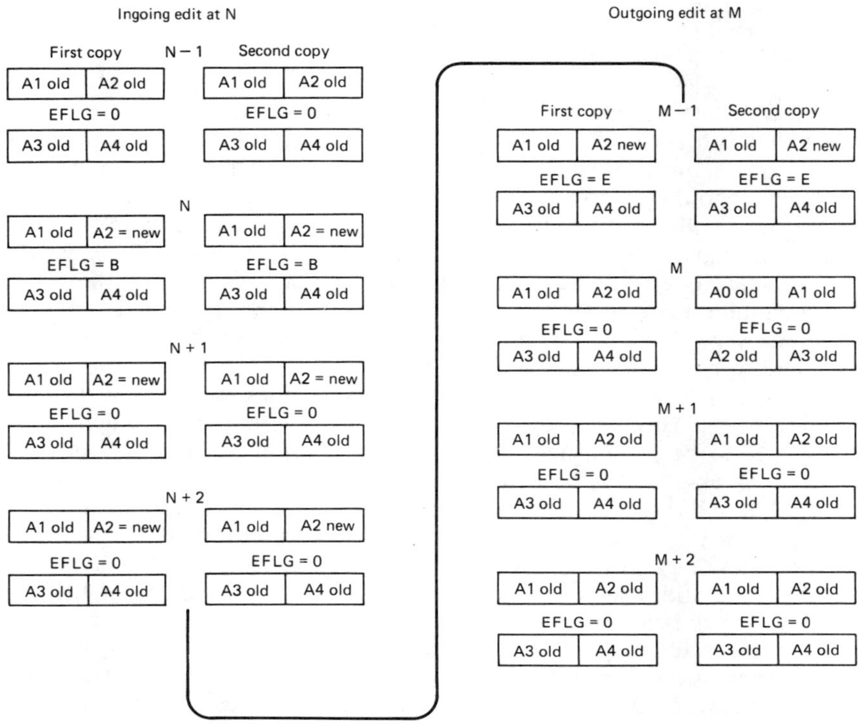

Figure 8.35 The use of the edit flag during an insert edit on channel 2. The change in the flag causes the replay circuitry to V-fade at the beginning and end of the edit area

existing recording in the area of the edit must first be played back, de-interleaved and corrected, and then digitally crossfaded at any point with the insert/assemble audio. The output of the crossfader has then to be re-interleaved, and the result has to be recorded in the blocks which were played back earlier to update them. The problem is that, until the block has been played, the contents cannot be de-interleaved, and by the time the correction, crossfading and re-interleave have been performed, the block has long gone. There are two solutions to this. The first is that the area of the edit point is played into a memory on a first pass of the tape; then the machine is backed up, and the crossfade to new material and the re-recording take place on a second pass. This is slow, but allows the use of a simple DVTR. The second method is to use an advanced replay head, separated by the decode/correction–encode delay period. In the DVTR this would require extra heads on the scanner and an additional amount of memory to delay the data from the advanced replay by the time between the heads minus the decode/encode delay. However, if the machine is fitted with track-following heads to allow variable-speed operation, these heads could be made to deflect so that they would become the advanced replay heads.

The sequence of events in such a system is shown in Figure 8.36. As the complete audio structure of the channel to be edited is rewritten, there is no loss of redundancy at the edit crossfade point, and the resistance to corruption will be the same as elsewhere.

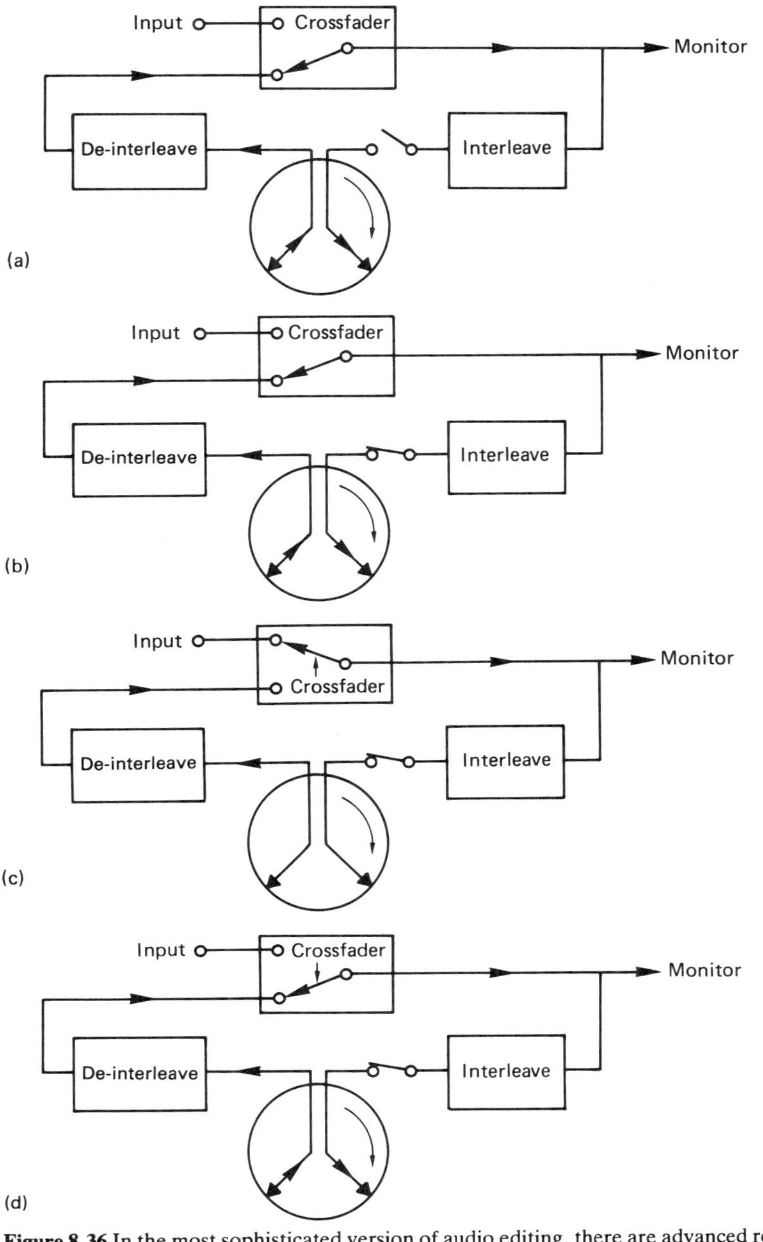

Figure 8.36 In the most sophisticated version of audio editing, there are advanced replay heads on the scanner, which allow editing to be performed on de-interleaved data. An insert sequence is shown. At (a) the replay-head signal is decoded and fed to the encoder which, after some time, will produce an output representing what is already on the tape. In (b), at a sector boundary, the write circuits are turned on, and the machine begins to re-record. In (c) the crossfade is made to the insert material. At (d) the insert ends with a crossfade back to the signal from the advanced replay heads. After this, the write heads will once again be recording what is already on the tape, and the write circuits can be disabled at a sector boundary. An assemble edit consists of the first three of these steps only

(5) In order to dub one channel to another (track bouncing), the de-interleave and correction process must precede the record interleave and, clearly, to avoid loss of lip-sync an advanced playback head is necessary as for the most sophisticated edit technique. Synchronous recording is similar in principle to track bouncing, except that a performer can listen to the playback of an existing recording and play along to it to record another track in synchronism.

As a DVTR can individually edit audio channels, it does not require a great leap of the imagination to see that a multitrack audio-only machine could be made by filling the tape tracks with audio sectors. The number of channels this would permit is staggering, and it may be more useful to incorporate a video channel using data reduction. The high recording density of helical scan means that the tape speed would be much lower than any other kind of multitrack audio recorder, and the apparent rewinding speed would be much higher, making possible a workstation approach as is currently implemented with hard disks.

References

1. ANON. AES recommended practice for professional digital audio applications employing pulse code modulation: preferred sampling frequencies. AES5-1984 (ANSI S4.28-1984). *J. Audio Eng. Soc.*, **32**, 781–785 (1984)
2. MURAOKA, T., IWAHARA, M. and YAMADA, Y. Examination of audio bandwidth requirements for optimum sound signal transmission. *J. Audio Eng. Soc.*, **29**, 2–9 (1982)
3. MURAOKA, T., YAMADA, Y. and YAMAZAKI, M. Sampling frequency considerations in digital audio. *J. Audio Eng. Soc.*, **26**, 252–256 (1978)
4. FINCHAM, L. R. The subjective importance of uniform group delay at low frequencies. Presented at the 74th Audio Engineering Society Convention (New York, 1983), preprint 2056(H-1)
5. ADAMS, R. W. Companded predictive delta modulation: a low-cost technique for digital recording. *J. Audio Eng. Soc.*, **32**, 659–672 (1984)
6. v.d. PLASSCHE, R. J. Dynamic element matching puts trimless converters on chip. *Electronics* (16 June 1983)
7. v.d. PLASSCHE, R. J. and GOEDHART, D. A monolithic 14 bit D/A converter. *IEEE J. Solid-State Circuits*, **SC-14**, 552–556 (1979)
8. v.d. PLASSCHE, R. J. and DIJKMANS, E. C. A monolithic 16 bit D/A conversion system for digital audio. In *Digital Audio*, ed. B. A. Blesser *et al.*, New York: Audio Engineering Society, pp. 54–60 (1982)
9. ADAMS, R. W. Design and implementation of an audio 18-bit A/D converter using oversampling techniques. Presented at the 77th Audio Engineering Society Convention (Hamburg, 1985), preprint 2182
10. Audio Engineering Society, AES recommended practice for digital audio engineering – serial transmission format for linearly represented digital audio data. *J. Audio Eng. Soc.,* **33**, 975–984 (1985)
11. Draft standard for a digital audio interface. IEC report 84/WG11
12. EIA RS-422A. Electronic Industries Association, 2001 Eye Street NW, Washington, DC 20006, USA
13. SMART, D. L. Transmission performance of digital audio serial interface on audio tie lines. *BBC Des. Dep. Tech. Memo.*, 3.296/84
14. European Broadcasting Union, Specification of the digital audio interface. *EBU Doc. Tech.*, 3250
15. LAGADEC, R. and McNALLY, G. J. Labels and their formatting in digital audio recording and transmission. Presented at the 74th Audio Engineering Society Convention (New York, 1983), preprint 2003
16. Ampex Corporation, Ampex digital format for video and audio tape recording of composite video signals using 19mm type D-2 cassette. Redwood City, CA (1988)

Index